Engineering and Scientific Computations in PASCAL

620.0028 H87E 7116507
HUELSMAN, LAWRENCE P.
ENGINEERING AND SCIENTIFIC
COMPUTATIONS IN PASCAL

Harper & Row
Computer Science and Technology Series

Bartee: *BASIC Computer Programming,* Second Edition

Bellin: *The Complete Computer Maintenance Handbook*

Bork: *Personal Computers for Education*

Chou: *Microcomputer Programming in BASIC with Business Applications,* Second Edition

Gallier: *Logic for Computer Science: Foundations of Automatic Theorem Proving*

Greenwood and Brodinski: *Enjoying BASIC: A Comprehensive Guide to Programming*

Halpern: *Microcomputer Graphics Using PASCAL*

Huelsman: *Engineering and Scientific Computations in PASCAL*

Jackson and Fischer: *Learning Assembly Language: A Guide for BASIC Programmers*

Jones: *PASCAL: Problem Solving and Programming with Style*

Lamprey, Macdonald, and Roberts: *Programming Principles Using PASCAL*

Lin: *Computer Organization & Assembly Language Programming for the PDP-11 and VAX-11*

Mason: *Learning APL: An Array Processing Language*

Newell: *Introduction to Microcomputing*

O'Shea and Eisenstadt: *Artificial Intelligence: Tools, Techniques and Applications*

Passafiume and Douglas: *Digital Logic Design: Tutorials and Laboratory Exercises*

Rafiquzzaman: *Microprocessors and Microcomputer Development Systems*

Shumate: *Understanding ADA*

Touretzky: *LISP: A Gentle Introduction to Symbolic Computation*

Weir: *Cultivating Minds: A LOGO Casebook*

Wood: *Theory of Computation: A Primer*

Engineering and Scientific Computations in PASCAL

Lawrence P. Huelsman
University of Arizona

HARPER & ROW, PUBLISHERS, New York
Cambridge, Philadelphia, San Francisco
London, Mexico City, São Paulo, Singapore, Sydney

Sponsoring Editor: John Willig
Project Editor: Ellen MacElree
Cover Design: Wanda Lubelska Design
Text Art: Fineline Illustrations, Inc.
Production: Willie Lane
Compositor: Waldman Graphics, Inc.
Printer and Binder: R.R. Donnelley & Sons Company

Engineering and Scientific Computations in PASCAL

Library of Congress Cataloging in Publication Data

Huelsman, Lawrence P.
Engineering and scientific computations in PASCAL.

Includes index.
1. Engineering—Data processing. 2. PASCAL (Computer program language) 3. Electronic digital computers—Programming. I. Title.
TA345.H84 1986 620'.0028'542 85-8476
ISBN 0-06-042994-1

85 86 87 88 9 8 7 6 5 4 3 2 1

To Jo and David

Contents

Preface

This book is designed to introduce the use of digital-computational techniques in the solution of engineering problems. The techniques presented are especially suitable for microcomputer implementation. The book has two main applications. The first of these is as a supplementary text for use in undergraduate engineering courses. In such courses it is desirable to provide the student with a computational capability for analyzing realistic systems for which the conventional ''hand'' analysis techniques would be hopelessly cumbersome. In illustrating this application, many of the example problems included in the book have been chosen from those likely to be found in a modern introductory course in circuit theory, since almost all engineering students take such a course. Obviously, the same analysis techniques can be used to solve problems in a wide range of other engineering subjects.

The second application of this book is for the engineer who desires to use a high-level programming language to implement the computations required in a microcomputer-based product. To make such an application as efficient as possible, each of the computational techniques presented is realized as a self-contained subprogram. Thus it is possible to reduce the amount of programming required in a given design by including only the techniques actually required.

The high-level language chosen to realize the digital-computational techniques is PASCAL. This language has shown itself to be well adapted to a wide range of microcomputer applications. In addition, it is a language capable of producing efficiently designed, well-structured programs. The presentation used assumes that the reader is already familiar with the PASCAL language. For the reader who does not have such a familiarity, Appendix A presents a self-contained introduction to the syntax of the language.

The following goals were used as a guide in the development of the material:

1. *The emphasis of the presentation should be on the application of numerical techniques to engineering rather than on the techniques themselves.* For an extended discussion of the numerical properties of the techniques, reference is easily made to one of the many currently available texts on numerical analysis.
2. *The reader should be motivated by being shown how numerical techniques make it possible to solve problems that cannot be solved by nonnumerical analysis methods.* For example, it is shown how numerical differential-equation-solving techniques may be used to provide solutions for the variables of systems that contain time-varying and nonlincar elements as well as for those containing only linear time-invariant ones.
3. *The book should provide readers with a set of computational tools that they can apply to a wide range of problems.* This goal has been implemented by developing all the numerical techniques as subprograms. Taken as a group, these subprograms form a software package that is readily applicable to a broad range of engineering and scientific applications.
4. *The programs and subprograms that are developed should be kept as simple as possible.* This was done so that emphasis could be placed on theory rather than on sophisticated programming. This makes the understanding of the techniques easier for the reader. In addition, it minimizes the amount of computer capabilities required, making the resulting programs readily usable in microcomputer environments. Because of the simplicity of the programs, modifications are readily made where necessary to satisfy the particular needs of specific applications.
5. *The problems and examples should be used to develop the reader's confidence in the results obtained from the numerical techniques and to emphasize the need to substantiate results.* To implement this goal, many of the problems have been chosen so that the answers can be verified by direct (nonnumerical) hand analysis.
6. *The programs should be usable on as many different types and makes of computers as possible.* To help ensure this, only the most basic capabilities of the PASCAL language (those most likely to be found on both ''full-size'' computers and microcomputers) have been utilized in implementing the digital-computational techniques.

A few words about the organization of the book may be of interest. It is divided into two main sections. The first includes Chaps. 2 to 6 and is concerned with the ways in which system elements and systems themselves may be treated in the time domain. These chapters deal with the use of numerical-integration techniques, solution of simultaneous sets of differential equations, state-variable formulation of systems, and so on. The techniques presented in these chapters are applicable to the time-varying and nonlinear cases as well as to the more usual linear time-invariant one. Also in this section of the text, a study of the solution of a set of simultaneous algebraic equations, such as might be encountered in the description of a resistance network, is given. An example of an approach to the solution of a nonlinear set of equations is given as an exercise.

The second main section includes Chaps. 7 to 12. It is concerned with the treatment of systems in the complex frequency domain. These chapters deal with the application of the digital computer to the determination of partial-fraction expansions, the inverse Laplace transformation, Nyquist plots, and root-locus plots. Some material on root solving is included. These chapters also cover the sinuosoidal steady-state analysis of systems. A treatment of Fourier-series analysis is given. Finally, in Chap. 13, some examples of the extension of the material presented in this text to larger-scale problems are given.

A set of exercises of varying difficulty is provided at the end of each chapter. These exercises fall into two main categories. Some of them are designed to provide students with additional material on numerical techniques. Others are included to teach them more about the application of the techniques. As an aid to the teacher who wishes to make use of the material, a complete instructors manual has been prepared that gives listings for the programs and copies of the resulting output for all the problems in the book. The book concludes with three appendices. Appendix A provides a review of the basic principles of PASCAL.This material is given in sufficient depth so that it may be used as an introductory treatment of the language for the reader who has not been exposed to it. Appendix B provides extra information on some of the more lengthy subprograms that are used in the text. Appendix C contains listings of programs that have been modified for microcomputer implementations.

One of the major features of the presentation of the material in this text is the inclusion of subprograms to plot the data resulting from the calculations for the problems. There are two such programs, and they are used extensively throughout the book. The first of these, the procedure Plot5, may be used to simultaneously plot several functions of an independent variable. It is most useful for plotting functions of time. A second procedure, Xyplt, provides a capability for making *x-y* plots as a function of a third variable. It is most useful for Nyquist and root-locus studies. As in the other programs to be found in this text, no attempt has been made to achieve high levels of sophistication in these plotting programs. Their purpose is to provide the reader with a useful and simple plotting capability that does not require a large amount of computer memory or large quantities of computer run time. They make it possible for the plotting of data to be done on the printer associated with the computer, thereby avoiding the need for the extra expense and delay frequently associated with the use of a more sophisticated plotting facility.

The material in this text has been included as part of a course in circuit theory taken by juniors in electrical computer engineering at the University of Arizona. The central processor used for the problem assignments was a Cyber 175. All the procedures described in this text were stored as a permanent file library in the computer. Thus, the students were able to access them directly, using a compiler ''include'' command, making it possible for them to submit relatively compact programs and also eliminating a considerable amount of printed output.

The programs described in this book have been run on a wide variety of computers ranging in size from a Cyber 175 main frame to a Kaypro (a CP/M microcomputer with 64K of core) and an IBM PC. The author hopes that this material will make it possible for an engineer to program a microcomputer effectively for many engineering applications. In addition, it should help the engineering instructor to demonstrate effectively to students the power and insight obtained from using the digital computer and a modern, structured, high-level language such as PASCAL to solve a wide variety of basic engineering problems.

I wish to acknowledge the support and encouragement given to this project by Dr. Roy H. Mattson, head of the Department of Electrical and Computer Engineering of the University of Arizona. In addition, the efforts and comments of many undergraduate students, especially Robert Olson, Andrew Piziali, and Charles Smith, who assisted in the development of earlier versions of some of the subprograms, are most gratefully acknowledged. I would also like to thank the reviewers of the manuscript for their many helpful comments: Louis Browning, University of Santa Clara; Frederic J. Mowle, Unicorn Technical Consultants, Inc. and Purduc University; Henry A. Etlinger, Rochester Institute of Technology; Franklin Prosser, Indiana University; Leon Levine, University of California at Los Angeles; and Lee Rosenthal, Fairleigh Dickinson University. Finally, my special thanks to Susan Kinsey and Joann Main for their excellent typing work.

Lawrence P. Huelsman

Engineering and Scientific Computations in PASCAL

Chapter 1

Introduction

The revolution brought about in the modern world by the invention of the digital computer has been paralleled and even surpassed with the introduction of the microcomputer. Although barely a decade has passed since the first microcomputer appeared, its uses today cover an unbelievably wide range of applications including programmable calculators, intelligent cash registers, talking microwave ovens, automobile pollution controls, and video arcade games. Concurrent with the development of these versatile devices has come the introduction of a new generation of high-level programming languages used to communicate with all types of digital computers. These new languages provide for more efficient programming, usually referred to as *structured* programming, than do their predecessors. In addition, these languages can usually be implemented with smaller amounts of computer resources than were required by the older high-level languages. As a result, they are far more suited to use in a microcomputer environment. One of the best known of such languages is PASCAL.[1] It readily lends itself to the development of well-structured programs that are easily used in microcomputer applications. In this book we show how the PASCAL language may be used to implement the computations most often required in engineering applications. Our discussion assumes that the reader is already familiar with the basic syntax of the language. For those readers who do not have such a familiarity, an introduction to PASCAL is given in Appendix A.

One of the most important techniques used in structured programming is the organization of a given programming task into a set of subtasks. Each of these subtasks is then implemented by a separate subprogram. The resulting modularity makes the overall program easier to write, read, maintain, and modify. In the remainder of this chapter we show how the subprogram concept can be applied to various types of engineering com-

[1]The PASCAL language (named after Blaise Pascal, the seventeenth century mathematician who is considered to be the inventor of the adding machine) was developed in the late 1960s by Niklaus Wirth, a professor at the Swiss Federal Technical Institute in Zurich.

putations. We also develop a consistent set of structures for these computations. By this means we preview the entire contents of this book, illustrating its goal of presenting a basic set of subprograms that can be used as building blocks to solve a wide variety of engineering and scientific problems. Note that the explanations of the programs given here are necessarily brief and introductory. The chapters that follow contain detailed treatments of both theory and implementation.

1.1 THE TIME DOMAIN

Engineering and scientific calculations in general take place in one of two environments. These environments are called the time domain and the frequency domain. In the *time domain,* computations are made on physical variables, that is, ones which are directly observable and measurable. Examples of such variables are voltage, current, position, and velocity. In the *frequency domain,* the physical variables are subjected to some type of mathematical transformation before the computations are made. As a result, even though the variables are no longer physically observable, general input-output relations are more readily derived, and properties of classes of systems can be explored and categorized.

The first part of this book is concerned with the time domain and the operations that various types of systems impose on sets of physical variables. In Chap. 2 we consider the first such operation, namely, integration. As an example, in electrical systems, this operation relates the voltage and current variables of an inductor or capacitor. As another example, in a translational mechanical system consisting only of a mass, this operation relates force and velocity. There are many other examples. Since the integration operation is common to so many physical systems, we implement it by a subprogram. The one chosen here uses a method called trapezoidal integration. To permit the application of this subprogram to as broad a range of systems as possible, the integrand is specified by another subprogram. The overall implementation of an integration problem can now be realized by the structure shown in Fig. 1.1.

One of the advantages of the structured modular approach to program development is the ease with which the resulting program can be modified. This advantage is illustrated in Chap. 2 in the implementation of a more efficient integration algorithm called the Romberg method. This method is also implemented as a subprogram. It uses the original integration method, the trapezoidal one, in making its computations. Thus, the new program structure includes the original method as one of its modules. The overall

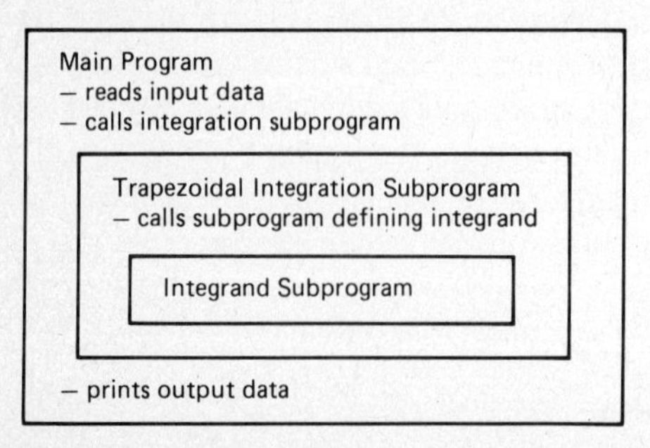

Figure 1.1 Program structure for trapezoidal integration.

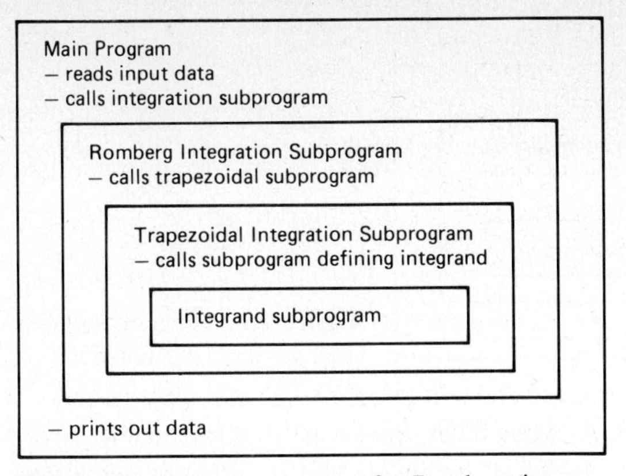

Figure 1.2 Program structure for Romberg integration.

implementation of the more efficient integration method has the structure shown in Fig. 1.2.

In many time-domain problems involving integration, the desired output is not the value of an integral at some specified time (a definite integral). Instead, what is desired is a set of values of the integral at a sequence of values of time over some specified range (an indefinite integral). Such a set of values is readily displayed in a plot. In Chap. 3 we introduce some techniques to implement the solution of this problem. The first of these is the use of iterative application of an integration algorithm to generate the required set of values of the integral. The second is the use of a plotting subprogram to display these values. The structure of the program that accomplishes this is shown in Fig. 1.3.

Frequently, in a problem involving integration, the integrand is specified by a set of data points rather than by an explicit mathematical function. The modularity of the structured programs we have been discussing makes it easy to accommodate this situation. The subprogram that is used to define the integrand is simply replaced by one that generates a piecewise-linear representation of the specified data points. This approach is discussed in Chap. 4. It works well with either the definite or the indefinite (plotted)

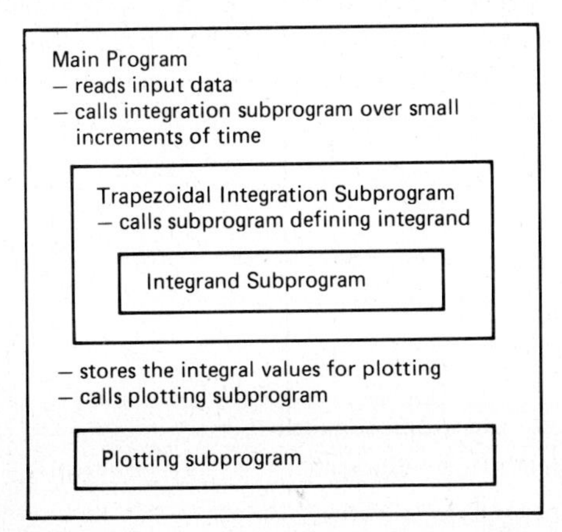

Figure 1.3 Program structure for indefinite integration.

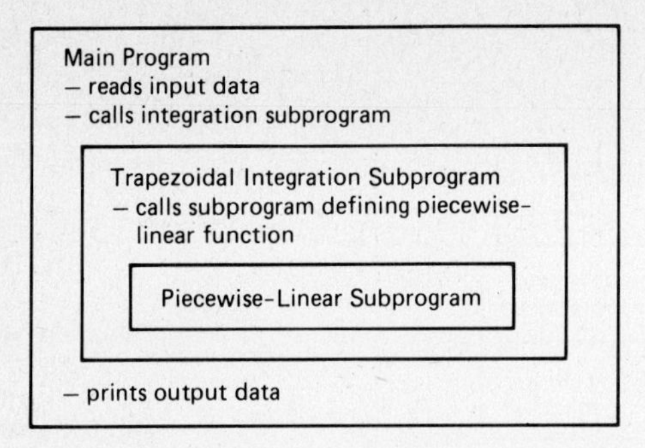

Figure 1.4 Program structure for integrating a piecewise-linear function.

integral case. The resulting form of the program is shown in Fig. 1.4. In this figure, note that only the block specifying the integrand is different from the structure shown in Fig. 1.1.

The second type of time-domain operation that systems impose on physical variables is one characterized by a differential equation. As an example, in an electrical system, such an equation relates the voltage and current variables for a network consisting of a resistor and a single energy-storage element such as an inductor or capacitor. As another example, in a translational mechanical environment, a differential equation relates force and velocity in a system consisting of mass and friction. There are many other examples. Since the operation of solving a differential equation is common to so many types of physical systems, in Chap. 5 it is implemented by a subprogram. The subprogram uses the Runge-Kutta method to solve the differential equation. To provide the necessary flexibility to allow the method to be applied to as many different types of systems as possible, the differential equation to be solved is specified by a separate subprogram. The method may be used to solve nonlinear and time-varying differential equations as well as the more usual linear and time-invariant ones. Applying the Runge-Kutta subprogram iteratively results in a set of sequential solution values which may be plotted. The structure of the program is illustrated in Fig. 1.5. A more efficient algorithm for the

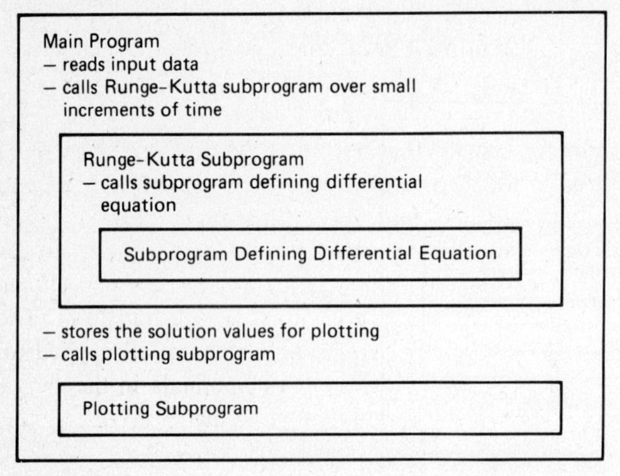

Figure 1.5 Program structure for solution of a differential equation.

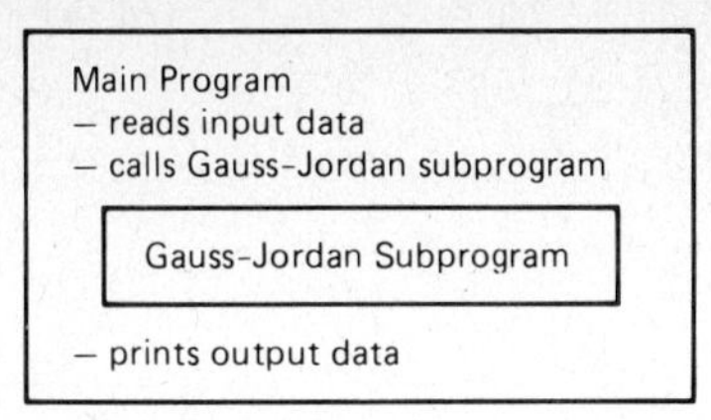

Figure 1.6 Program structure for solving a set of simultaneous equations.

solution of a differential equation is the Adams-Bashforth method. This is also presented in Chap. 5. It is called a predictor-corrector method. Here again, we see one of the advantages of the modular programming approach, since the Runge-Kutta subprogram is used to provide the initial solution values that the Adams-Bashforth method requires for its operation. In Chap. 6, the Runge-Kutta method is extended to the solution of sets of differential equations involving more than a single variable. The techniques used are similar to those introduced in Chap. 5, but arrays of variables are used rather than single variables. This, of course, greatly expands the generality of the approach.

The final type of time-domain operation that systems impose on physical variables is one characterized by a set of linear algebraic equations, usually referred to as a simultaneous set of equations. An example is the mesh or node equations relating the voltage and current variables of a network of resistors. In Chap. 7 a subprogram is provided to implement the solution of such a set of equations. It uses the Gauss-Jordan method. The overall implementation of a program applying this method is shown in Fig. 1.6.

1.2 THE FREQUENCY DOMAIN

The second part of this book is concerned with the frequency domain. It treats operations on mathematical transformations of physical variables. In general, these transformed variables are complex rather than real. Since few PASCAL implementations provide a complex variable type, in Chap. 8 a user-defined complex type is introduced. A set of subprograms is presented for implementing the various algebraic operations on these complex variables. The availability of these subprograms is assumed throughout the second part of the book.

Almost all computations in the frequency domain require operations on polynomials of the complex frequency variable s. Two of the most important of these are the differentiation of a polynomial and the evaluation of it for some complex value of its argument. In Chap. 8 these operations are implemented by two subprograms. Using these subprograms, one of the most important computations in the frequency domain can be programmed. This is the determination of the inverse Laplace transform. The program for doing this is modularized into three subprograms which are all applied to the data representing a rational function, that is, a ratio of polynomials in the complex frequency variable s. The first subprogram is a root-solving one that finds the poles of the rational function, that is, the zeros of its denominator polynomial. The second subprogram determines the residues associated with each of the poles; thus it finds the partial-fraction expansion. It uses the polynomial differentiation and evaluation subprograms referred to

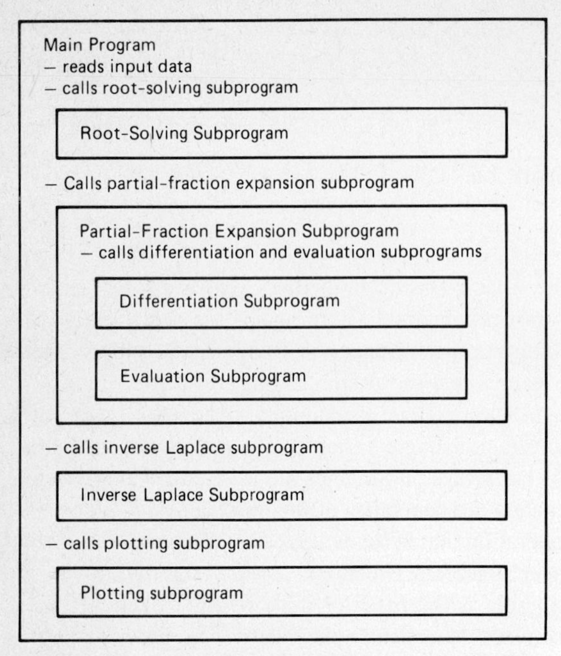

Figure 1.7 Program structure for inverse Laplace transform.

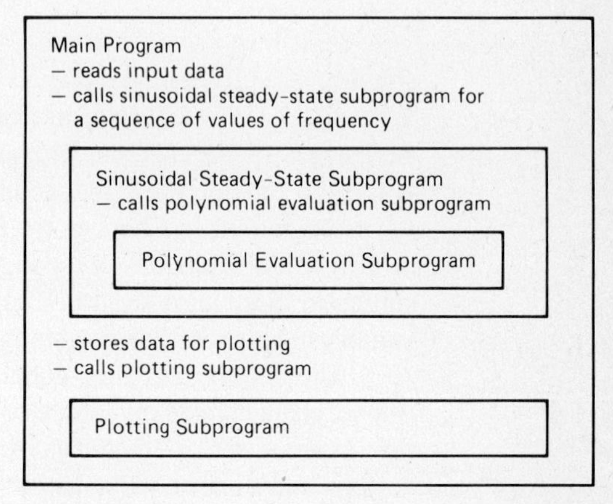

Figure 1.8 Program structure for sinusoidal steady-state analysis.

above. The third subprogram uses the poles and the residues to compute the inverse transform for a sequence of values of time. It also stores this data for plotting. The overall program has the structure shown in Fig. 1.7. The details are given in Chap. 8. Note that the plotting subprogram used to display the results is the same one used in the first part of the book.

Another important frequency-domain computation required for various types of engineering systems is the determination of sinusoidal steady-state response. This is treated in Chap. 9. Just as an iterative sequence of time values was used in Chaps. 2 to 6, here an iterative sequence of sinusoidal frequencies is required. These frequencies are used as input to the rational function describing the system, the same function used in Chap. 8. Here, however, the function is evaluated for the argument $j\omega$ rather than the complex argument s. The evaluation of the rational function is performed by separately evaluating the numerator and denominator polynomials, using the polynomial-evaluation subprogram introduced in Chap. 8. The resulting data is displayed as a plot. The overall program structure has the form shown in Fig. 1.8. Both linear and logarithmic (Bode) plots may be generated by varying the method used to generate the sequence of frequency values.

The advantages of the modular approach to program design are well illustrated by the treatment of Fourier series methods given in Chap. 10. The determination of the Fourier coefficients is accomplished using the integration subprogram presented in Chap. 2. The subprogram that defines the integrand is used to multiply the periodic function, for which the series is being determined, by the appropriate sine and cosine terms. The periodic function is, in turn, specified by its own subprogram. The form of the structure

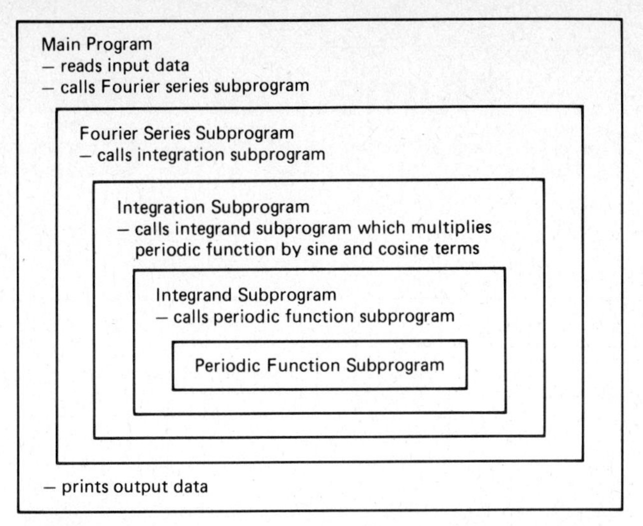

Figure 1.9 Program structure for finding Fourier coefficients.

of the program is shown in Fig. 1.9. After the Fourier coefficients have been determined, their modification, representing the effect of passing the periodic input signal through a specific system, is accomplished by applying the subprograms discussed in Chap. 9. Finally, the harmonic terms and their modified Fourier coefficients are summed to obtain a representation of the output waveform. The computations required in the overall process are generally prohibitive for ''hand'' processing. Thus, the use of the programs described here is especially attractive.

The subprograms introduced in Chap. 7 to obtain the solution of a set of simultaneous equations are extended to the sinusoidal steady-state case in Chap. 11. The subprograms used are similar to those of Chap. 7; however, they have been modified to use complex variables rather than real ones. These modified subroutines permit the use of sinusoidal steady-state analysis methods without the necessity of deriving the rational function for the system. As such, they provide an alternative approach to sinusoidal steady-state analysis which is useful in many situations.

Some final applications of frequency-domain analysis are given in Chap. 12 with the presentation of methods for determining Nyquist plots and root-locus plots. Finally, in Chap. 13, a description is given of the ways in which the subprogram modules developed in earlier chapters of this book can be combined to provide other types of engineering-system analyses.

1.3 CONCLUSION

In this chapter we have presented an introduction to one of the most important concepts of this book—the development of modular, well-structured programs for performing the computations most frequently encountered in analyzing engineering systems. In the following chapters we present detailed discussions of these techniques.

Chapter 2

Numerical Integration: The Inductor and the Capacitor

In this chapter we discuss the application of digital-computation techniques to the solution of integrals. As an example of such an application, we use the terminal relations for linear time-invariant inductors and capacitors. The techniques of numerical integration[1] that are introduced here are readily applicable to many other physical situations.

2.1 TERMINAL RELATIONS FOR THE INDUCTOR AND THE CAPACITOR

The inductor and the capacitor are network elements which may be referred to as *two-terminal* elements (the resistor is another two-terminal element). The terminal behavior of these elements may be characterized by defining a voltage variable $v(t)$ and a current variable $i(t)$, with the respective reference polarities shown in Fig. 2.1*a*. For the inductor, these variables are related by the integral equation

$$i(t_b) = \frac{1}{L}\int_{t_a}^{t_b} v(t)\,dt + i(t_a) \tag{2.1}$$

The variables, their reference polarities, and the symbol for an inductor are shown in Fig. 2.1*b*. Similarly, for a capacitor, the voltage and current variables are related by the equation

$$v(t_b) = \frac{1}{C}\int_{t_a}^{t_b} i(t)\,dt + v(t_a) \tag{2.2}$$

[1]This is also called *numerical quadrature*.

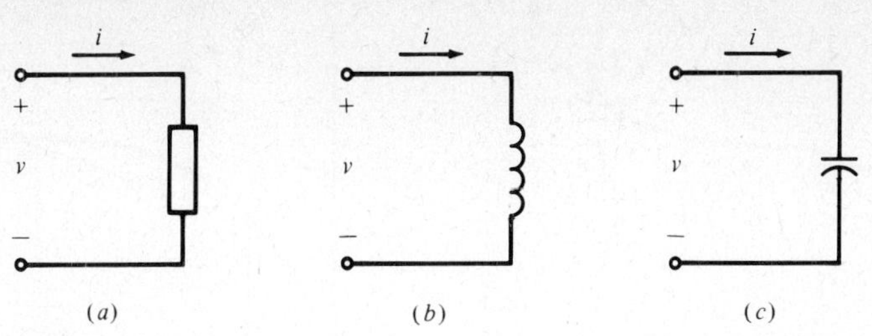

Figure 2.1 Voltage and current variables for two-terminal elements.

The variables, their reference polarities, and the symbol for a capacitor are shown in Fig. 2.1*c*.

If we compare (2.1) and (2.2), we see that both equations are examples of an integral relationship of the form

$$f(t_b) = K\int_{t_a}^{t_b} y(t)\,dt + f(t_a) \tag{2.3}$$

The integral in (2.3) is sometimes referred to as a *definite integral* since both limits of integration are fixed. With respect to (2.3), in the remainder of this chapter, we investigate the following problem:

Problem. Given $y(t)$ and $f(t_a)$ as shown in (2.3), find $f(t_b)$.

It is readily apparent that this can apply to the determination of the current through an inductor if the voltage applied across the terminals of the inductor is specified and if the current through the inductor at some time t_a is known. Similarly, the problem defined above can apply to the determination of the voltage that appears across the terminals of a capacitor if the current through the capacitor is specified and if the voltage across the capacitor at some time t_a is known. Many other physical applications will suggest themselves to the reader.

2.2 NUMERICAL INTEGRATION

The solution of the problem posed in Sec. 2.1 requires some means of integrating the function $y(t)$ between the limits t_a and t_b. Of course, if $y(t)$ is some simple function such as t, t^2, or $\sin t$, the integration can be performed directly through the use of well-known integration relations, integral tables, etc. These, however, are of little use when $y(t)$ is not defined by an easily integrable mathematical relation. In such a case, it is frequently more advantageous to use techniques of numerical integration.

As an example of a numerical-integration technique, consider the function $y(t)$ shown in Fig. 2.2*a*. The definite integral of this function between the limits of t_a and t_b is equal to the shaded area A under the curve. An approximation to this area may be

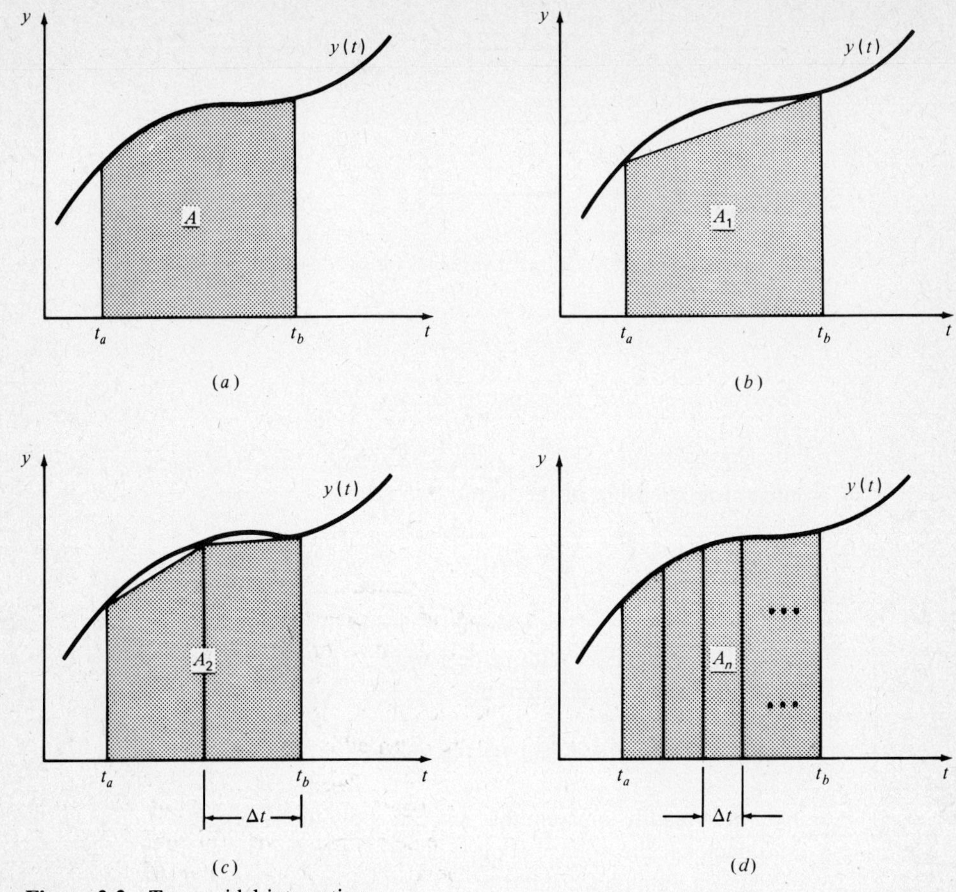

Figure 2.2 Trapezoidal integration.

made by forming the trapezoid shown in Fig. 2.2*b*. If we define the area of the trapezoid as A_1 and define $\Delta t = t_b - t_a$, we see that

$$A_1 = \left[\frac{y(t_a) + y(t_b)}{2}\right]\Delta t \tag{2.4}$$

An even better approximation to the area A can be found by dividing the interval from t_a to t_b into two equal segments and forming two trapezoids, as shown in Fig. 2.2*c*. If we call the area found by this approximation A_2, we see that

$$A_2 = \left[\frac{y(t_a) + y(t_a + \Delta t)}{2}\right]\Delta t + \left[\frac{y(t_a + \Delta t) + y(t_b)}{2}\right]\Delta t \tag{2.5}$$

where $\Delta t = (t_b - t_a)/2$. The above relation may be rewritten in the form

$$A_2 = \left[\frac{y(t_a)}{2} + y(t_a + \Delta t) + \frac{y(t_b)}{2}\right]\Delta t \tag{2.6}$$

The above approach is easily extended to the case where n trapezoidal sections are used

n	A_n
2	7.00000
5	6.72000
10	6.68000
20	6.67000
50	6.66720
100	6.66680

Figure 2.3 The effect of increasing the number of iterations.

to approximate area A, as shown in Fig. 2.2*d*. If we let A_n be the total area defined by the n trapezoids shown in this figure, then

$$A_n = \left[\frac{y(t_a)}{2} + \sum_{i=1}^{n-1} y(t_a + i\,\Delta t) + \frac{y(t_b)}{2}\right]\Delta t \tag{2.7}$$

where $\Delta t = (t_b - t_a)/n$. In the limit, as n becomes large, area A_n approaches area A. As an illustration of this, consider the function $y(t) = 1 + t + t^2$. If we let $t_a = 0$ and $t_b = 2$, the result of applying the above process for various values of n is shown in Fig. 2.3. The actual value of the integral is easily seen to be $6\frac{2}{3}$.

The procedure described above is usually referred to as *trapezoidal integration*.[2] The error in determining the integral is called *truncation error*. It is proportional to $(\Delta t)^2$. Thus, doubling the number of iterations reduces the error by approximately one-quarter. As an example of this, in Fig. 2.3, the error for $n = 10$ is 0.01333 while the error for $n = 20$ is 0.00333. Note that $0.00333/0.01333 \approx \frac{1}{4}$. If a very large number of trapezoidal sections is used, or if the computer word length is small, then another error, called *round-off error,* may need to be considered. An additional discussion of these errors may be found in any of the books on numerical analysis methods listed in the bibliography. Although there are many more sophisticated numerical-integration schemes, trapezoidal integration has the advantage of extreme simplicity both in concept and in implementation. In the next section we describe a means for performing trapezoidal integration on the digital computer.

2.3 THE PROCEDURE IntegTrpz

In this section we describe a digital-computer program for performing trapezoidal integration as defined by (2.7). To permit us to make as general a usage as possible of this program, we prepare it as a procedure, which can be "called" by the main program.[3] Examination of (2.7) shows that the procedure must be supplied with the following information:

1. The initial value t_a of the independent variable t. This input to the procedure will be given the variable name timeA in the parameter listing for the procedure.
2. The final value t_b of the independent variable t. We name this variable timeB.
3. The number of trapezoidal sections n that it is desired to use between the values t_a and t_b. We name this variable n.

[2]This is the simplest of a series of integration methods specified by the Newton-Cotes formulas.

[3]For a review of the properties of subprograms and their usage in PASCAL programming see Appendix A.

4. The value of the integral as found by the procedure. We name this output variable area.
5. The function $y(t)$. This function may take any of several forms; for example, it may be a piecewise-linear function, or it may be discontinuous. In order to accommodate as many forms of $y(t)$ as possible, we simply define it as a separate subprogram of the "function" type. Thus, in our program we specify a subprogram FUNCTION Y (t : real) : real;, which will be called by the trapezoidal-integration procedure whenever needed. In addition, we call the function by using a function parameter in the procedure heading.[4]

From the above considerations, our parameter list for the trapezoidal-integration procedure must be of the form

```
(timeA, timeB : real; n : integer; VAR area : real; FUNCTION Y
  (t : real) : real)
```

Let us use the name IntegTrpz (for *integ*ration by *trap*ezoidal method) for the procedure. Thus, the procedure will be declared by the PASCAL heading

```
PROCEDURE IntegTrpz (timeA, timeB : real; n : integer; VAR area : real;
                     FUNCTION Y (t : real) : real);
```

A listing and a flowchart of the procedure are given in Fig. 2.4. A comparison of the statements given in this figure with (2.7) should make the purpose of the various PASCAL statements clear.

For future reference, a summary of the important features of the procedure IntegTrpz is given in Table 2.1. Example 2.1 illustrates its use.

TABLE 2.1 SUMMARY OF THE CHARACTERISTICS OF THE PROCEDURE IntegTrpz

Identifying Statement PROCEDURE IntegTrpz (timeA, timeB : real; n : integer; VAR area : real; FUNCTION Y (t : real) : real);

Purpose To find the definite integral of a specified function $y(t)$, that is, to find a, where

$$a = \int_{ta}^{tb} y(t)\,dt$$

by using trapezoidal integration.

Additional Subprograms Required This procedure calls the function identified by the statement

```
FUNCTION Y (t : real) : real;
```

This function must be used to define $y(t)$, that is, the integrand.

Input Arguments

timeA	The lower limit t_a of the variable of integration t
timeB	The upper limit t_b of the variable of integration t
n	The number n of trapezoidal segments used to approximate the area, that is, the number of iterations used in performing the trapezoidal integration

Output Argument

area The value of the integral

[4]If the implementation of PASCAL on which these programs are being run does not permit the use of function parameters, this parameter may be omitted from the parameter list and a globally defined function $y(t)$ used in its place. The process is illustrated in Appendix C.

```
PROCEDURE IntegTrpz (timeA, timeB : real; n : integer; VAR area : real;
                    FUNCTION Y (t : real) : real);

(* Trapezoidal integration procedure
     timeA - Initial value of independent variable t
     timeB - Final value of independent variable t
     n - Numoer of trapezoidal sections used
     area - Value of integral from timeA to timeB
   A function Y(t) must be used to define the integrand. *)

   VAR
      t, dt : real;
      i     : integer;

   BEGIN

      (*   Initialize the value of time t to the value timeA
           and calculate dt, the increment in time    *)

      t := timeA;
      dt := (timeB - timeA) / n;

      (*   Initialize the value of the integral    *)

      area := Y(timeA) / 2.0;

      (*   Compute the n - 1 terms in the summation    *)

      IF N <> 1 THEN
         BEGIN
            FOR i := 1 to n-1 DO
               BEGIN
                  t := t + dt;
                  area := area + Y(t)
               END
         END;

      (*   Finisn tne computation of the integral    *)

      area := (area + Y(timeB) / 2.0) * dt
   END;
```

(*a*)

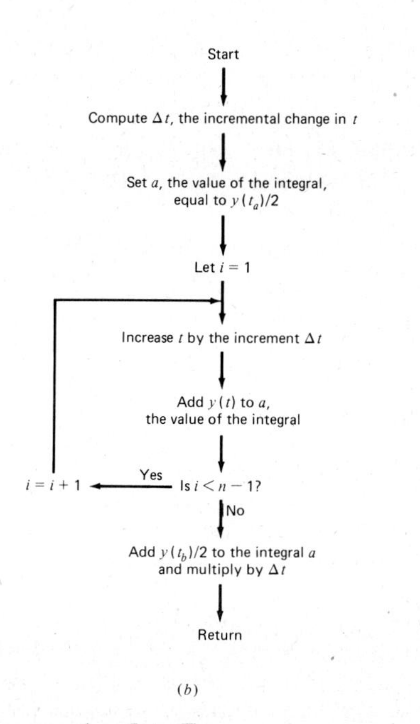

(*b*)

Figure 2.4 The procedure IntegTrpz.

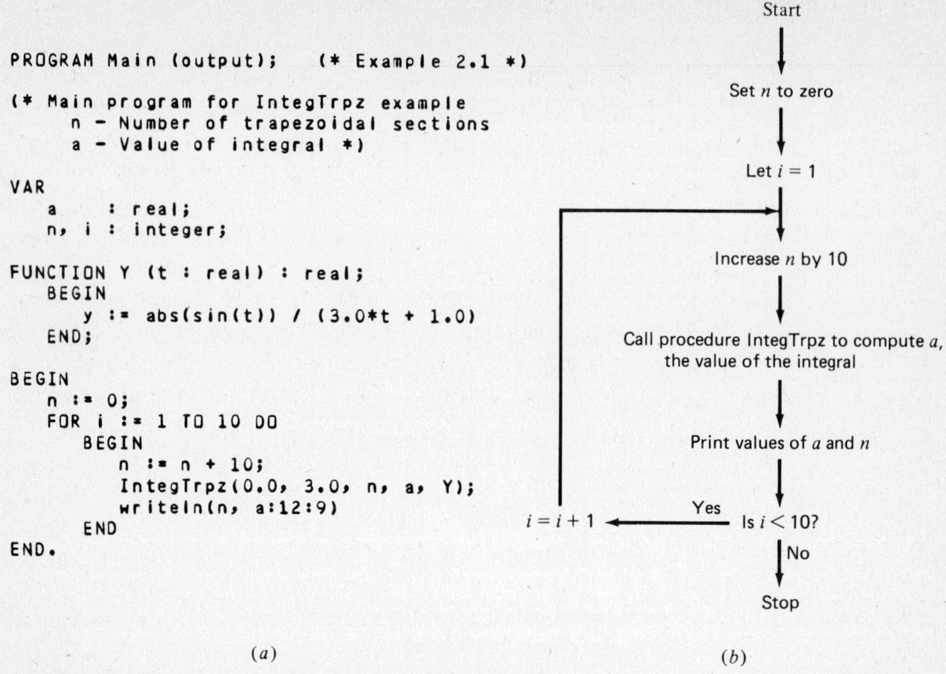

```
PROGRAM Main (output);    (* Example 2.1 *)

(* Main program for IntegTrpz example
     n - Number of trapezoidal sections
     a - Value of integral *)

VAR
    a     : real;
    n, i  : integer;

FUNCTION Y (t : real) : real;
    BEGIN
       y := abs(sin(t)) / (3.0*t + 1.0)
    END;

BEGIN
    n := 0;
    FOR i := 1 TO 10 DO
       BEGIN
          n := n + 10;
          IntegTrpz(0.0, 3.0, n, a, Y);
          writeln(n, a:12:9)
       END
END.
```


(*a*) (*b*)

Figure 2.5 Function and main program for Example 2.1.

EXAMPLE 2.1

If a voltage defined as $v(t) = |(\sin t)/(3t + 1)|$ V is applied to the terminals of a 1-H inductor, and if the current $i(t)$ through the inductor is zero at $t = 0$, that is, if $i(0) = 0$, we may use IntegTrpz to find the current $i(t)$ at $t = 3$ sec. To do this, we need a PASCAL function to compute the term $|(\sin t)/(3t + 1)|$. In addition, a main program must be constructed to set up values for ta (=0.0), tb (=3.0), and n, and to call the procedure. A listing and a flowchart of this function and main program are given in Fig. 2.5. For brevity, the listing of IntegTrpz has not been repeated in this figure. The program has been set up to provide a series of solutions for different values of the constant n. The output from this program, showing the values of a provided by the subroutine for various values of the iteration constant n, is given in Fig. 2.6. It is evident that at least n = 100 is needed for four-digit accuracy.

n	a
10	0.407907241
20	0.413690167
30	0.414812223
40	0.415209653
50	0.415394489
60	0.415495135
70	0.415555904
80	0.415595380
90	0.415622460
100	0.415641837

Figure 2.6 Output listing for Example 2.1.

2.4 THE ROMBERG INTEGRATION METHOD

One of the disadvantages of the trapezoidal-integration method presented in Sec. 2.2 is the lack of provision for the user to specify the accuracy with which the value of the integral is to be obtained. In this section we describe a method of numerical integration which solves this problem. The method also has the advantage that it converges to the correct answer more rapidly, that is, using fewer evaluations of the function which is being integrated, than does the trapezoidal one. It is called the Romberg method.

The *Romberg integration method* is based on two different computations. The first of these is the calculation of a sequence of trapezoidal integrations starting with a single section ($n = 1$) and continuing with other integrations in which the number of sections is doubled in each succeeding one. The results are stored in the first column of a two-dimensional array with elements R_{ij}, that is, in the elements R_{i1} (note that the second subscript is 1). Thus, from (2.4) through (2.7), using the terminology of Sec. 2.2, we obtain

$$\begin{aligned} R_{11} = A_1 &= \left[\frac{y(t_a)}{2} + \frac{y(t_b)}{2}\right]\Delta t & \Delta t &= t_b - t_a \\ R_{21} = A_2 &= \left[\frac{y(t_a)}{2} + y(t_a + \Delta t) + \frac{y(t_b)}{2}\right]\Delta t & \Delta t &= \frac{t_b - t_a}{2} \\ R_{31} = A_4 &= \left[\frac{y(t_a)}{2} + \sum_{i=1}^{3} y(t_a + i\,\Delta t) + \frac{y(t_b)}{2}\right]\Delta t & \Delta t &= \frac{t_b - t_a}{4} \end{aligned} \tag{2.8}$$

More generally

$$R_{i1} = A_m = \left[\frac{y(t_a)}{2} + \sum_{i=1}^{m-1} y(t_a + i\,\Delta t) + \frac{y(t_b)}{2}\right]\Delta t \tag{2.9}$$

where

$$m = 2^{i-1} \qquad \Delta t = \frac{t_b - t_a}{m} \tag{2.10}$$

The second computation used in the Romberg method is an extrapolation called the *Richardson extrapolation* which is used to speed the convergence process. This is applied to determine the elements R_{ij} of the rows of the two-dimensional array to the right of the first column elements R_{i1}. The Richardson extrapolation is specified as

$$R_{ij} = \frac{4^{j-1}R_{i,j-1} - R_{i-1,j-1}}{4^{j-1} - 1} \qquad \text{for } i = 2, 3, 4, \ldots, n \quad j = 2, 3, \ldots, i \tag{2.11}$$

The combination of the two computational steps is performed on a row-by-row basis and thus generates the elements of a triangular array having the form

Number of trapezoidal sections	Result of trapezoidal integration	Result of Richardson extrapolation			
1	R_{11}				
2	R_{21}	R_{22}			
4	R_{31}	R_{32}	R_{33}		
8	R_{41}	R_{42}	R_{43}	R_{44}	
.	.	.	.	.	
.	.	.	.	.	
.	.	.	.	.	
2^{n-1}	R_{n1}	R_{n2}	R_{n3}	R_{n4}	$\cdots R_{nn}$

(2.12)

The diagonal terms R_{jj} of this array can be shown to converge to the value of the integral more rapidly than the terms resulting from trapezoidal integration alone. Thus, for a given value of n, that is, a given number of rows of the array of elements R_{ij} shown in (2.12), the best value of the integral is given by the quantity R_{nn}. The error of the elements in the first column of the array, as already shown for trapezoidal integration, is proportional to $(\Delta t)^2$. The error of the elements in succeeding columns of the array is proportional to $(\Delta t)^{2j}$, where j is the number of the column.

In implementing the Romberg method, the specification of an error tolerance is readily made by testing the values R_{nn} and $R_{n-1,n-1}$ and stopping the integration process when

$$|R_{nn} - R_{n-1,n-1}| < \text{errmax} \tag{2.13}$$

where errmax is some specified maximum error tolerance. In the next section we describe a means for implementing the Romberg integration method on the digital computer.

2.5 THE PROCEDURE Romberg

In this section we describe a PASCAL procedure for performing Romberg integration as defined by Eqs. 2.9, 2.10, and 2.11. The procedure must be supplied with the following inputs:

1. The initial value t_a of the independent variable t. This input to the procedure is given the variable name timeA in the parameter listing for the procedure.
2. The final value t_b of the independent variable. We define this variable as timeB.
3. The maximum number of rows n to be generated in the array of elements R_{ij}. This input provides one of the limits used to terminate the integration process. We name this variable nmax. Its maximum value is 20.
4. The value of the integral as computed by the procedure. This is the element R_{jj} of the two-dimensional array for the largest value of j used in the computation. We name this variable area.

5. The name of the PASCAL subprogram defining the function $y(t)$ which is to be integrated.
6. The maximum value of the error that is desired in performing the integration. This input provides the second limit used to terminate the integration process. We name this variable errmax.
7. A parameter allowing the user to determine whether the integration process terminated because the maximum value of n was reached, or because the maximum-error criterion was satisfied. The parameter is implemented by the variable test. If test is set to 1, an output message will be printed giving the final values of n and the final error as determined by the left member of (2.13). If test is set to 0, no output message will be given.

Using the name Romberg for the PASCAL procedure, and using the input variable names listed above, we may use the following heading to identify the procedure:

```
PROCEDURE Romberg (timeA, timeB : real; nmax : integer;
                   VAR area : real; FUNCTION Y (t : real) : real;
                   errmax : real; test : integer);
```

A listing and a flowchart of the procedure are given in Fig. 2.7. Note that it calls the procedure IntegTrpz defined in Sec. 2.3. A summary of the important features of the procedure Romberg is given in Table 2.2. Example 2.2 illustrates its use.

TABLE 2.2 SUMMARY OF THE CHARACTERISTICS OF THE PROCEDURE Romberg

Identifying Statement PROCEDURE Romberg (timeA, timeB : real; nmax : integer; VAR area: real; FUNCTION Y (t : real) : real; errmax : real; test : integer);

Purpose To find the definite integral of a specified function $y(t)$, that is, to find a where

$$a = \int_{ta}^{tb} y(t)\,dt$$

by using Romberg integration.

Additional Subprograms Required This procedure calls the procedure IntegTrpz and the function $Y(t)$.

Input Arguments

timeA	The lower limit t_a of the variable of integration t
timeB	The upper limit t_b of the variable of integration t
nmax	The maximum number of rows n generated in the **R** matrix
errmax	The maximum value of error between successive computations of the integral required before stopping the integration
test	A parameter that produces (test = 1) an output statement giving the final values of rows n and error. For test = 0 no output is given.

Output Argument

area	The value of the integral

Note: The maximum value permitted for nmax is 20.

```
PROCEDURE Romberg (timeA, timeB : real; nmax : integer; VAR a : real;
                  FUNCTION Y (t : real) : real; errmax : real;
                  test : integer);

(* Romberg integration procedure
     timeA - Initial value of independent variable t
     timeB - Final value of independent variable t
     nmax - Maximum number of rows of R array
     a - Value of integral from timeA to timeB
     errmax - Maximum value of error used as limit
     test - Output parameter, set to 1 to output
        stopping conditions, otherwise set to 0
   A function Y(t) must be used to define the integrand.
   This procedure calls the procedure IntegTrpz.             *)

   VAR
      error, fac : real;
      i, j, m    : integer;
      r          : ARRAY[1..20,1..20] OF real;

   BEGIN

      (*   Use the procedure IntegTrpz to find the array element r11 *)

      IntegTrpz(timeA, timeB, 1, r[1,1], Y);
      i := 1;  m := 1;

      (*   Find the other rows of the r array  *)

      REPEAT
         i := i + 1;  m := m * 2;
         fac := 1.0;

         (*   Use IntegTrpz to find the first element in row *)

         IntegTrpz(timeA, timeB, m, r[i,1], Y);

         (*   Use extrapolation to find other elements in row *)

         FOR j := 2 TO i DO
            BEGIN
               fac := fac * 4.0;
               r[i,j] := (fac * r[i,j-1] - r[i-1,j-1]) / (fac - 1.0)
            END;
         error := abs(r[i,i] - r[i-1,i-1]);

      (* Check terminating conditions for REPEAT/UNTIL loop *)

      UNTIL (i >= nmax) OR (error < errmax);

      (* Check to see if terminating conditions are to be printed *)

      IF test > 0 THEN
         writeln(' Romberg, n =', i:3, '    error =', error);

      (* Store output from R array in variable 'a' *)

      a := r[i,i]
   END;
```

(*a*)

Figure 2.7 Listing and flowchart for the procedure Romberg.

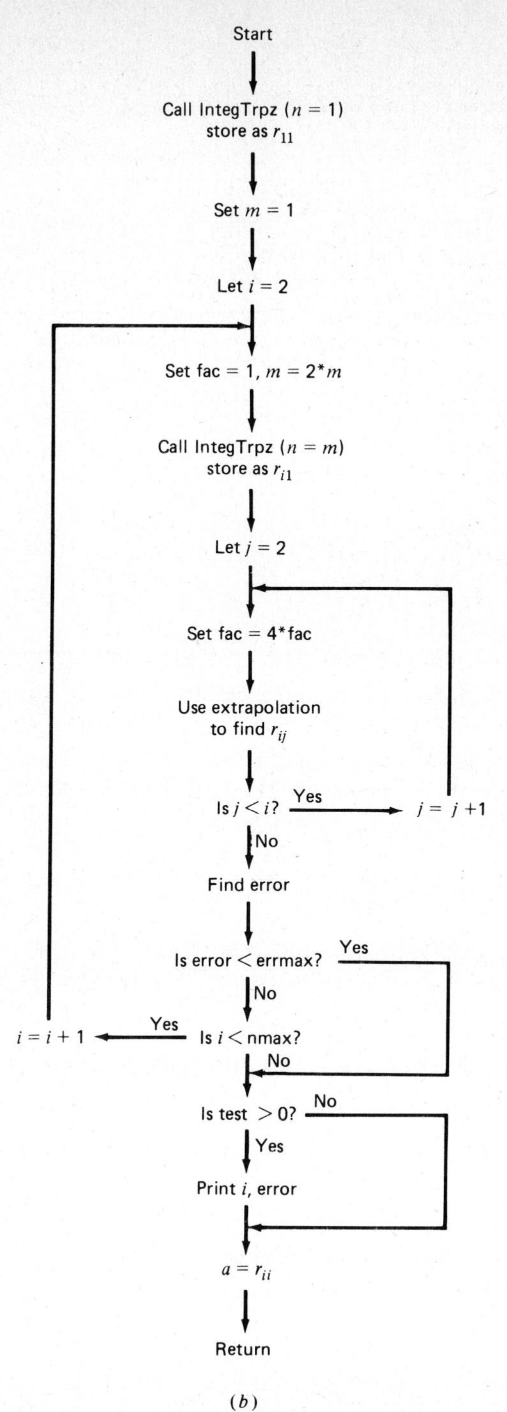

(*b*)

Figure 2.7 (*Continued*)

```
PROGRAM Main (output);     (* Example 2.2 *)

(* Main program for Romberg example
    n - Number of trapezoidal sections
    a - Value of integral *)

VAR
   a : real;
   i : integer;

FUNCTION Y (t : real) : real;
   BEGIN
      y := abs(sin(t)) / (3.0*t + 1.0)
   END;

BEGIN
   FOR i := 2 TO 5 DO
      BEGIN
         writeln;
         Romberg(0.0, 3.0, i, a, Y, 1.0E-04, 1);
         writeln('            n =', i:3, '      area =', a)
      END
END.
```

Figure 2.8 Function and main program for Example 2.2

EXAMPLE 2.2

Here we use the procedure Romberg to solve the problem given in Example 2.1. A listing of the program is given in Fig. 2.8. A value of nmax of 5 and a maximum error of 10^{-4} were used. The program stopped on the maximum value of n. The output is shown in Fig. 2.9. The **R** matrix is shown in Fig. 2.10. The number of calls of the function $y(t)$ necessary to obtain the final answer was 36. The data given in Fig. 2.6 shows that almost 100 calls of $y(t)$ were required to obtain the same answer using IntegTrpz alone. This readily verifies the improved convergence obtained by using the Romberg method.

2.6 CONCLUSION

In this chapter we have introduced the concept of calculating the definite integral of a specified function $y(t)$ [as defined by a PASCAL function $Y(t)$] through the use of numerical-integration procedures. There are many other approaches to the integration prob-

```
ROMBERG, N =  2    ERROR =  3.48613448868822E-001
         N =  2      AREA =  3.69781450077720E-001

ROMBERG, N =  3    ERROR =  3.68634209716031E-002
         N =  3      AREA =  4.06644871048800E-001

ROMBERG, N =  4    ERROR =  7.87529896187071E-003
         N =  4      AREA =  4.14520170010671E-001

ROMBERG, N =  5    ERROR =  1.11383909317221E-003
         N =  5      AREA =  4.15634009105841E-001
```

Figure 2.9 Output listing for Example 2.2.

```
0.021168001
0.282626088 0.369781450
0.373912702 0.404340907 0.406644871
0.403804630 0.413768605 0.414397118 0.414520170
0.412572623 0.415495238 0.415610400 0.415629658 0.415634009
```

Figure 2.10 **R** matrix for Example 2.2.

lem than the trapezoidal and Romberg ones discussed here. For example, gaussian quadrature allows the use of unequally spaced increments of time (another method of treating this case is given in Chap. 4). Trapezoidal and Romberg integrations, however, will prove satisfactory for the majority of engineering and scientific applications. Because the integration processes are implemented by the use of PASCAL procedures, any application requiring a more sophisticated integration routine is easily accommodated by the direct substitution in the main program of a procedure that implements the new integration routine. Thus, the subprogram concept provides the user with considerable flexibility in the design of programs.

PROBLEMS

2.1. **(a)** A current $i(t) = 20 \sin 2\pi t$ A is applied to a capacitor of value 0.26 F. If $v(t)$ is the voltage across the terminals of the capacitor and if $v(0) = 0$, determine the value of $v(0.5)$, that is, the value of $v(t)$ evaluated at $t = 0.5$, by using the procedure IntegTrpz. Determine the result for the following values of the iteration constant n: 10, 20, 50, 100.

(b) Calculate the error in the resulting value of $v(t)$ for each of the values of n by computing the integral by nonnumerical methods.

2.2. Across a two-terminal element of the type shown in Fig. 2.1*a*, the voltage and current are defined by the relations

$$v(t) = 10 + 5t \quad \text{V}$$
$$i(t) = 60 \sin 3\pi t \quad \text{A}$$

If the energy stored in the element is defined as $w(t)$ and if $w(0) = 0$, use the procedure IntegTrpz to find the energy in joules stored in the circuit at $t = 10$ sec; that is, find $w(10)$ [use the relation $w(t) = \int_0^t v(t)i(t)\,dt$].

2.3. A voltage $v(t) = 10e^{-t} \cos 4\pi t$ V is applied to a 10,000-Ω linear time-invariant resistor. If the total energy dissipated in the resistor is defined as $w(t)$, use the procedure IntegTrpz to find the amount of energy in joules dissipated in the resistor from $t = 1$ sec to $t = 4$ sec. [Use the relation

$$w(t) = 1/R \int_{ta}^{tb} v^2(t)\,dt \qquad \text{[Assume } w(1) = 0.]$$

2.4. When trapezoidal-integration routines are used on discontinuous functions, the use of larger values of the iteration constant n may actually produce poorer results. As an example of this, compute the integral of the function $y(t)$ shown in Fig. P2.4 for the following values of n: 10, 20, 50, and 100. Explain the results obtained.

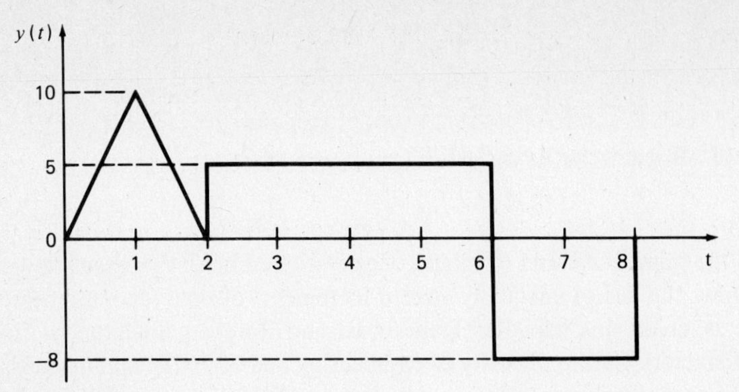

Figure P2.4

2.5. A more accurate integral formula (and also one that requires more computing time) is provided by Simpson's rule. This algorithm uses a parabolic approximation between successive values of the function being integrated. One form of this rule is

$$f(t_b) = \int_{ta}^{tb} y(t)\,dt$$

$$\cong (y_1 + 4y_2 + 2y_3 + 4y_4 + \cdots + 2y_{n-1} + 4y_n + y_{n+1})\frac{\Delta t}{3}$$

where $y_i = y[t_a + (i - 1)\,\Delta t]$, $\Delta t = (t_b - t_a)/n$, and n is even. Write a PASCAL procedure Isimp with the same argument listing as that used for the procedure IntegTrpz to perform integration by means of Simpson's rule. Check the operation of your program by applying it to any of the problems given for this chapter and comparing the results with those obtained by trapezoidal integration.

2.6. A voltage $v(t) = 40e^{-t/4}|\sin \pi t|$ V is applied to the terminals of a 1-H inductor. If $i(t)$ is the current through the inductor and if $i(0) = -50$ A, use the procedure IntegTrpz to find the current $i(5)$ in amperes. Determine the result for the following values of the iteration constant n: 10, 20, 50, and 100.

2.7. It may be shown that the integral of two sinusoidal functions, one of which is a harmonic of the other, is zero, in other words, that

$$\int_0^{2m\pi} \sin t \sin nt\,dt = 0 \qquad \begin{array}{l} n = 2, 3, 4, \ldots \\ m = 1, 2, 3, 4, \ldots \end{array}$$

Verify this result numerically by using the integration procedure IntegTrpz to find

$$\int_0^{2\pi} \sin t \sin 2t\,dt$$

Use values of 5, 10, 20, and 50 for the iteration constant n.

2.8. A series of sinusoids may be used to represent an arbitrary periodic function. Such a series is known as a *Fourier* series. For a representation of a square-wave function $f(t)$ defined by the relations $f(t) = 40$, $0 \le t \le 50$, and $f(t) = -40$, $50 < t \le 100$, the Fourier series will have the form

$$f(t) \cong g(t) = c_1 \sin\frac{2\pi}{100}t + c_2 \sin\frac{4\pi}{100}t + c_3 \sin\frac{6\pi}{100}t + \cdots$$

where the coefficients c_n are determined by the relations

$$c_n = \frac{1}{50}\int_0^{100} f(t) \sin \frac{2\pi}{100} nt \, dt$$

Use the trapezoidal-integration procedure IntegTrpz to find the coefficients c_1 to c_3. Use a value of 100 for the iteration constant n.

2.9. A time-varying resistor has a value of resistance given by the relation $R(t) = 1000 + 200 \sin 2000\pi t \ \Omega$. If a voltage $v(t) = 100 \sin 4000\pi t$ V is applied over a range of t from 0 to 0.0005 sec, find the energy in joules dissipated in the resistance. Use values of n of 10, 20, 50, and 100 {use the relation

$$w(t) = \int_0^t [v^2(t)/R(t)] \, dt\}.$$

2.10. In many physical situations is necessary to integrate a given function twice. For example, consider the determination of the energy stored in a capacitor by a specified input current. The voltage across the capacitor is given as $1/C$ times the integral of the current, and the stored energy is given as the integral of the product of the current and the voltage. Thus, if $w(t)$ is the energy, it may be expressed as

$$w(t) = \int_0^t i(t)\left[\left(\frac{1}{C}\right)\int_0^t i(t)\, dt\right] dt$$

where $w(t)$ is in joules. Using the PASCAL procedure IntegTrpz and taking advantage of the function parameter in its parameter list, implement this equation to find the energy stored in a 0.01-F capacitor over the range of time from 0 to 1 sec by a current $i(t) = 0.0001t$ A. Check your result using the relation $w(t) = 0.01\ [v^2(t)]/2$.

2.11. The rms (root-mean-square) value of a periodic function is found by the relation

$$F_{\text{rms}} = \sqrt{\frac{1}{T}\int_0^T f^2(t)\, dt}$$

where $f(t)$ is an arbitrary periodic function of period T sec. Use the procedure IntegTrpz to find the rms value of a function, one cycle of which is shown in Fig. P2.11.

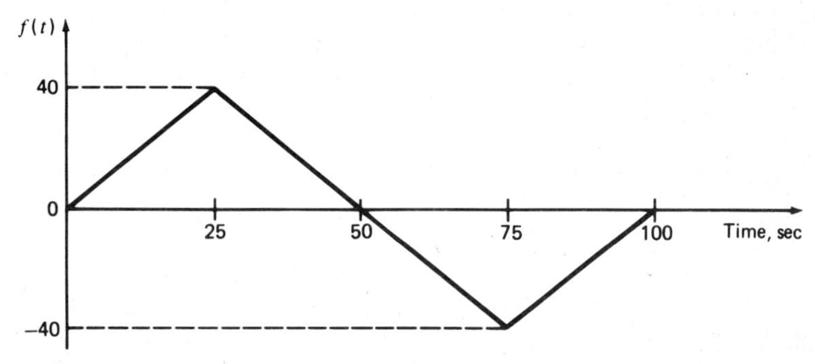

Figure P2.11

2.12. Repeat Prob. 2.1 using the procedure Romberg. Use values of 2, 3, 4, and 5 for nmax. For each of the values of nmax compare the approximate number of calls of the integrand required to obtain the same answer using IntegTrpz.

2.13. If the function being integrated has multiple nulls which coincide with the boundaries of the trapezoids, the low-order trapezoidal integrations used in the Romberg method may give inconsistent results. As an example of this, repeat Prob. 2.2 using the procedure Romberg.

2.14. Repeat Prob. 2.3 using the procedure Romberg.

2.15. Repeat Prob. 2.6 using the procedure Romberg. Use values of 2, 3, 4, and 5 for nmax. For each of the values of nmax compare the approximate number of calls of the integrand to obtain the same answer using IntegTrpz.

Chapter 3

Solution of Indefinite Integrals by Numerical Integration: Plotting

The type of problem that was formulated in the preceding chapter entailed the solution of definite integrals by numerical integration. In solving this problem, we are usually interested only in one piece of data—the value of the integral at the upper limit specified for the variable of integration. In this chapter we consider the more general problem of the indefinite integral, where, by the term ''indefinite,'' we imply that we are interested not only in the end point of the integration process but in the intermediate points as well.

3.1 DEFINITE AND INDEFINITE INTEGRALS

In Chap. 2 we discussed the use of digital-computation techniques in the evaluation of definite integrals. The problem discussed was, given $y(t)$ and $f(t_a)$, to find $f(t_b)$, where

$$f(t_b) = \int_{t_a}^{t_b} y(t)\, dt + f(t_a) \tag{3.1}$$

To solve this problem we employed digital-computation techniques, namely, the integration procedures IntegTrpz and Romberg to evaluate the definite integral given in (3.1). Frequently, in physical problems, we are concerned not only with determining the value of a function at some specific value of its argument but also with determining its behavior over a range of values of the independent variable. Thus we are concerned with the problem of finding $f(t)$ over some range of t, assuming $y(t)$ and $f(t_a)$ are known, where

$$f(t) = \int_{t_a}^{t} y(t)\, dt + f(t_a) \qquad t > t_a \tag{3.2}$$

We may now formulate the following problem:

Problem. Given $y(t)$ and $f(t_a)$ as shown in (3.2), find $f(t)$ for a specified range of t, where $t > t_a$.

The above formulation can apply to the problem of finding the current through an inductor if the voltage applied to the inductor is specified and if the current at some value of time is known. Similarly, it can apply to the problem of finding the voltage across a capacitor if the current through the capacitor is specified and if the voltage at some value of time is known. There are many other physical situations to which this problem applies.

3.2 TREATING THE INDEFINITE INTEGRAL AS A SERIES OF DEFINITE INTEGRALS

An indefinite integral of the form given in (3.2) may be solved explicitly only if $y(t)$ is a function which is easily mathematically integrable. For example, if $y(t) = t^2$ and $t_a = 0$, then $f(t_a) = 0$ and $f(t) = t^3/3$. Since digital-computation techniques, of necessity, deal with discrete data, that is, numbers, it is not possible, in general, to use them to obtain solutions in such a closed form. It is, however, possible to approximate such a solution by generating a sequence of numbers giving the value of $f(t)$ at successive values of the independent variable t. Such an approach has the advantage that it is

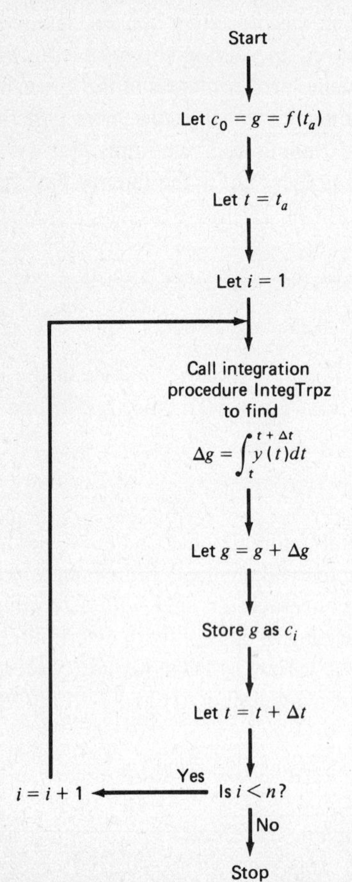

Figure 3.1 Flowchart for indefinite-integral problems.

applicable to all types of functions $y(t)$, not just to those which are in easily integrable mathematical form.

To provide such a sequence of values of $f(t)$, we need only make repeated application of the techniques of numerical integration described in Chap. 2. Thus we can solve problems involving indefinite integrals by the use of techniques for solving definite integrals. The detailed procedure for doing this is illustrated by the flowchart given in Fig. 3.1. This flowchart illustrates the procedure for finding $n + 1$ values of $f(t)$, for $t > t_a$, and storing these values in a one-dimensional array of variables $c_i (i = 0, 1, 2, \ldots, n)$ where

$$c_i = f(t_a + i\,\Delta t)$$

and where Δt is the desired spacing of the independent variable for which successive values of $f(t)$ are desired. The loop indicated in the figure, which makes successive computations for all values of i from 1 to n, is easily programmed by a FOR statement. Note that, since the successive increments Δt over which the function is integrated are, in general, small, the procedure IntegTrpz is used rather than the procedure Romberg. The general method is best illustrated by an example.

EXAMPLE 3.1

If a voltage defined as $v(t) = 20(1 - \cos 2\pi t)$ V is applied to the terminals of a 1-H inductor and if the current $i(t)$ through the inductor is zero at $t = 0$, that is, if $i(0) = 0$, we may use the procedure IntegTrpz described in Chap. 2 and the flowchart shown in Fig. 3.1 to find the values of $i(t)$ (in amperes) for 51 successive values of t, spaced 0.1 sec apart, starting from $t = 0$ and finishing at $t = 5$. A listing of a PASCAL program for accomplishing this is given in Fig. 3.2. A listing of the output values of t and $i(t)$ is

```
PROGRAM Main (output);      (* Example 3.1 *)

(* Main program for indefinite integral example
      t - time
      amps - value of inductor current at each time
      c - array of values of inductor current     *)

CONST twoPi = 6.2831853;
VAR
   t, dt, amps, damps : real;
   c                  : ARRAY[0..50] OF real;
   i                  : integer;

FUNCTION Volts (t : real) : real;
   BEGIN Volts := 20.0 * (1.0 - cos(twoPi * t)) END;

BEGIN
writeln('1');
   amps := 0.0; t := 0.0; dt := 0.1; c[0] := 0.0;
   writeln(t:5:1, c[0]);
      FOR i := 1 TO 50 DO
         BEGIN
            IntegTrpz(t, t+dt, 2, damps, Volts);
            amps := amps + damps;
            t := t + dt;
            c[i] := amps;
            writeln(t:5:1, c[i])
         END
END.
```

Figure 3.2 Main program and function for Example 3.1.

t	$i(t)$
0.0	0.0000000000000E+000
0.1	1.4443498619544E-001
0.2	9.9763273776292E-001
0.3	2.9976327342610E+000
0.4	6.1444349770272E+000
0.5	9.9999999886674E+000
0.6	1.3855565004636E+001
0.7	1.7002367258735E+001
0.8	1.9002367269241E+001
0.9	1.9855565032141E+001
1.0	2.0000000022665E+001
1.1	2.0144435004532E+001
1.2	2.0997632744767E+001
1.3	2.2997632727257E+001
1.4	2.6144434958691E+001
1.5	2.9999999966003E+001
1.6	3.3855564986300E+001
1.7	3.7002367251732E+001
1.8	3.9002367276245E+001
1.9	3.9855565050477E+001
2.0	4.0000000045330E+001
2.1	4.0144435022868E+001
2.2	4.0997632751770E+001
2.3	4.2997632720253E+001
2.4	4.6144434940354E+001
2.5	4.9999999943336E+001
2.6	5.3855564967962E+001
2.7	5.7002367244726E+001
2.8	5.9002367283248E+001
2.9	5.9855565068813E+001
3.0	6.0000000067995E+001
3.1	6.0144435041204E+001
3.2	6.0997632758774E+001
3.3	6.2997632713247E+001
3.4	6.6144434922015E+001
3.5	6.9999999920669E+001
3.6	7.3855564949624E+001
3.7	7.7002367237721E+001
3.8	7.9002367290251E+001
3.9	7.9855565087149E+001
4.0	8.0000000090659E+001
4.1	8.0144435059540E+001
4.2	8.0997632765776E+001
4.3	8.2997632706241E+001
4.4	8.6144434903676E+001
4.5	8.9999999898001E+001
4.6	9.3855564931285E+001
4.7	9.7002367230715E+001
4.8	9.9002367297253E+001
4.9	9.9855565105485E+001
5.0	1.0000000011332E+002

Figure 3.3 Listing of output for Example 3.1.

given in Fig. 3.3 (these are the variables t and c_i in the flowchart of Fig. 3.1, and the PASCAL variables t and c[i] in the program shown in Fig. 3.2). It should be noted that the number of iterations performed by IntegTrpz in determining each new value of $f(t)$ is 2, which is the value of the third argument in the actual parameter list in the statement calling IntegTrpz.

3.3 GRAPHICAL DISPLAY OF DATA: PLOTTING

The presentation of a set of values of a function in a tabular form, as illustrated in Fig. 3.3, is not very satisfactory from the viewpoint of indicating the manner in which the function varies. Thus, it is usually desirable to provide some means for plotting the computed data. Actually, it is even more advantageous to be able to plot more than one set of data simultaneously. As an illustration of this, in Example 3.1 it might be desirable to plot both the applied voltage and the resulting current. Other variables, such as power or energy, might also be of interest. In this section, therefore, we discuss a simple technique for simultaneously plotting several sets of data. It should be noted that plotting equipment is usually available as a facility of any modern computing center. The use of such equipment, however, can be undesirable for many applications, since the user may be penalized by any or all of the following: (1) an increase of "turnaround" time (since the turnaround time required for the use of the plotter will be added to the turnaround time for using the computer), (2) a requirement for additional programming to prepare data for the plotter, and (3) additional charges for the use of the plotting equipment. As an alternative to the use of such facilities, in the next section we present a method for using the printer associated with the computer to make a plot as part of the output data from the program being processed. The additional computer time required for this is small, and this type of plot is usually quite sufficient for the simple output-display purposes we are considering here.

In order to use the plotting procedure, the data to be plotted must be stored in a two-dimensional array. We may use a PASCAL array to provide such storage. For example, let us use a[i,j] for the storage of the data. If we let i be an index identifying the variable and j be an index which indicates the storage location for the various values of each variable, then for the first variable that it is desired to plot, the first value of such a variable must be stored in a[1,0], the second value in a[1,1], the third value in a[1,2], etc. Similarly, for the second variable that it is desired to plot, the first value must be stored in a[2,0], the second in a[2,1], the third in a[2,2], etc. The pattern is easily extended. We see that the maximum value permitted by the dimensioning for the index i gives the maximum number of variables that may be stored, while the maximum value permitted for j gives the maximum number plus 1 of values of each of the variables that may be stored. We limit ourselves to a maximum of five variables and a maximum of 101 values of each variable. Thus, defining the type plotArray, we require that the array a be declared as follows:

```
TYPE plotArray = ARRAY[1..5,0..100] OF real;
VAR a:plotArray;
```

As an example of storing a set of data in such an array, the 51 values of current $i(t)$ which were computed in Example 3.1 could easily have been stored for future plotting by replacing the statements

```
c[0]:= 0.0; and c[i]:= amps;
```

in the program shown in Fig. 3.2 with the statements

```
a[1,0]:= 0.0; and a[1,i]:= amps;
```

In the next section we discuss a procedure which provides a plot from such a two-dimensional array of data.

3.4 THE PROCEDURE Plot5

This procedure is designed to plot a two-dimensional array of data which has been prepared according to the description given in the preceding section. It is based on the use of equally spaced values of the independent variable, and it has the following features:

1. *Number of variables—number of points*. From one to five variables may be plotted simultaneously. Thus, if we consider plotting the array a[i,j], the maximum value of i is 5. From 1 to 101 points may be plotted for each variable; thus, starting from 0, the maximum value of j is 100. These limits are the result of the dimensioning in the plotting subroutine; that is, the two-dimensional variable to be plotted is dimensioned [1..5,0..100].
2. *Identification of different variables*. All the data points relating to the first variable to be plotted are identified by printing the letter *A* on the plot at each data point. Thus, if we are plotting the array a[i,j], all the values of a[1,j] are plotted with the letter *A*. (It should be noted that the *A* which is printed has nothing to do with the a which is the name of the array.) Similarly, data points for the second variable, which are stored in the array as a[2,j], are plotted with a *B*. In the same manner, data stored in the array as a[3,j], a[4,j], and a[5,j], representing the third, fourth, and fifth variables that it is desired to plot, are identified in the resulting plot by the letters *C, D,* and *E*, respectively.
3. *Orientation of the plot*. In order to simplify the plotting routine, successive values of the independent variable, as represented by successive values of the index j, are plotted on successive *lines* by the printer; thus the coordinate, which we should normally refer to as the abscissa, or x axis, increases in a downward direction on the plot, while the scale for the values of the dependent functions, that is, the variables that are being plotted, which constitutes the ordinate, or y axis, increases from left to right across the printed line. This is illustrated in Fig. 3.4. The more conventional orientation of the plot is easily achieved by rotating it 90° counterclockwise.
4. *Scaling the ordinate*. Frequently, in the computation and plotting of the values of functions, the problem of correctly scaling the data before plotting it becomes a critical one. To aid the user in this matter, an automatic scaling option has been included in the Plot5 procedure. This option is called by requesting a maximum ordinate value of 999. In this case, Plot5 will automatically scale each of the functions being plotted so that it fills the entire ordinate range of the plot. If the automatic scaling option is not used, the user may select any desired range of 100 units for the ordinate. This is done by specifying the maximum ordinate value. The minimum ordinate value is automatically set 100 units lower. For example, choosing 100 as a maximum ordinate value, the minimum is 0; similarly, choosing 50 as a maximum ordinate value, the minimum is -50, etc. For such a user-chosen range of ordinate values, it is usually desirable to scale the data to be plotted by multiplying it and/or adding or subtracting suitable constants before plotting. This is readily done by a main program statement. Values of the user-chosen ordinate range are printed along the ordinate every 10 units.

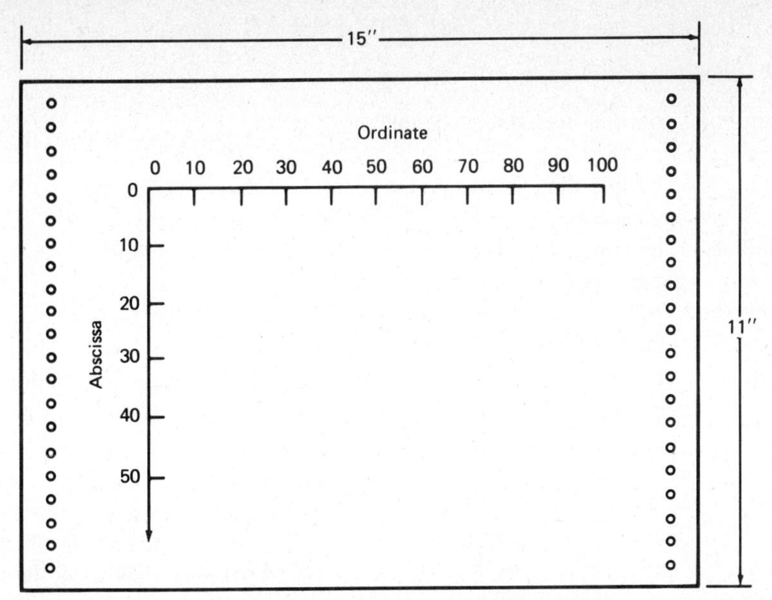

Figure 3.4 Orientation of plot produced by procedure Plot5.

5. *Scaling the abscissa.* The abscissa of the plot is scaled from 0 to the maximum number minus 1 of values that are entered into the two-dimensional data array for each of the variables that it is desired to plot. Values of this scale are printed on the plot every 10 units.
6. *Coordinate lines.* Vertical and horizontal lines are automatically drawn on the resulting plot every 10 units in both the abscissa direction and the ordinate direction.
7. *Printout of data.* In the space at the right of the plot, as a convenience, the values of the first variable, that is, the one stored as a[1,j], are printed using an *E* (or exponential) format.
8. *Data that exceed scale.* If any of the data stored in the array that is being plotted exceed the upper or lower limits of the ordinate scale that has been chosen, the symbol $ will be printed on the upper and lower edge of the plot, indicating that the value of the particular number was not within the specified range.
9. *Coinciding data points.* If, for a given value of the independent variable, two of the dependent variables have values that are so close they would normally be plotted at the same point, the procedure will plot only the letter farthest in the alphabet at that point. Thus, if the second and fifth variables both had the same value for some value of the independent variable, an *E* (not a *B*) would be printed.
10. *Averaging variable values.* All values of a variable that are within 0.5 unit of an integer value are rounded off to that value. For example, a value 96.6 would be plotted as 97, as would a value of 97.499.

The procedure Plot5 is called by the PASCAL statement

Plot5(a, numPlots, numPoints, ordMx);

where the various arguments are defined as:

a — The two-dimensional array containing the values to be plotted {a must be dimensioned [1..5,0..100] (see Sec. 3.3)}.

numPlots — The number of variables that it is desired to plot [numPlots must have a value from 1 to 5 (see 1 and 2 above)].

numPoints — The number of points minus 1 of each variable that it is desired to plot [numPoints must have a value from 0 to 100 (see 1 and 5 above)].

ordMx — The maximum value desired for the ordinate scale. The minimum value for the ordinate will be ordMx − 100 (see 4 above). If ordMx is given the value of 999, automatic scaling will be used.

A summary of the important features of the procedure Plot5 is given in Table 3.1.

We shall not take time at this point in our development of the text material to analyze the inner workings of the procedure Plot5. A brief description of the subprogram, together with a flowchart and a listing of its PASCAL statements, is given in Appendix B. This information is sufficient to permit interested readers to produce their own copy of the subprogram. From this point on, however, we assume that this subprogram is available as one of our tools for the efficient use and application of digital-computation techniques to basic engineering computations. Example 3.2 illustrates the use of the procedure Plot5.

TABLE 3.1 SUMMARY OF THE CHARACTERISTICS OF THE PROCEDURE Plot5

Identifying Statement PROCEDURE Plot5 (VAR a : plotArray; numPlots, numPoints, ordMx : integer);

Purpose To provide a simultaneous plot of several functions $y_i(t)$ for discrete evenly spaced values of the independent variable t.

Additional Subprograms Required None

Input Arguments

a — The two-dimensional array of variables a[i,j] giving the value of the function $y_i(t)$ at the value $j \Delta t$ of the independent variable t (where Δt is the spacing between adjacent values of t)

numPlots — The number of functions $y_i(t)$ which are to be plotted

numPoints — The number of increments of t to be used in plotting the functions (including the value at $t = 0$, the actual number of values which are plotted will be n + 1)

ordMx — The maximum value of the ordinate scale used in plotting the functions (the minimum value is ordMx − 100)

Output A plot is provided of the functions, with the positive direction of the independent variable taken as downward on the plot. The letter A is used to indicate the points for y_1, the letter B for y_2, etc. The values of the function y_1 are printed along the edge of the plot. If a function exceeds the range of the ordinate scale, the symbol $ is printed along the right or left edge of the plot depending on whether the function was greater than or less than the permitted range.

Note: The dimensioning used for the array a must be declared in the TYPE declarations of the main program by a statement

TYPE plotArray = ARRAY[1..5,0..100] OF real;

This permits plotting 1, 2, 3, 4, or 5 functions for up to 100 equal-valued increments of the independent variable t.

```
PROGRAM Main (output);     (* Example 3.2 *)

(* Main program for example of plotting
      t - time
      amps - value of inductor current
      a - plotting array    *)

CONST twoPi = 6.2831853;
TYPE plotArray = ARRAY[1..5,0..100] OF real;
VAR
   t, dt, amps, damps : real;
   a                  : plotArray;
   i                  : integer;

FUNCTION Volts (t : real) : real;
   BEGIN Volts := 20.0 * (1.0 - cos(twoPi * t)) END;

BEGIN
   amps := 0.0; t := 0.0; dt := 0.1; a[1,0] := 0.0; a[2,0] := 0.0;
   FOR i := 1 TO 50 DO
      BEGIN
         IntegTrpz(t, t+dt, 2, damps, Volts);
         amps := amps + damps;
         t := t + dt;
         a[1,i] := amps;
         a[2,i] := Volts(t)
      END;
   Plot5(a, 2, 50, 100)
END.
```

Figure 3.5 Main program and function for Example 3.2.

EXAMPLE 3.2

If a voltage defined as $v(t) = 20(1 - \cos 2\pi t)$ V is applied to the terminals of a 1-H inductor and if the current $i(t)$ through the inductor is zero at $t = 0$, that is, if $i(0) = 0$ (this is the problem used in Example 3.1), we may use the procedure IntegTrpz and the procedure Plot5 to obtain a plot of the values of $i(t)$ (as A) and $v(t)$ (as B) for 50 successive values of t spaced 0.1 sec apart, starting from $t = 0$ sec and finishing at $t = 5$ sec. The general procedure follows that given in the flowchart in Fig. 3.1. A listing of a PASCAL program for accomplishing this is given in Fig. 3.5. The resulting plot is shown in Fig. 3.6. For convenience in interpreting the plot, lines have been drawn linking the data points for the two variables, and some supplementary information has been lettered on the plot.

3.5 THE PROCEDURE Print5

Frequently it is difficult to anticipate the actual range of a set of variables that it is desired to plot. In such a case, the use of the Plot5 procedure described in Sec. 3.4 may not provide enough information to rescale the data for another attempt at plotting. When users are uncertain of the actual range of the variables they desire to plot, a preliminary test of the program is desirable. For such a test, the output variables may be stored in the same two-dimensional array that was described in Sec. 3.3; however, instead of a plot of the variables, an output listing of the values of the variables is needed. As a convenience in providing such a listing, in this section we present a procedure Print5,

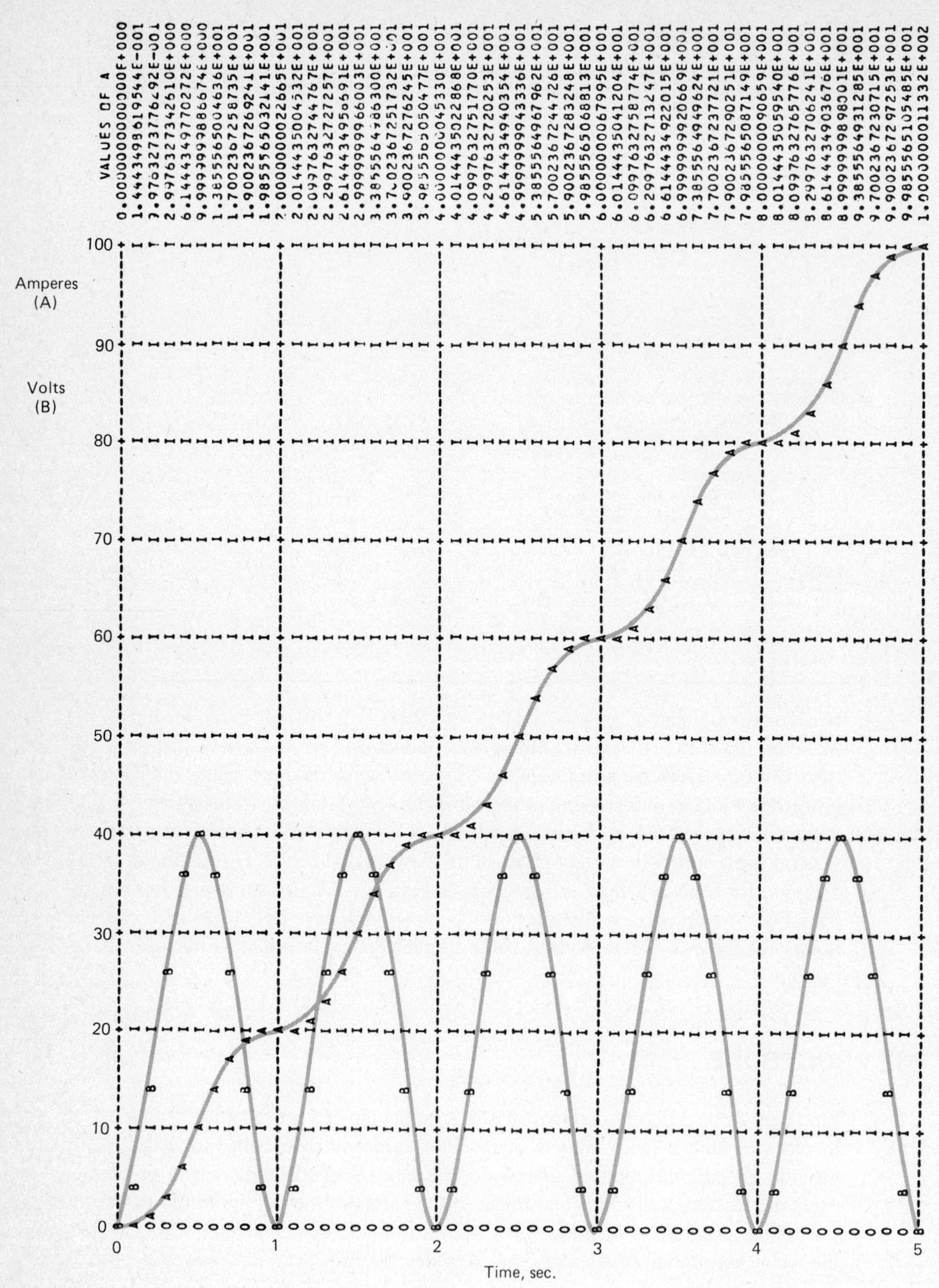

Figure 3.6 Output plot for Example 3.2.

```
PROCEDURE Print5 (VAR a : plotArray; numFuncts, numValues : integer);

(* Procedure for printing the values of the elements
   of an array
     a - Array of user-defined type plotArray specified
         as [1..5,0..100] of type real
     numFuncts - Number of variables (1 - 5)
     numValues - Number (minus one) of values of each variable
         which are to be printed (0 - 100)
     i - Index for variable number (range 1 - numFuncts)
     j - Index for number of values of each variable
         (range 0 - numValues)                                *)

VAR
   i, j : integer;

BEGIN

   (*    Print the heading    *)

   write('  J  ');
   FOR i := 1 TO numFuncts DO
      write('         A(',i:1,',J)          ');
   writeln;

   (*    Print the data stored in the array    *)

   FOR j := 0 TO numValues DO
      BEGIN
         writeln;
         write(j:3);
         FOR i := 1 TO numFuncts DO
            write(a[i,j])
      END
END;
```

Figure 3.7 Listing of the procedure Print5.

which uses the same [1..5,0..100] dimensioned array as the procedure Plot5 does but provides a listing of the values of the dependent variables at all the values of the independent variable. Examination of such a listing readily provides the necessary information for scaling the variables, which can then be plotted when the program is rerun. The procedure is called by the PASCAL statement

TABLE 3.2 SUMMARY OF THE CHARACTERISTICS OF THE PROCEDURE Print5

Identifying Statement PROCEDURE Print5 (VAR a : plotArray; numFuncts, numValues : integer);

Purpose To provide an output listing of the values of several functions $y_i(t)$ for discrete evenly spaced values of the independent variable t.

Additional Subprograms Required None

Input Arguments

a	The two-dimensional array of variables a[i,j] giving the value of the function $y_i(t)$ at the value $j\ \Delta t$ of the independent variable t (where Δt is the spacing between adjacent values of t)
numFuncts	The number of functions $y_i(t)$ that are to have their values printed
numValues	The number of increments of t for which the values of the functions are to be printed (including the value at $t = 0$, the actual number of values that are printed will be numValues + 1)

Output The procedure provides a listing of the values of the quantities a[1,j], a[2,j], etc., on each line. It continues this for numValues + 1 lines, that is, for the numValues + 1 values of j.

Note: The dimensioning used for the array a is the same as that used in the procedure Plot5 (see Table 3.1).

Print5(a, numFuncts, numValues);

where a, numFuncts, and numValues have the same definitions as the variables a, numPlots, and numPoints which were used in the procedure Plot5 and which are explained in Sec. 3.4. A listing of the PASCAL statements for the procedure Print5 is shown in Fig. 3.7. A summary of the important features of this procedure is given in Table 3.2. The result of using it to obtain a listing for the values of $v(t)$ and $i(t)$ for Example 3.2, corresponding to the plotted data shown in Fig. 3.6, is shown in Fig. 3.8.

```
 J          A(1,J)                     A(2,J)

 0   0.0000000000000E+000   0.0000000000000E+000
 1   1.4443498619544E-001   3.8196601040609E+000
 2   9.9763273776292E-001   1.3819660085188E+001
 3   2.9976327342610E+000   2.6180339846530E+001
 4   6.1444349770272E+000   3.6180339653739E+001
 5   9.9999999886674E+000   4.0000000000000E+001
 6   1.3855565004636E+001   3.6180339938139E+001
 7   1.7002367258735E+001   2.6180339983093E+001
 8   1.9002367269241E+001   1.3819660221751E+001
 9   1.9855565032141E+001   3.8196601884616E+000
10   2.0000000022665E+001   0.0000000000000E+000
11   2.0144435004532E+001   3.8196600195606E+000
12   2.0997632744767E+001   1.3819659948626E+001
13   2.2997632727257E+001   2.6180339709967E+001
14   2.6144434958691E+001   3.6180339769338E+001
15   2.9999999966003E+001   4.0000000000000E+001
16   3.3855564986300E+001   3.6180340022540E+001
17   3.7002367251732E+001   2.6180340119655E+001
18   3.9002367276245E+001   1.3819660358313E+001
19   3.9855565050477E+001   3.8196602728614E+000
20   4.0000000045330E+001   0.0000000000000E+000
21   4.0144435022868E+001   3.8196599352597E+000
22   4.0997632751770E+001   1.3819659812061E+001
23   4.2997632720253E+001   2.6180339573403E+001
24   4.6144434940354E+001   3.6180339684936E+001
25   4.9999999943336E+001   4.0000000000000E+001
26   5.3855564967962E+001   3.6180340106943E+001
27   5.7002367244726E+001   2.6180340256225E+001
28   5.9002367283248E+001   1.3819660494883E+001
29   5.9855565068813E+001   3.8196603572665E+000
30   6.0000000067995E+001   0.0000000000000E+000
31   6.0144435041204E+001   3.8196598508542E+000
32   6.0997632758774E+001   1.3819659675491E+001
33   6.2997632713247E+001   2.6180339436830E+001
34   6.6144434922015E+001   3.6180339600530E+001
35   6.9999999920669E+001   4.0000000000000E+001
36   7.3855564949624E+001   3.6180340191348E+001
37   7.7002367237721E+001   2.6180340392795E+001
38   7.9002367290251E+001   1.3819660631453E+001
39   7.9855565087149E+001   3.8196604416714E+000
40   8.0000000090659E+001   0.0000000000000E+000
41   8.0144435059540E+001   3.8196597664478E+000
42   8.0997632765776E+001   1.3819659538919E+001
43   8.2997632706241E+001   2.6180339300260E+001
44   8.6144434903676E+001   3.6180339516124E+001
45   8.9999999898001E+001   4.0000000000000E+001
46   9.3855564931285E+001   3.6180340275754E+001
47   9.7002367230715E+001   2.6180340529365E+001
48   9.9002367297253E+001   1.3819660768026E+001
49   9.9855565105485E+001   3.8196605260777E+000
50   1.0000000011332E+002   0.0000000000000E+000
```

Figure 3.8 Output from procedure Print5.

3.6 CONCLUSION

In this chapter we have considerably enlarged our capability for obtaining useful results from digital-computation techniques, by describing and implementing a simple approach to the problem of simultaneously plotting one or more functions of an independent variable. The procedure Plot5 is a relatively compact program that is economical from the viewpoints of overall program size, compiling time, and execution time. It will be a most useful tool in our future studies of the implementation of various digital-computation techniques.

PROBLEMS

3.1. **(a)** A current $i(t) = 20 \sin 2\pi t$ A is applied to a capacitor of value 0.26 F, at $t = 0$. For the range of time from 0 to 2 sec, plot $i(t)$ and the resulting voltage across the capacitor $v(t)$. Assume that $v(0) = 0$. Use a time increment of 0.04 sec so as to provide 51 points on the resulting plot. In the procedure IntegTrpz, set n to 2.
(b) What is the phase difference between $v(t)$ and $i(t)$?

3.2. **(a)** Find the maximum error in $v(t)$ as found in Prob. 3.1 by mathematical (nonnumerical) integration.
(b) If the number of iterations used in each operation of the procedure IntegTrpz is increased from 2 to 5, find the new value of the maximum error.

3.3. For the situation described in Prob. 3.1, plot $v(t)$, the voltage across the capacitor, and $w(t)$, the energy in joules stored in the capacitor, using the same range of time and the same time increment specified in that problem [use the relation $w(t) = (\frac{1}{2})Cv^2(t)$].

3.4. Across a two-terminal element of the type shown in Fig. 2.1*a*, the voltage and current are defined by the relations

$$v(t) = 10 + 5t \qquad \text{V}$$

$$i(t) = 50 \sin\left(\frac{\pi t}{2}\right) \qquad \text{A}$$

If the energy stored in the element is defined as $w(t)$ and if $w(0) = 0$, plot $v(t)$, $i(t)$ (scaled by $\frac{1}{5}$), and $w(t)$ (in joules, scaled by 0.02) for a range of t from 0 to 10 sec. Use a time increment of 0.2 sec to produce a plot of 51 points.

3.5. A voltage $v(t) = 50e^{-t} \cos \pi t$ V is applied to a 10-kΩ (10,000-Ω) linear time-invariant resistor. If the energy dissipated in the resistor is defined as $w(t)$ and the power as $p(t)$, plot $p(t)$ (in watts) and $w(t)$ (in joules) over the range of time from 0 to 4 sec. Use 51 points.

3.6. A voltage $v(t) = 40e^{-t/4} |\sin \pi t|$ V is applied to the terminals of a 1-H inductor. For the range of time from 0 to 5 sec, plot $v(t)$ and $i(t)$, the resulting current in amperes through the inductor, assuming that $i(0) = -50$. Use a time increment of 0.1 sec so as to provide 51 points on the resulting plot. In the procedure IntegTrpz, set n to 2.

3.7. The first two nonzero coefficients of the Fourier series (see Prob. 2.8) for the sawtooth wave $g(t)$ shown in Fig. P3.7 are given by the expression

$$h(t) = \frac{320}{\pi^2}\left(\sin \frac{2\pi t}{100} - \frac{1}{9} \sin \frac{6\pi t}{100}\right)$$

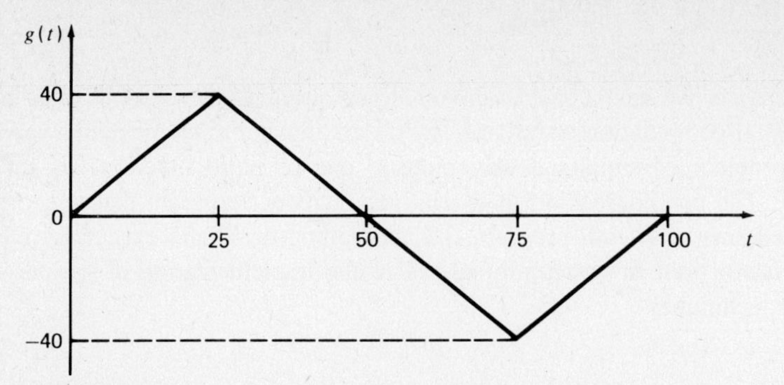

Figure P3.7

Use the procedure Plot5 to plot one half-cycle of the sawtooth wave $g(t)$ and one half-cycle of the approximation given as $h(t)$ above. Plot 51 points, corresponding to values of t from 0 to 50 sec.

3.8. A time-varying resistor has a value of resistance given by the relation $R(t) = 1000 + 200 \sin 2000\pi t\ \Omega$. If a voltage $v(t) = 100 \sin 4000\pi t$ V is applied over a range of time from 0 to 0.001 sec, plot $w(t)$, the energy in joules dissipated in the resistance (scaled by 10^4), and the voltage $v(t)$ (scaled by $\frac{1}{4}$). Use 51 points, and use a value of n for the procedure IntegTrpz of 2.

3.9. Integrate the function $f(x)$ given below by using the trapezoidal-integration procedure IntegTrpz, over a range of time from 0 to 1 sec for integer values of the iteration constant n from 4 to 50. Use the procedure Plot5 to plot the absolute value of the error scaled by 100 as a function of the iteration constant.

$$f(x) = 1 + x^2 + x^3 + x^4$$

Chapter 4

Piecewise-Linear Representation

In the preceding chapters we have seen how to integrate and plot functions that can be easily described mathematically. Frequently, however, the engineer must deal with functions described by graphical or tabular data, for which no explicit mathematical representation is available. In this chapter we present a means of dealing with such functions through the use of piecewise-linear representations.

4.1 PIECEWISE-LINEAR REPRESENTATION OF DATA

Let us assume that some experiment has generated a waveshape of a physical quantity that has the form shown in Fig. 4.1. If this waveshape represents a voltage or a current as a function of time and if we desire to investigate what happens when such a voltage or current is applied to a circuit, then we must have some analytic way of representing the waveshape. In general, it is not feasible to derive an explicit mathematical expression for such an arbitrary function. As an alternative procedure, which is quite useful in practice, we present a means of representing this function by choosing a set of points on the waveform, recording the values of t and y, where $y = y(t)$, and using these values as the defining information for the function. If the function is to be integrated, care must be exercised to define accurately the portions of the waveform that have large magnitudes, since a small relative error in large values of the function will have a larger effect on the integral than a similar error in the function where its magnitude is small.

An example set of points that might be used to define the function shown in Fig. 4.1 is given in Fig. 4.2. The straight-line segments joining adjacent points in such a set are said to form a *piecewise-linear model* of the function. Obviously, many other sets of points are possible.

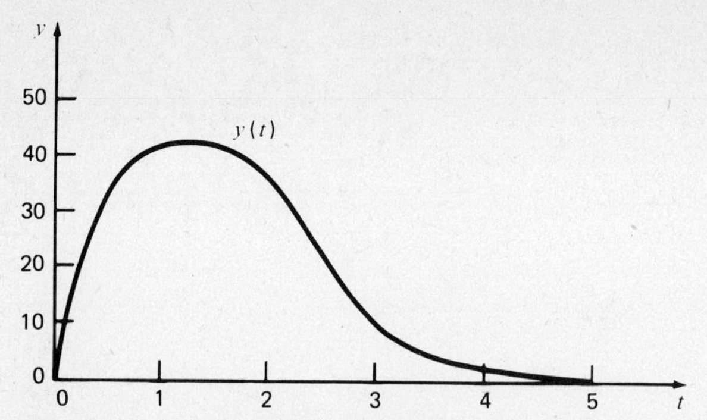

Figure 4.1 An arbitrary waveshape.

Once the points defining the piecewise-linear model for the given waveshape have been selected, we require some means of evaluating the function $y(t)$ for values of x that may not be directly specified by the data points. Thus we pose the following problem:

Problem. Given a set of n data points (t_i, y_i) defining a piecewise-linear representation for a given function $y(t)$, where $y_i = y(t_i)$, find the value of y for any point t which is not one of the given values t_i.

This is obviously a simple problem in linear geometry, since if t lies between the points t_{i-1} and $t_i (i = 2, 3, \ldots, n)$, then y lies between the values y_{i-1} and y_i and is determined by the relation

$$y = y_{i-1} + \frac{y_i - y_{i-1}}{t_i - t_{i-1}}(t - t_{i-1}) \tag{4.1}$$

The above equation simply computes the slope of the line segment between the points (t_{i-1}, y_{i-1}) and (t_i, y_i) and determines the value of y by simple proportion. In the next section we describe a digital-computer subprogram for performing this computation.

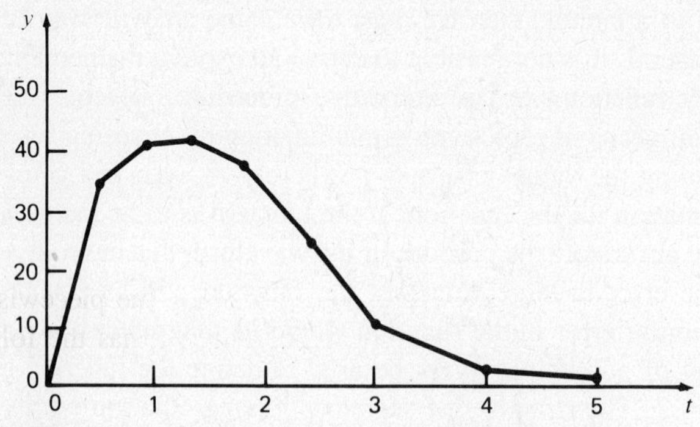

Figure 4.2 A piecewise-linear representation of a waveshape.

4.2 THE FUNCTION PiecewiseLin

In this section we describe a PASCAL function which can be used to provide a piecewise-linear representation of a function specified by a set of data points. From an examination of the problem stated in Sec. 4.1 and the linear interpolation equation given in (4.1), we see that the function must be supplied with the following information:

1. The values t_i of the independent variable t corresponding to the known data points.
2. The values y_i of the dependent variable y, where $y_i = y(t_i)$, corresponding to the data points given in 1.
3. The total number n of data points that have been selected for the piecewise-linear representation.
4. The value of the independent variable t at which it is desired to evaluate the function.

The output of the PASCAL function gives the value y of the function as determined by the piecewise-linear approximation. From the above considerations, our argument listing for the piecewise-linear function must be of the form

```
(VAR tdata, ydata : array50; ndata
 : integer; t : real) : real;
```

where array50 is defined in the main program by the type statement

```
TYPE array50 = ARRAY[1..50] OF real;
```

and where the various arguments are defined as follows:

tdata	The one-dimensional array containing (in ascending order) the values t_i of the independent variable t at the chosen data points
ydata	The one-dimensional array containing the values y_i of the dependent variable y corresponding to the t_i
ndata	The number n of data points
t	The value of the independent variable t at which it is desired to evaluate the function

If we define this function as PiecewiseLin, then the heading for it is

```
FUNCTION PiecewiseLin (VAR tdata, ydata : array50; ndata
                       : integer; t : real) : real;
```

A listing of the PASCAL statements of a function providing the piecewise-linear representation described above is given in Fig. 4.3. The function has the following characteristics:

1. The value of t with which the function is supplied is first checked to make certain that it lies within the range of data specified by the data points. Since

```
FUNCTION PiecewiseLin (VAR tdata, ydata : array50;
                       ndata : integer; t : real) : real;

VAR i, im      : integer;
    outRange : boolean;

BEGIN
   outRange := false;
   IF t = tdata[1] THEN PiecewiseLin := ydata[1]
      ELSE IF t < tdata[1] THEN outrange := true
         ELSE IF t = tdata[ndata] THEN PiecewiseLin := ydata[ndata]
            ELSE IF t > tdata[ndata] THEN outRange := true
               ELSE
                  BEGIN
                     i := 2;
                     WHILE t >= tdata[i]
                        DO i := i + 1;
                     im := i - 1;
                     PiecewiseLin := ydata[im] + (ydata[i] - ydata[im])
                        * (t - tdata[im]) / (tdata[i] - tdata[im])
                  END;
   IF outRange = true THEN
      BEGIN
         PiecewiseLin := 999999.0;
         writeln;
         writeln(' T = ', t,' OUT OF RANGE');
         writeln
      END
   END;
```

Figure 4.3 Listing of the function PiecewiseLin.

TABLE 4.1 SUMMARY OF THE CHARACTERISTICS OF THE FUNCTION PiecewiseLin

Identifying Statement FUNCTION PiecewiseLin (VAR tdata, ydata : array50; n data : integer; t : real) : real;

Purpose To construct a piecewise-linear approximation for a function $y(t)$ from a set of n points (t_i, y_i), and to use this approximation to give a numerical value for the function y for a given value of the independent variable t

Additional Subprograms Required: None

Input Arguments

tdata	The one-dimensional array of variables tdata[i] in which are stored (in ascending order) the values of the independent variable t_i used in constructing the piecewise-linear approximation to $y(t)$
ydata	The one-dimensional array of variables ydata[i] in which are stored the values y_i corresponding to the values t_i, where $y_i = y(t_i)$
ndata	The number of points (t_i, y_i) of data
t	The value of the independent variable t for which the approximate value of $y(t)$ is desired

Output

PiecewiseLin	Approximate value of the function $y(t)$

the values of t_i are given in increasing order, the following inequality must be satisfied:

$$t_1 \leq t \leq t_n \quad (4.2)$$

If t does not lie within this range, the function stops further execution of the program.

2. The second action of the function is to determine the smallest value of t_i that is greater than t. It does this by comparing t with the values t_2, t_3, . . . until it finds one, say t_k, that is greater than t.
3. If t_k is the first value of the t_i that is greater than t, then t must lie between t_k and t_{k-1}. These values and the values of y_k and y_{k-1} are inserted in (4.1) to determine the value of y.

As an aid in understanding the operation of the function PiecewiseLin, a flowchart for it is given in Fig. 4.4. A summary of the significant features of this function is given in Table 4.1.

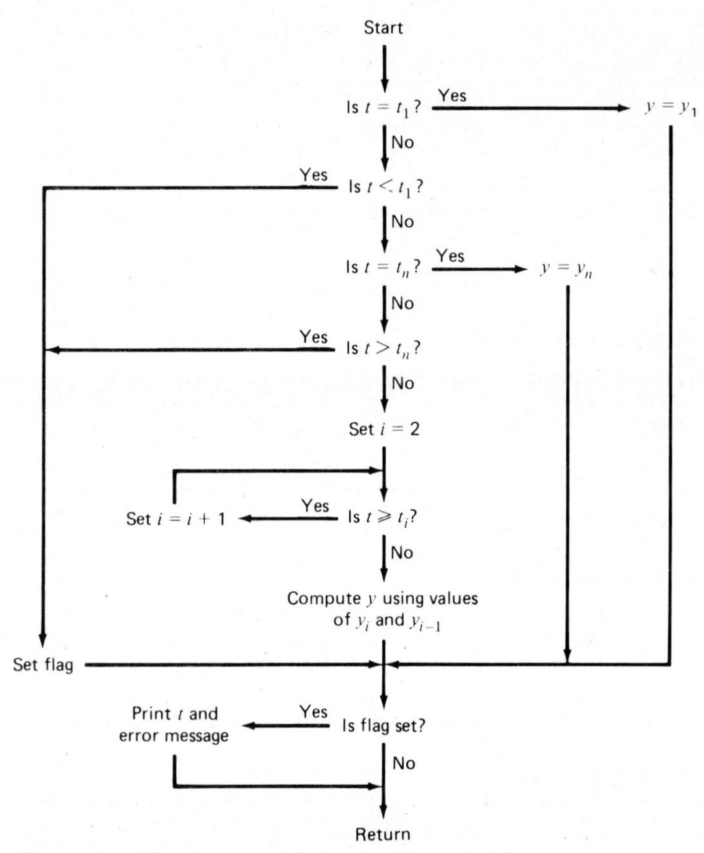

Figure 4.4 Flowchart for the function PiecewiseLin.

```
PROGRAM Main (input,output);    (* Example 4.1 *)

(* Main program for PiecewiseLin example
      ndata - number of data points
      ydata - array of inductor voltage values
      tdata - array of time values
      t     - time
      amps  - inductor current
   Program requires IntegTrpz, PiecewiseLin (with
   modified argument list), and Plot5 subprograms *)

CONST ndata = 9;
TYPE
   plotArray = ARRAY[1..5,0..100] OF real;
   array50   = ARRAY[1..50] OF real;
VAR
   t, dt, amps, damps : real;
   tdata, ydata       : array50;
   a                  : plotArray;
   i                  : integer;

FUNCTION PiecewiseLin (t : real) : real;

VAR i, im     : integer;
    outRange : boolean;

BEGIN
   outRange := false;
   IF t = tdata[1] THEN PiecewiseLin := ydata[1]
      ELSE IF t < tdata[1] THEN outrange := true
         ELSE IF t = tdata[ndata] THEN PiecewiseLin := ydata[ndata]
            ELSE IF t > tdata[ndata] THEN outRange := true
               ELSE
                  BEGIN
                     i := 2;
                     WHILE t >= tdata[i]
                        DO i := i + 1;
                     im := i - 1;
                     PiecewiseLin := ydata[im] + (ydata[i] - ydata[im])
                        * (t - tdata[im]) / (tdata[i] - tdata[im])
                  END;
   IF outRange = true THEN
      BEGIN
         PiecewiseLin := 999999.0;
         writeln;
         writeln(' T = ', t,' OUT OF RANGE');
         writeln
      END
   END;

BEGIN
   t := 0.0; dt := 0.1; amps := 0.0; a[1,0] := 0.0; a[2,0] := 0.0;
   FOR i := 1 TO ndata DO
      readln(tdata[i], ydata[i]);
   FOR i := 1 TO 50 DO
      BEGIN
         IntegTrpz(t, t+dt, 2, damps, PiecewiseLin);
         amps := amps + damps;
         t := t + dt;
         a[1,i] := amps;
         a[2,i] := PiecewiseLin(t)
      END;
   Plot5(a, 2, 50, 100)
END.
```

(*a*)

Figure 4.5 Listing and flowchart for Example 4.1.

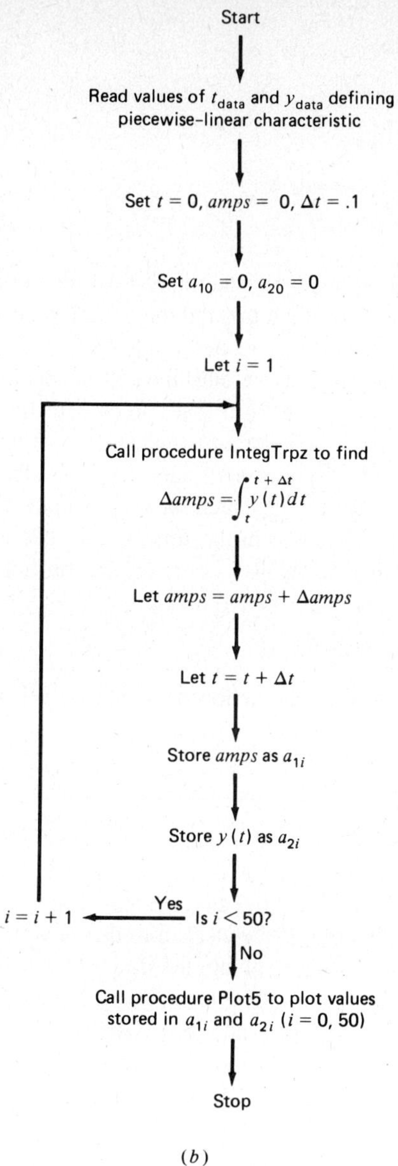

(*b*) **Figure 4.5** (*Continued*)

4.3 INTEGRATING THE FUNCTION PiecewiseLin

In Sec. 4.2 we discussed the use of the function PiecewiseLin to solve the problem posed in Sec. 4.1, that is, to give us a value of a dependent variable $y(t)$ for every value of an independent variable t, within some range specified by a set of values t_i and the corresponding values y_i, where $y_i = y(t_i)$. Now let us suppose that we desire to integrate a

```
0.0    00.0
0.5    36.0
0.9    41.4
1.3    42.6
1.8    38.3
2.4    25.1
3.0    11.0
4.0     3.0
5.0     1.0
```

Figure 4.6 Input data for Example 4.1.

function that is defined by such a set of points, using the trapezoidal-integration procedure IntegTrpz discussed in Chap. 2. It will be recalled that the integration procedure required the use of a function parameter Y(t : real) : real to define the integrand. Therefore, if we desire to integrate a piecewise-linear function, we must have some means of relabeling our piecewise-linear function PiecewiseLin in a form that agrees with this function parameter. To do this, we must remove the arrays tdata and ydata and the integer variable ndata from the parameter list. To provide the necessary data values for these quantities, we can simply treat them as global variables, in which case the values established for them in the main program will also be available in the function. In this case, the only change needed in the subprogram given in Fig. 4.3 is to replace the function heading shown there with the following one:

```
FUNCTION PiecewiseLin (t:real):real;
```

As an illustration of the integration of a function defined by a set of data points, consider the following:

EXAMPLE 4.1

A voltage function $v(t)$ is defined by the piecewise-linear representation given in Fig. 4.2. If this voltage is applied to the terminals of a 1-H inductor and if the current $i(t)$ through the inductor at $t = 0$ is zero, that is, if $i(0)$, then we may find $i(t)$ over the range of time from 0 to 5 sec by using the function PiecewiseLin to define $v(t)$. We may then make repeated use of the procedure IntegTrpz to obtain the desired values of the current $i(t)$, where $i(t) = \int_0^t v(t)\,dt$ for $0 \le t \le 5$ sec. The results may be displayed by using the subroutine Plot5 defined in Chap. 3. A listing and a flowchart of a program for doing this are given in Fig. 4.5. The input data for the variables tdata and ydata are given in Fig. 4.6. The values of the variable tdata represent time (in seconds), and the values of the variable ydata represent voltage (in volts). The output plot showing the input voltage $v(t)$ and the resulting current $i(t)$ is shown in Fig. 4.7.

4.4 CONCLUSION

In this chapter we have discussed the use of the PASCAL function PiecewiseLin to provide a representation for functions that are specified by graphical means or by tables of data. The representation is based on a piecewise-linear model for the specified function. We also demonstrated the application of the IntegTrpz trapezoidal-integration procedure for integrating such a function.

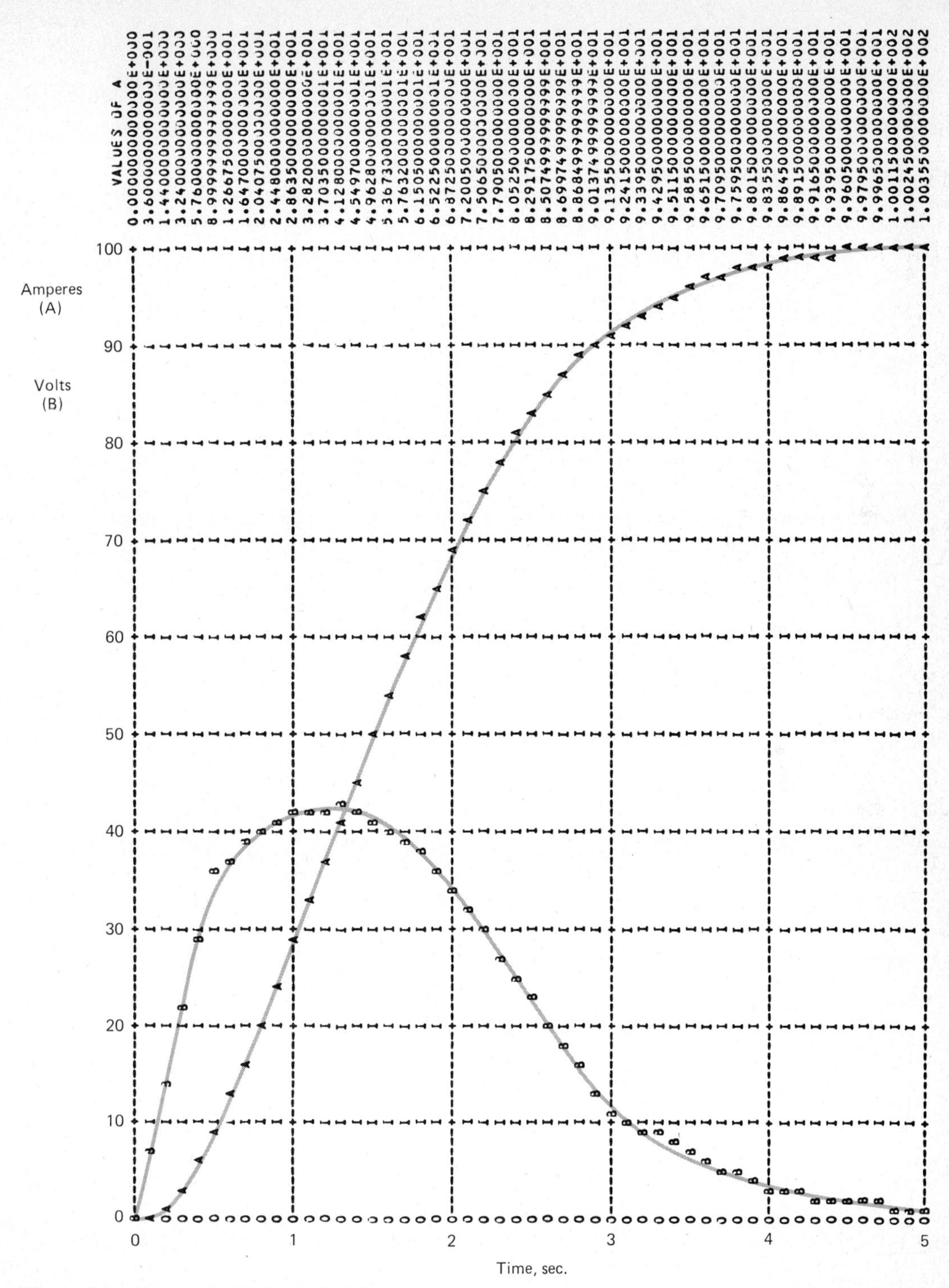

Figure 4.7 Output plot for Example 4.1.

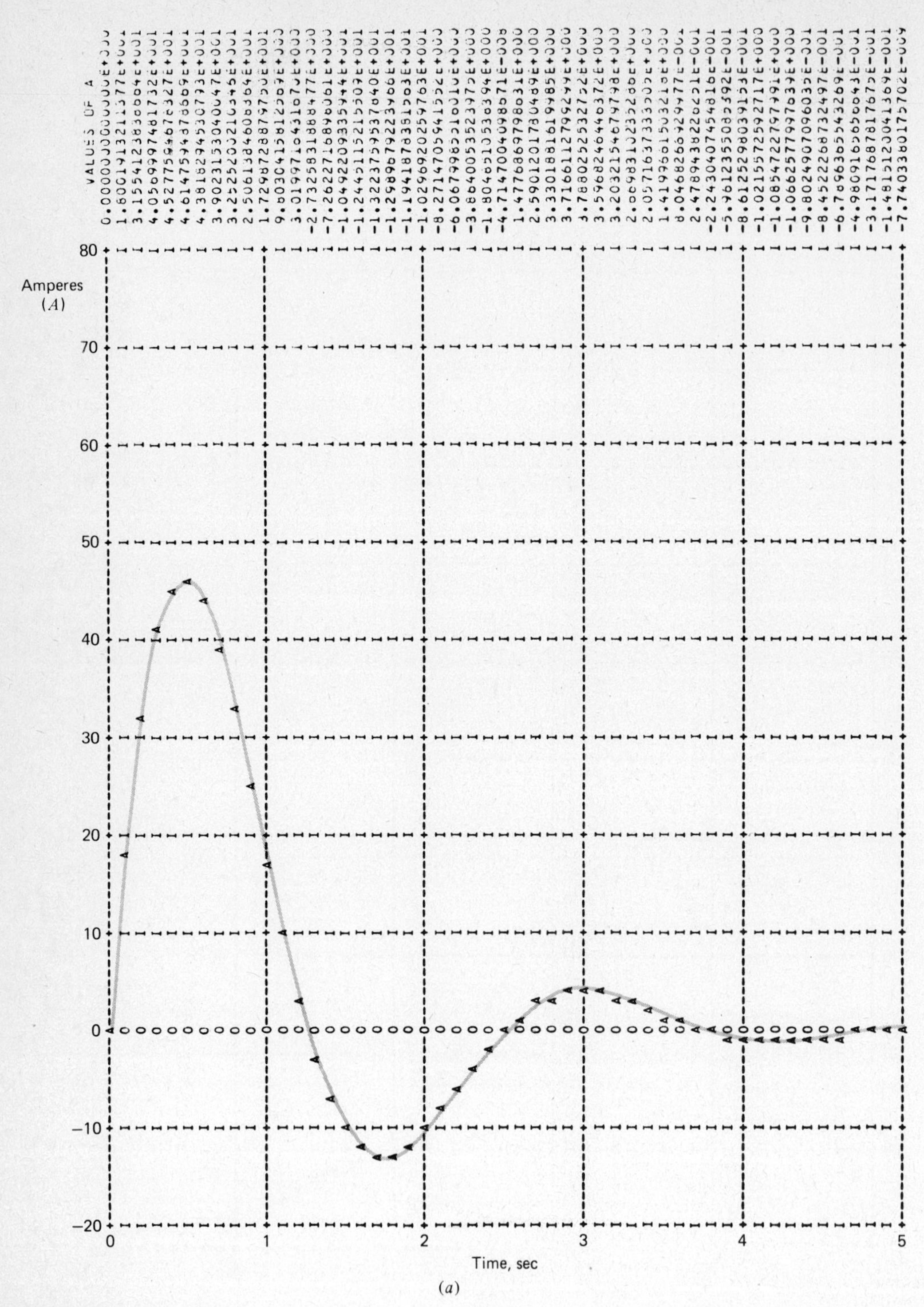

(a)

Figure P4.1

PROBLEMS

4.1. **(a)** A current $i(t)$ (in amperes) is defined by the function plotted in Fig. P4.1*a*, with the data points listed in Fig. P4.1*b*. If the current is applied to the terminals of a 1-F capacitor, use the data points to determine a piecewise-linear representation for the current, and use the procedure IntegTrpz to determine the resulting voltage $v(t)$ in volts that appears across the terminals of the capacitor. Plot $v(t)$ and $i(t)$ for a range of time from 0 to 5 sec. Use 51 points. Assume that $v(0) = 0$.

t	$i(t)$
0.00	0.000E+000
0.10	1.800E+001
0.20	3.155E+001
0.30	4.057E+001
0.40	4.528E+001
0.50	4.615E+001
0.60	4.382E+001
0.70	3.902E+001
0.80	3.253E+001
0.90	2.506E+001
1.00	1.730E+001
1.10	9.803E+000
1.20	3.020E+000
1.30	-2.733E+000
1.40	-7.262E+000
1.50	-1.049E+001
1.60	-1.245E+001
1.70	-1.322E+001
1.80	-1.299E+001
1.90	-1.194E+001
2.00	-1.030E+001
2.10	-8.271E+000
2.20	-6.058E+000
2.30	-3.864E+000
2.40	-1.805E+000
2.50	-4.715E-008
2.60	1.478E+000
2.70	2.590E+000
2.80	3.330E+000
2.90	3.717E+000
3.00	3.768E+000
3.10	3.597E+000
3.20	3.203E+000
3.30	2.670E+000
3.40	2.057E+000
3.50	1.420E+000
3.60	8.047E-001
3.70	2.479E-001
3.80	-2.243E-001
3.90	-5.961E-001
4.00	-8.613E-001
4.10	-1.022E+000
4.20	-1.085E+000
4.30	-1.066E+000
4.40	-9.802E-001
4.50	-8.452E-001
4.60	-6.790E-001
4.70	-4.981E-001
4.80	-3.172E-001
4.90	-1.482E-001
5.00	-7.740E-009

(*b*)

Figure P4.1 (*Continued*)

(b) Make an estimate as to the actual mathematical expression that is represented by the plot given in Fig. P4.1*a*, and, using this as an explicit function for $i(t)$, determine the voltage $v(t)$ that appears across the terminals of the capacitor. Compare the results obtained with the two methods.

4.2. Repeat Prob. 4.1, using every second data point from Fig. P4.1*b* to determine the piecewise-linear approximation to the current $i(t)$. Compare the results with those obtained from Prob. 4.1.

4.3. A nonlinear resistor whose actual value of resistance is a function of the current through it is defined by the set of values of current (in amperes) and resistance (in ohms) given in Fig. P4.3. If a current $i(t) = 0.01t$ A is passed through this resistor for a range of time from 0 to 0.5 sec, find and plot the total energy $w(t)$ (using 51 points) for the specified time range.

i	$R(i)$
0.000	0.0
0.002	2004.8
0.004	2019.2
0.006	2043.2
0.008	2076.8
0.010	2120.0
0.012	2172.8
0.014	2235.2
0.016	2307.2
0.018	2388.8
0.020	2480.0
0.022	2580.8
0.024	2691.2
0.026	2811.2
0.028	2940.8
0.030	3080.0
0.032	3228.8
0.034	3387.2
0.036	3555.2
0.038	3732.8
0.040	3920.0
0.042	4116.8
0.044	4323.2
0.046	4539.2
0.048	4764.8
0.050	5000.0

Figure P4.3

4.4. In Fig. P4.4*a*, a listing of values of time (in seconds) and voltage (in volts) defining a voltage $v(t)$ is given. Similarly, in Fig. P4.4*b*, a listing of values of time and current (in amperes) defining a current $i(t)$ over the same total range of time is given. If $v(t)$ and $i(t)$, as defined by the data, are the variables associated with a two-terminal element, find the energy $w(t)$ in

t	$v(t)$
0.00	0.0000
0.10	0.1980
0.20	0.3840
0.30	0.5460
0.40	0.6720
0.46	0.7253
0.54	0.7651
0.58	0.7698
0.62	0.7533
0.68	0.7311
0.74	0.6695
0.80	0.5760
0.90	0.3420
1.00	0.0000

(*a*)

t	$i(t)$
0.00	0.0000
0.14	0.0388
0.30	0.1719
0.40	0.2944
0.60	0.5904
0.70	0.7399
0.80	0.8704
0.86	0.9322
0.92	0.9764
1.00	1.0000

(*b*)

Figure P4.4

joules and the power $p(t)$ in watts and plot them for 51 points over the specified time range. (*Hint:* In this problem, the function PiecewiseLin must be used in its general form with the argument list given in Sec. 4.2. The function used as the integrand of the procedure IntegTrpz should be defined as the product of two calls to the function PiecewiseLin.)

4.5. The data listed in Fig. P4.5*a* defines a set of values of current (in amperes) and voltage (in volts) to specify the *v-i* characteristic of a nonlinear two-terminal resistor. If such a resistor is inserted in the circuit shown in Fig. P4.5*b*, find the steady-state operating point for the circuit. (*Hint:* Choose successive values of current, and find the voltage v both from the data and from the source and linear resistance. When the difference between the two values reaches a minimum, that value of current determines the best guess at the operating point. A more sophisticated approach is given in the following problem.)

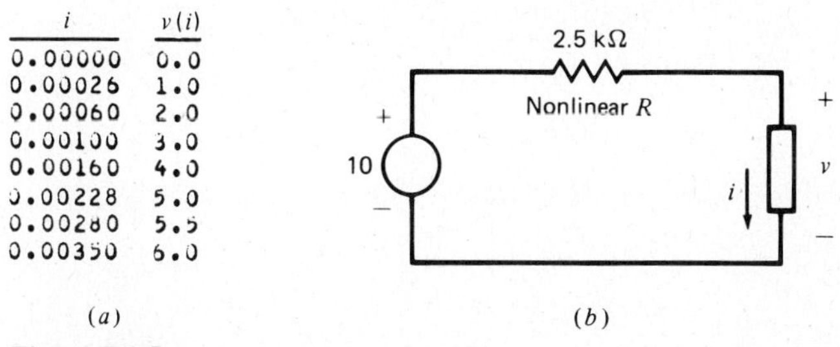

i	$v(i)$
0.00000	0.0
0.00026	1.0
0.00060	2.0
0.00100	3.0
0.00160	4.0
0.00228	5.0
0.00280	5.5
0.00350	6.0

Figure P4.5

4.6. Problem 4.5 can also be solved by a one-dimensional search. To do this, the range of the current variable is divided into two half-ranges, and the error function giving the difference between the values of voltage found by the data of Fig. P4.5*a* and the circuit of Fig. P4.5*b* is calculated for small deviations on either side of the median current value. The lower value of the error implies that the minimum (we assume that there is only one minimum) lies in that half-range of current. This half-range is now subdivided into two parts, and the process is repeated. Write a program that will perform this search for the conditions defined in Prob. 4.5.

Chapter 5

Solution of Differential Equations: Time-Varying and Nonlinear Inductors and Capacitors

In the previous chapters we applied some basic numerical techniques to solve the first-order *integrals* that define the relations between a pair of variables. We illustrated these techniques by considering the terminal variables of lumped linear time-invariant inductors and capacitors. In this section we consider the first-order *differential* equations that may relate a pair of variables. We discover that the numerical methods for solving such differential equations may be applied not only to lumped linear time-invariant inductors and capacitors but also to the case where these elements are time-varying and/or non-linear.

5.1 DIFFERENTIAL EQUATIONS FOR THE INDUCTOR AND THE CAPACITOR

In (2.1) and (2.2) we presented the integral equations relating the terminal variables of voltage and current for typical two-terminal elements, namely, the lumped linear time-invariant inductor and capacitor. In this section we discuss typical differential relations as encountered for these elements. By differentiating with respect to time the equations referenced above, we obtain (for the case where L and C are constant) the differential equations

$$v(t) = Li'(t) \tag{5.1}$$

$$i(t) = Cv'(t) \tag{5.2}$$

where the prime indicates differentiation with respect to the independent variable t. Both of the relations given in (5.1) and (5.2) are examples of first-order linear ordinary differential equations with constant coefficients. Such equations may always be written in the form

$$y'(t) = g(t) \tag{5.3}$$

where g is an arbitrary function.

In the following sections of this chapter we present numerical techniques for the solution of (5.3), that is, for finding $y(t)$. Actually, the same techniques can be used to solve a more general problem, which includes the non-constant-coefficient case and the nonlinear case. A general expression for a first-order ordinary differential equation that is not necessarily linear and whose coefficients need not be constant is

$$y'(t) = g(t, y) \tag{5.4}$$

where the prime indicates differentiation with respect to t. As an example of such an equation, let us consider the case of a time-varying inductor, that is, one in which the value of the inductance varies with time. Thus, $L = L(t)$. For this case, (5.1) becomes

$$v(t) = \frac{d}{dt}[L(t)i(t)] = L(t)i'(t) + L'(t)i(t) \tag{5.5}$$

Rearranging the terms of (5.5) to put it in the form given in (5.4), we obtain

$$i'(t) = g(t, i) = \frac{v(t) - i(t)L'(t)}{L(t)} \tag{5.6}$$

where it is assumed that the variation of the inductance with time is given; that is, $L(t)$ and thus $L'(t)$ are known quantities. Similarly, for a time-varying capacitance, we obtain

$$v'(t) = g(t, v) = \frac{i(t) - v(t)C'(t)}{C(t)} \tag{5.7}$$

Since the relations given in (5.4), (5.6), and (5.7) are first-order differential equations, one initial condition is required for their solution. In view of the above, we may pose the following problem:

Problem. Given an ordinary first-order differential equation defined by (5.4), and some initial condition y_i, where $y_i = y(t_i)$, find $y(t)$, for $t > t_i$.

In the following section we discuss the application of numerical techniques to solve this problem.

5.2 NUMERICAL TECHNIQUES FOR THE SOLUTION OF FIRST-ORDER DIFFERENTIAL EQUATIONS

In Fig. 5.1 the locus of a function $y(t)$ with respect to its independent variable t is shown. Let us assume that y_i, the value of the function $y(t)$ when $t = t_i$, is known. We may make an estimate of the value of y at some nearby value of t, which we call t_{i+1}, by

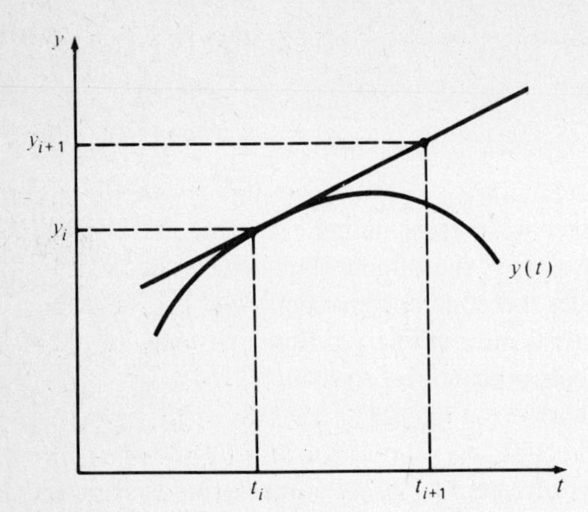

Figure 5.1 Euler's method.

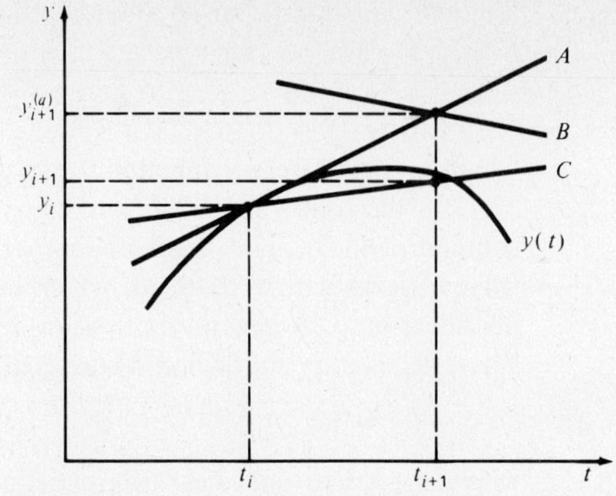

Figure 5.2 The improved Euler method.

simply drawing a tangent to the curve $y(t)$ at the point (t_i, y_i) and evaluating the function at the point on the tangent corresponding to the value of t_{i+1}. The process is illustrated in Fig. 5.1. From this figure we may write

$$y_{i+1} = y_i + y_i'(t_{i+1} - t_i) \tag{5.8}$$

The term y_i' is simply found from some known relation of the form of (5.4), that is,

$$y_i' = g(t_i, y_i) \tag{5.9}$$

Such a method is frequently referred to as *Euler's method*. It involves one evaluation of the function $g(t, y)$; therefore, it may be called a *first-order method*. The *truncation error* which results from using only a single evaluation, and thus "truncating" the series representation for $y(t)$, is proportional to $(\Delta t)^2$, where $\Delta t = t_{i+1} - t_i$. Other sources of error common to all numerical-analysis techniques are *round-off error* caused by the limited number of significant digits used by the computer in performing the computations, and *propagation error* due to the accumulation of previous errors. The references given in the bibliography provide additional information on errors. Euler's method is too crude to be of much practical use, but it does illustrate the basic approach which is applicable to other methods, namely, given y_i, the value of the function $y(t)$ for $t = t_i$, use (5.4) to *estimate* the value y_{i+1}, that is, the value of $y(t)$ for $t = t_{i+1}$. Obviously, the process can be repeated to produce a sequence of values of $y(t)$, and the resulting set of values can be plotted to define the function.

A more accurate method of solution for the first-order differential equation given in (5.4) is provided by the *improved Euler method* illustrated in Fig. 5.2. To locate the point (t_{i+1}, y_{i+1}) shown in this figure we simply average the slopes of the tangents at the points (t_i, y_i) and $(t_{i+1}, y_{i+1}^{(a)})$ and use this value to determine the slope of the line determining the new point (t_{i+1}, y_{i+1}). In the figure, the tangent at (t_i, y_i) is illustrated by line A, and the tangent at the point $(t_{i+1}, y_{i+1}^{(a)})$ is illustrated by line B. If we let r be the average of the two slopes as indicated by line C, we see that, from (5.4),

$$r = \tfrac{1}{2}[g(t_i, y_i) + g(t_{i+1}, y_{i+1}^{(a)})] \tag{5.10}$$

If we let $\Delta t = t_{i+1} - t_i$, then $y_{i+1}^{(a)} = y_i + \Delta t\, y'_i$. Thus, we may write (5.10) in the form

$$r = \tfrac{1}{2}[g(t_i, y_i) + g(t_i + \Delta t, y_i + \Delta t\, y'_i)] \tag{5.11}$$

From Fig. 5.2, we see that an approximation for y_{i+1} is given by the relation

$$y_{i+1} = y_i + \Delta t\, r \tag{5.12}$$

Since r is a function of t_i, y_i, and Δt, we see that we may predict a new value of $y(t)$, namely, y_{i+1}, using only our knowledge of the previous value of $y(t)$, namely, y_i, and the function $g(t, y)$ defining the differential equation.

The improved Euler method described above is frequently referred to as a *second-order method,* since two evaluations of the function $g(t, y)$ are required to find the function r of (5.11) for each determination of a new value of $y(t)$. The error at each step is proportional to $(\Delta t)^3$. The relations given abovc are easily programmed for the digital computer. The development of such a program is left to the reader as a problem assignment (see Prob. 5.7).

5.3 THE PROCEDURE DiffEqn

Although the second-order method for the solution of a first-order ordinary differential equation given in the preceding section is adequate for many purposes, we use a more sophisticated method to provide a subprogram "building block" to solve the problem posed in Sec. 5.1 and to be used for our subsequent solution of such differential equations. The method is based on a fourth-order approximation; that is, it requires four evaluations of the function $g(t, y)$ to determine the next value of the dependent variable $y(t)$. It is usually referred to as the *Runge-Kutta method.* A derivation of the method would take us too far from our primary goal in this text; so in this section we present without proof the actual relations for the implementation of the method. The reader will note their similarity to the general approach for the solution of differential equations described in Sec. 5.2.

Given the function $g(t, y)$ defined in (5.4), and a known value y_i of the dependent variable $y(t)$ at some value t_i of the independent variable t, we define the four functions

$$\begin{aligned}
g_1 &= g(t_i, y_i) \\
g_2 &= g\left(t_i + \frac{\Delta t}{2}, y_i + \frac{\Delta t\, g_1}{2}\right) \\
g_3 &= g\left(t_i + \frac{\Delta t}{2}, y_i + \frac{\Delta t\, g_2}{2}\right) \\
g_4 &= g(t_i + \Delta t, y_i + \Delta t\, g_3)
\end{aligned} \tag{5.13}$$

where $\Delta t = t_{i+1} - t_i$. It should be noted that the expressions of (5.13) must be calculated in the order given, since the calculation of g_2 requires that the value of g_1 be known, etc. After the functions described above have been calculated, the next value of $y(t)$ is found by the relation

$$y_{i+1} = y_i + (g_1 + 2g_2 + 2g_3 + g_4)\frac{\Delta t}{6} \tag{5.14}$$

This is a fourth-order method, and the error is proportional to $(\Delta t)^5$. The relations described above are easily programmed for the digital computer. Obviously, we need a function subprogram [similar to the function subprogram $y(t)$ that was used in Sec. 2.3] to define the relation $y' = g(t, y)$ given in (5.4). Let us define $g(t, y)$ as our function. It will be identified by the PASCAL function heading

```
FUNCTION G (t, y:real):real;
```

plus the necessary statements used to define the function. The statements defining the function will, of course, be different for each different problem that we may desire to solve.

In addition to having available a function $G(t, y)$ defined as given above, a procedure designed to compute a solution to a first-order differential equation must be provided with the following information:

1. The value t_i of the independent variable t at which we desire to start the solution process
2. The value y_i of the dependent variable $y(t)$ corresponding to the value $t = t_i$
3. The value of the independent variable t at which it is desired to stop the solution process
4. The number of iterations, that is, the number of intermediate calculations that it is desired to make in going from the starting value of t to the final value of t.

As an output, the procedure must provide us with a value of the dependent variable $y(t)$ at the final value of t. Consideration of the above inputs and outputs for the procedure leads us to a parameter list for it that has the form

```
(tInitial, yInitial, tstop:real; iter:integer; VAR y:real;
 FUNCTION G (t, y:real):real);
```

where tInitial is the variable name for t_i, etc. Let us use the name DiffEqn (for *diff*erential *eq*uatio*n* solution by Runge-Kutta method) for the procedure. It will be identified by the heading

```
PROCEDURE DiffEqn (tInitial, yInitial, tstop:real; iter:integer;
                   VAR y:real; FUNCTION G (t, y:real):real);
```

A listing and a flowchart of a set of PASCAL statements defining a procedure to perform the operations listed in (5.13) and (5.14) and calling a function $g(t, y)$ are given in Fig. 5.3. The operation can be readily understood by consulting the referenced equations. A summary of the information concerning this procedure is given in Table 5.1.

```
PROCEDURE DiffEqn (tInitial, yInitial, tstop : real; iter : integer;
                   VAR y : real; FUNCTION G(t, y : real) : real);

(* Runge-Kutta differential equation solving procedure
     tInitial - Initial value of time
     yInitial - Initial value of dependent variable y(t)
     tstop - Final value of time
     iter - Number of iterations from tInitial to tstop
     y - Value of y(t) at tstop
   A function G(t, y) must be used to define the derivative
   of y(t).                                                  *)

   VAR
     t, dt, g1, g2, g3, g4 : real;
     i                     : integer;

   BEGIN

     (*   Initialize the values of t and y and compute
          the value of dt                            *)

     t := tInitial;
     y := yInitial;
     dt := (tstop - tInitial) / iter;

     (*   Perform iterations to find the value of y at
          tstop                                      *)

     FOR i := 1 TO iter DO
        BEGIN
          g1 := G(t, y);
          g2 := G(t + dt/2.0, y + dt*g1/2.0);
          g3 := G(t + dt/2.0, y + dt*g2/2.0);
          g4 := G(t + dt, y + dt*g3);
          y := y + (g1 + 2.0*g2 + 2.0*g3 + g4) * dt / 6.0;
          t := t + dt
        END
   END;
```

(a)

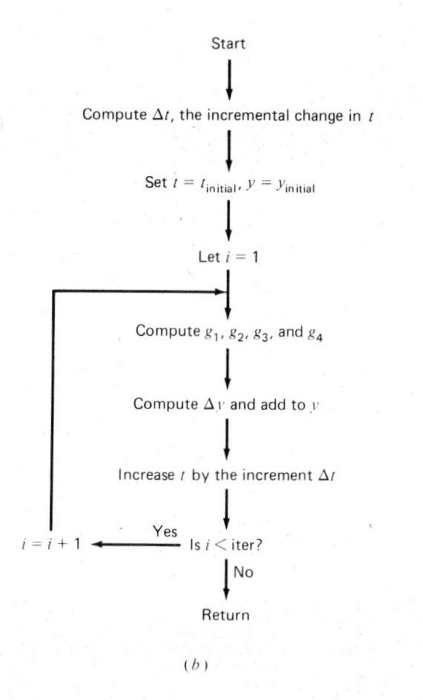

(b)

Figure 5.3 The procedure DiffEqn.

TABLE 5.1 SUMMARY OF THE CHARACTERISTICS OF THE PROCEDURE DiffEqn

Identifying Statement PROCEDURE DiffEqn (tInitial, yInitial, tstop : real; iter : integer; VAR y : real; FUNCTION G (t, y : real) : real);

Purpose To solve the differential equation

$$y' = g(t, y)$$

and thus to find the value of $y(t)$ at the value of t specified as t_{stop}, starting from t_i, a known value of t, and y_i, where $y_i = y(t_i)$.

Additional Subprograms Required This procedure calls the function identified by the statement

FUNCTION G (t, y : real) : real;

The function must be used to define the differential equation, that is, to specify $g(t, y)$.

Input Arguments

tInitial	The initial value t_i of the independent variable t
yInitial	The initial value y_i of $y(t)$, that is, $y(t_i)$
tstop	The final value t_{stop} of t for which the value of $y(t)$ is desired
iter	The number of iterations used by the procedure in going from t_i to t_{stop}

Output Argument

y	The value of $y(t)$ at the final value of the independent variable t, that is, $y(t_{stop})$

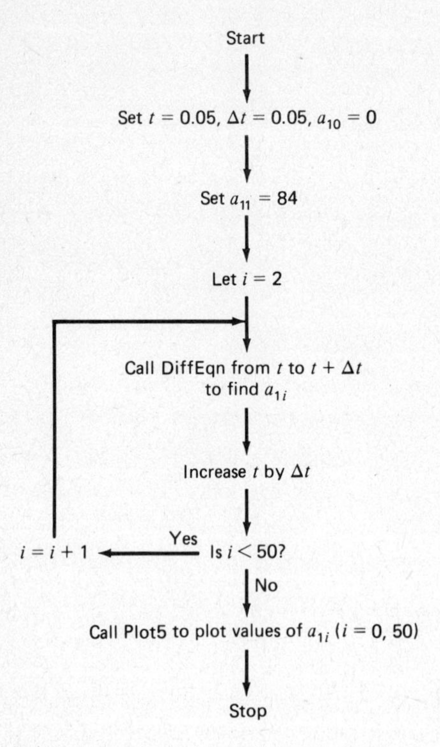

Figure 5.4 Flowchart for Example 5.1

```
PROGRAM Main (output);     (* Example 5.1 *)

(* Main program for linear time-varying DiffEqn example
      iter  - Number of iterations between plotted points
      amps  - Plotting array                                    *)

TYPE
   plotArray = ARRAY[1..5,0..100] OF real;

VAR
   i, iter : integer;
   t, dt   : real;
   amps        : plotArray;

FUNCTION G (t, amps : real) : real;
   BEGIN
      G := (4.0 - amps) / t
   END;

BEGIN

   (* Initialize values of current and time at t = 0.05
      Set unused plotting array element to zero              *)

   amps[1,1] := 84.0;
   amps[1,0] := 0.0;
   t := 0.05;
   dt := 0.05;
   iter := 2;

   (* Compute and plot the values of the inductor current *)

   FOR i := 2 TO 50 DO
      BEGIN
         DiffEqn(t, amps[1,i-1], t+dt, iter, amps[1,i], G);
         t := t + dt
      END;
   Plot5(amps, 1, 50, 100)
END.
```

Figure 5.5 Main program and function for Example 5.1.

As an example of the use of the procedure DiffEqn, consider the following problem:

EXAMPLE 5.1

A time-varying inductor is defined over a range of time from 0.05 to 2.5 sec by the relation $L(t) = t$ H. Let the voltage applied to such an inductor be constant and equal to 4 V. Let $i(t)$ be the current through the inductor, and let $i(0.05) = 84$ A. We may use the procedure DiffEqn to find the current $i(t)$ for the specified time range. The function $G(t$, amps) is defined by (5.6). The procedure Plot5 is used to display the results. A flowchart of the computational process used is shown in Fig. 5.4. A listing of the function $G(t$, amps) and the main program is given in Fig. 5.5. A plot of the output is given in Fig. 5.6. Although it is difficult, in general, to determine explicitly the solution to a differential equation with time-varying coefficients, for this particular example it is easily verified, by direct substitution in (5.6), that the solution is $i(t) = 4(1 + 1/t)$ A. A comparison of this result with the numerical values of the function plotted in Fig. 5.6 readily establishes the excellent agreement between the numerical results and the explicit mathematical solution.

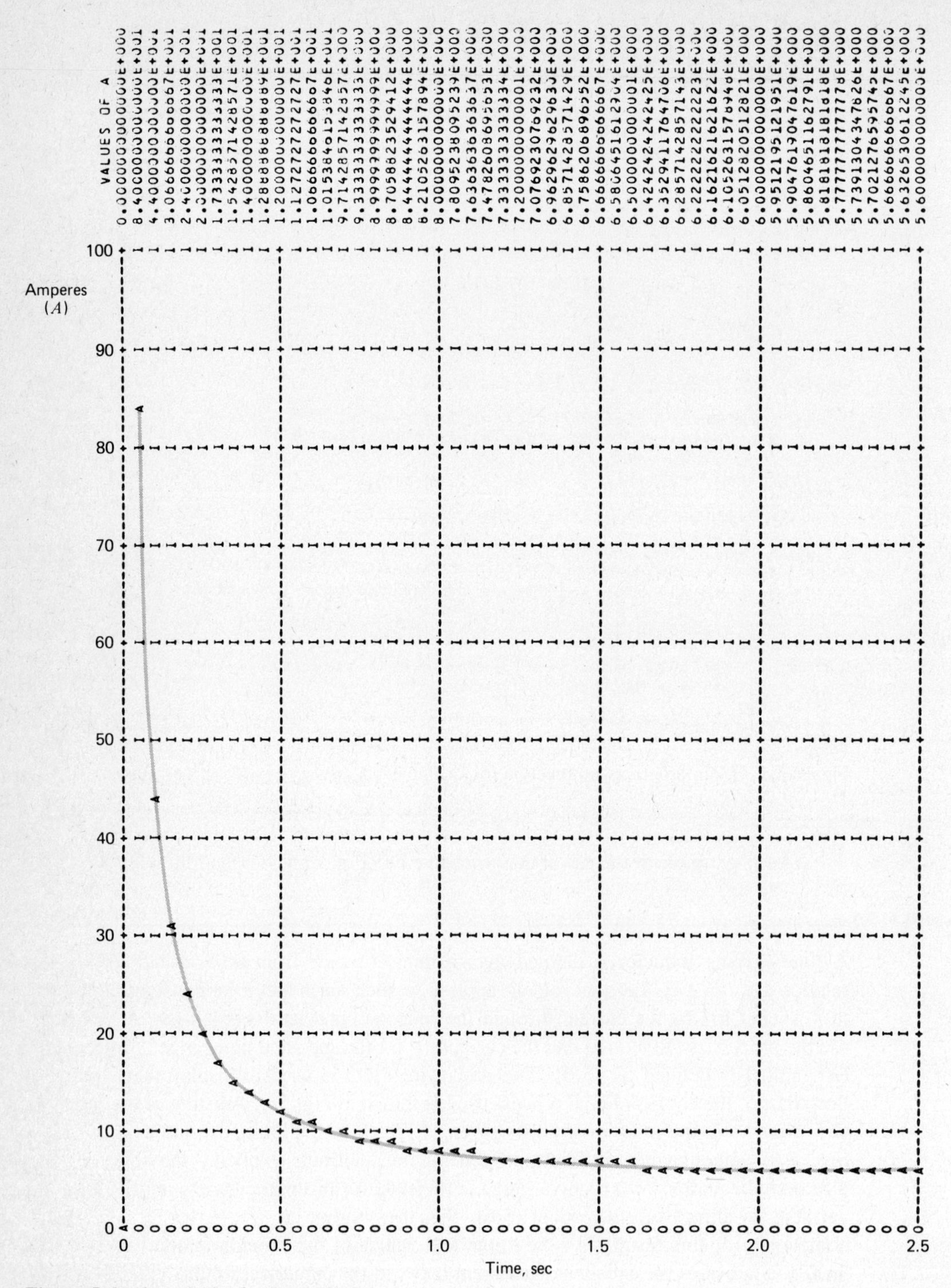

Figure 5.6 Output plot for Example 5.1.

5.4 SOLUTION OF NONLINEAR DIFFERENTIAL EQUATIONS

The problem posed in Sec. 5.1 not only includes cases where the coefficients of the first-order differential equations are functions of the independent variable, that is, where they are *not* constant, but also includes cases where the first-order differential equations are nonlinear, that is, those in which at least one term is of higher than first degree in the dependent variable and/or its derivative. For example, the equation $y'(t)y(t) = k$ is a nonlinear first-order differential equation. Such equations arise frequently in engineering situations. As an example, consider the case where the inductance L of a two-terminal inductor, as a result of saturation of the core, is a function of the current $i(t)$ flowing through it. Thus we may write $L = L(i)$. From the basic differential equation relating the voltage $v(t)$ and the current $i(t)$ for a two-terminal inductor we know that

$$v(t) = \frac{d}{dt}[Li(t)] \tag{5.15}$$

Since L is a function of $i(t)$, we obtain

$$v(t) = L(i)i'(t) + i(t)i'(t)\frac{d}{di}L(i) \tag{5.16}$$

where the prime indicates differentiation with respect to t. If we rearrange the above equation to put it in the form given in (5.4), we obtain

$$i'(t) = g(t, i) = \frac{v(t)}{L(i) + i(t)(d/di)L(i)} \tag{5.17}$$

Since $L(i)$ and thus $(d/di)L(i)$ are known functions, (5.17) may be solved for $i(t)$ if a function $v(t)$ is specified and if some initial condition on the current $i(t)$ is given.

A similar situation exists for the case of a nonlinear capacitor, that is, one in which the capacitance C is a function of the voltage appearing across the terminals of the capacitor. Such a dependence is found in modeling the capacitive effects associated with many semiconductor devices. In such a case we may write $C = C(v)$. The first-order differential equation written in the form of (5.4) for such a capacitor is

$$v'(t) = g(t, v) = \frac{i(t)}{C(v) + v(t)(d/dv)C(v)} \tag{5.18}$$

Example 5.2 shows the solution of a first-order nonlinear differential equation.

EXAMPLE 5.2

A nonlinear inductor is defined over a range of current from 0 to 0.1 A by the relation $L(i) = 3 - 6i^2(t)$ H, where $i(t)$ is the current through the inductor. Let a voltage $v(t) = 6 - 144t^2$ V be applied to the inductor over the range of time from 0 to 0.2 sec, and let $i(0) = 0$. We may use the procedure DiffEqn to find the current $i(t)$ for the specified time range, using (5.17) to define the function $G(t$, amps) and the procedure Plot5 to display the resulting variation of $i(t)$. The flowchart used is similar to that shown in Fig. 5.4 for Example 5.1. A listing of the function $G(t$, amps) and the main program is given in Fig. 5.7. A plot of $i(t)$, $v(t)$, and $L[i(t)]$ is shown in Fig. 5.8. Although it is

```
PROGRAM Main (output);     (* Example 5.2 *)

(* Main program for nonlinear DiffEqn example
      iter - Number of iterations between plotted points
      a - plotting array                                    *)

CONST   (* Values of scale factors used in plotting *)
   iscale = 200.0;
   vscale = 10.0;
   Lscale = 10.0;

TYPE
   plotArray = ARRAY[1..5,0..100] OF real;

VAR
   t, dt, oldamps, amps : real;
   i, iter              : integer;
   a                    : plotArray;

FUNCTION G (t, amps : real) : real;
   BEGIN
      G := (6.0 - 144.0*t*t) / (3.0 - 18.0*amps*amps)
   END;

BEGIN

   (* Initialize values of variables *)

   oldamps := 0.0;
   t := 0.0;
   dt := 0.005;
   iter := 2;
   a[1,0] := 0.0;
   a[2,0] := 6.0 * vscale;
   a[3,0] := 3.0 * Lscale;

   (* Compute and plot inductor current, voltage,
      and inductance                                        *)

   FOR i := 1 TO 40 DO
      BEGIN
         DiffEqn(t, oldamps, t+dt, iter, amps, G);
         a[1,i] := amps * iscale;
         oldamps := amps;
         a[2,i] := (3.0 - 6.0*amps*amps) * vscale;
         t := t + dt;
         a[3,i] := (6.0 - 144.0*t*t) * Lscale;
      END;
   Plot5(a, 3, 40, 100)
END.
```

Figure 5.7 Main program and function for Example 5.2.

difficult, in general, to determine explicitly the solution to a nonlinear differential equation, for this particular example it is easily verified by direct substitution in (5.17) that the solution is $i(t) = 2t$. Comparison of this result with the numerical values of the function plotted in Fig. 5.8 readily establishes the validity of the numerical results.

5.5 CIRCUITS CONTAINING RESISTORS AND A SINGLE INDUCTOR OR A SINGLE CAPACITOR

The same techniques that have been used in this chapter to solve for the relations between the terminal variables of a single inductor or capacitor can also be applied to networks

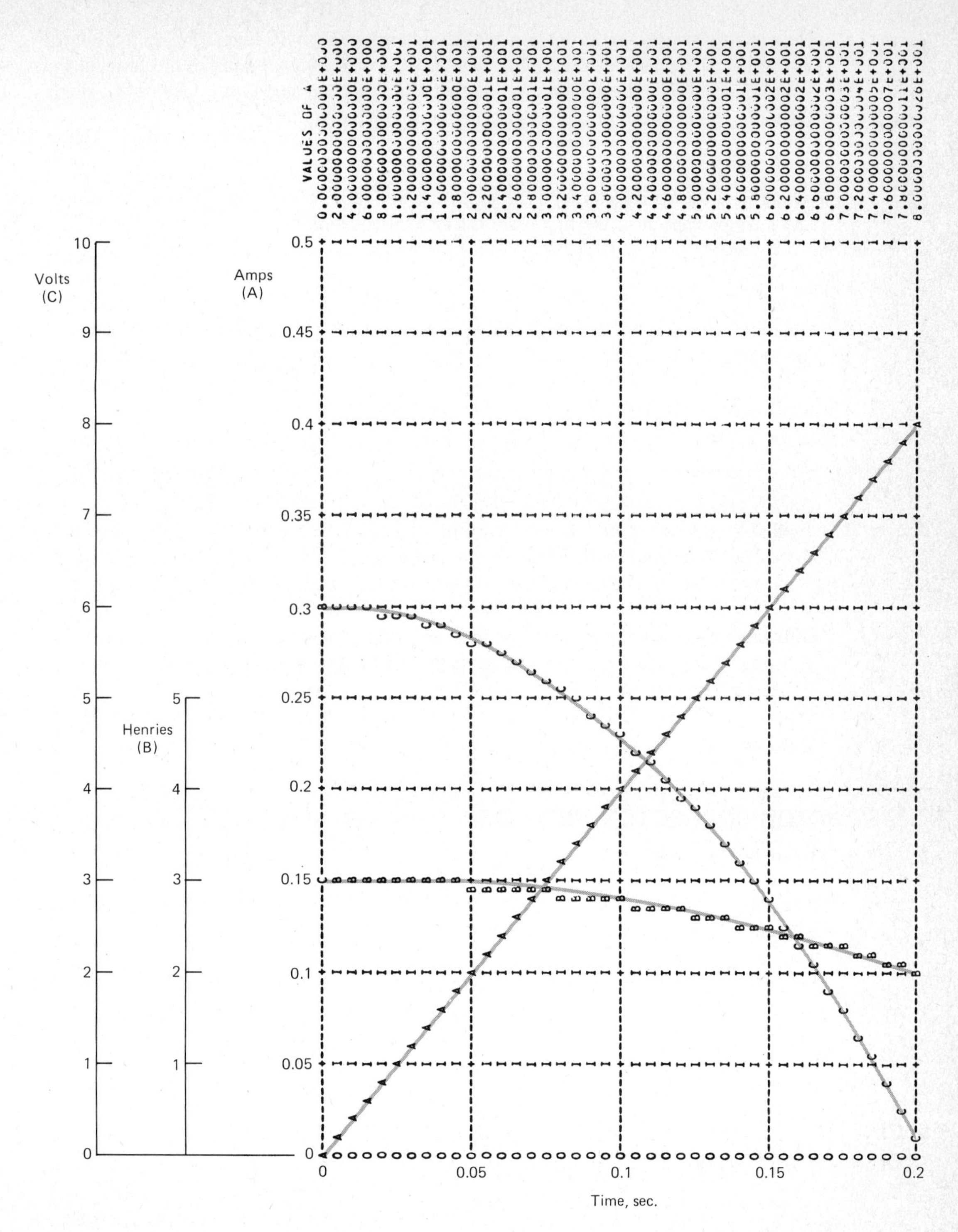

Figure 5.8 Output plot for Example 5.2.

containing resistive elements *plus* a single inductor or a single capacitor. The reason for this is that the equations describing such circuits can always be put in the form of the first-order differential equation given in (5.4). If the circuit contains resistors and a single inductor, the first-order differential equation will have the form

$$i_L'(t) = g(t, i_L) \tag{5.19}$$

where i_L is the current through the inductor. Similarly, if the circuit contains resistors and a single capacitor, the first-order differential equation will have the form

$$v_C'(t) = g(t, v_C) \tag{5.20}$$

where v_C is the voltage across the capacitor. The form of the first-order differential equations given in (5.19) and (5.20) will also apply to the case where the resistive and the inductive or capacitive elements are time-varying and/or nonlinear. Rearranging the circuit equations to put them in the form of (5.19) or (5.20) is easily accomplished by application of Kirchhoff's laws (see Example 5.3).

EXAMPLE 5.3

An *RL* circuit containing a single inductor is shown in Fig. 5.9. Let the voltage $v(t)$ applied to this circuit be a square pulse of 80 V from 0 to 0.003 sec as shown in Fig. 5.10. The current $i(t)$ in the circuit is assumed to be zero before application of the pulse. We may use the procedure DiffEqn to determine the current $i(t)$ for a range of time from 0 to 0.005 sec and the procedure Plot5 to display the results. The flowchart for the computational process is the same as that used for Example 5.1. It is shown in Fig. 5.4. A listing of the main program and the function $G(t$, amps) that is called by the procedure DiffEqn is given in Fig. 5.11. A plot of the resulting output current and the input voltage is shown in Fig. 5.12. The resulting values of the current $i(t)$ are easily verified by direct computation.

5.6 PREDICTOR-CORRECTOR METHODS: THE PROCEDURE AdBash

The numerical methods for the solution of a differential equation introduced in the earlier sections of this chapter are frequently referred to as *one-step methods*. They use information about a single previous value of $y(t)$ — $y(t_i)$ and $y'(t_i)$ — to find the next value $y(t_{i+1})$. Thus they are said to be *self-starting*. In general they are based on a Taylor

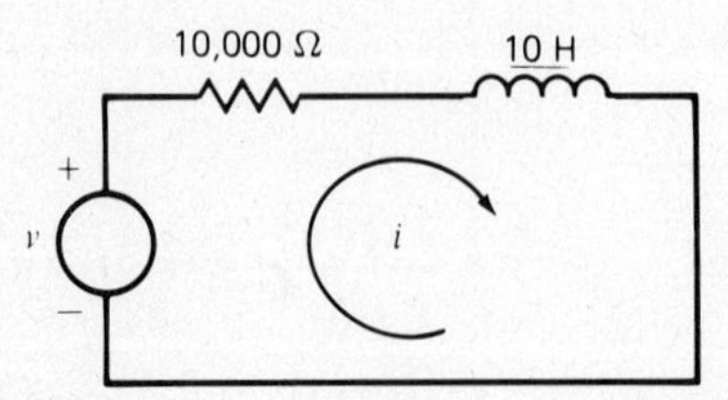

Figure 5.9 Circuit with a single inductor.

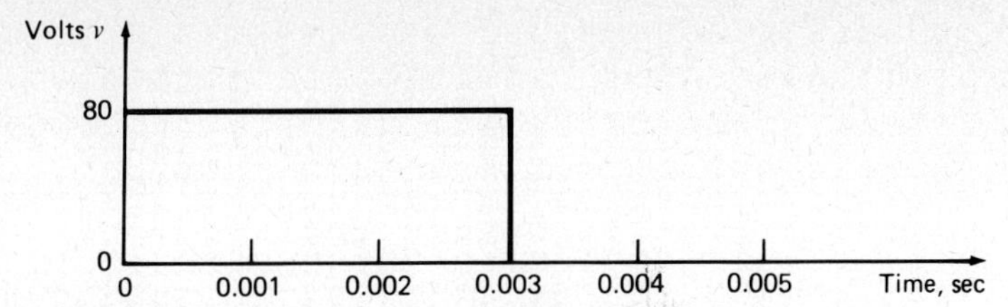

Figure 5.10 Input-voltage waveform for Example 5.3.

```
PROGRAM Main (output);     (* Example 5.3 *)

(* Main program for resistor-inductor DiffEqn example
      iter - Number of iterations between plotted points
      a - Plotting array                                    *)

CONST   (* Value of scale factor used in plotting *)
   iscale = 10000.0;

TYPE
   plotArray = ARRAY[1..5,0..100] OF real;

VAR
   t, dt, oldamps, amps : real;
   i, iter              : integer;
   a                    : plotArray;

FUNCTION G (t, amps : real) : real;
   BEGIN
      IF t <= 0.003
         THEN G := (80.0 - amps*10000.0) / 10.0
         ELSE G := (-amps*10000.0) / 10.0
   END;


BEGIN

   (* Initialize values of variables *)

   t := 0.0;
   dt := 1.0E-04;
   oldamps := 0.0;
   iter := 10;
   a[1,0] := 0.0;
   a[2,0] := 80.0;

   (* Compute and plot the inductor current and the
      input voltage                                    *)

   FOR i := 1 TO 50 DO
      BEGIN
         DiffEqn(t, oldamps, t+dt, iter, amps, G);
         a[1,i] := amps * iscale;
         oldamps := amps;
         t := t + dt;
         IF t <= 0.003
            THEN a[2,i] := 80.0
            ELSE a[2,i] := 0.0
      END;
   Plot5(a, 2, 50, 100)
END.
```

Figure 5.11 Main program and function for Example 5.3.

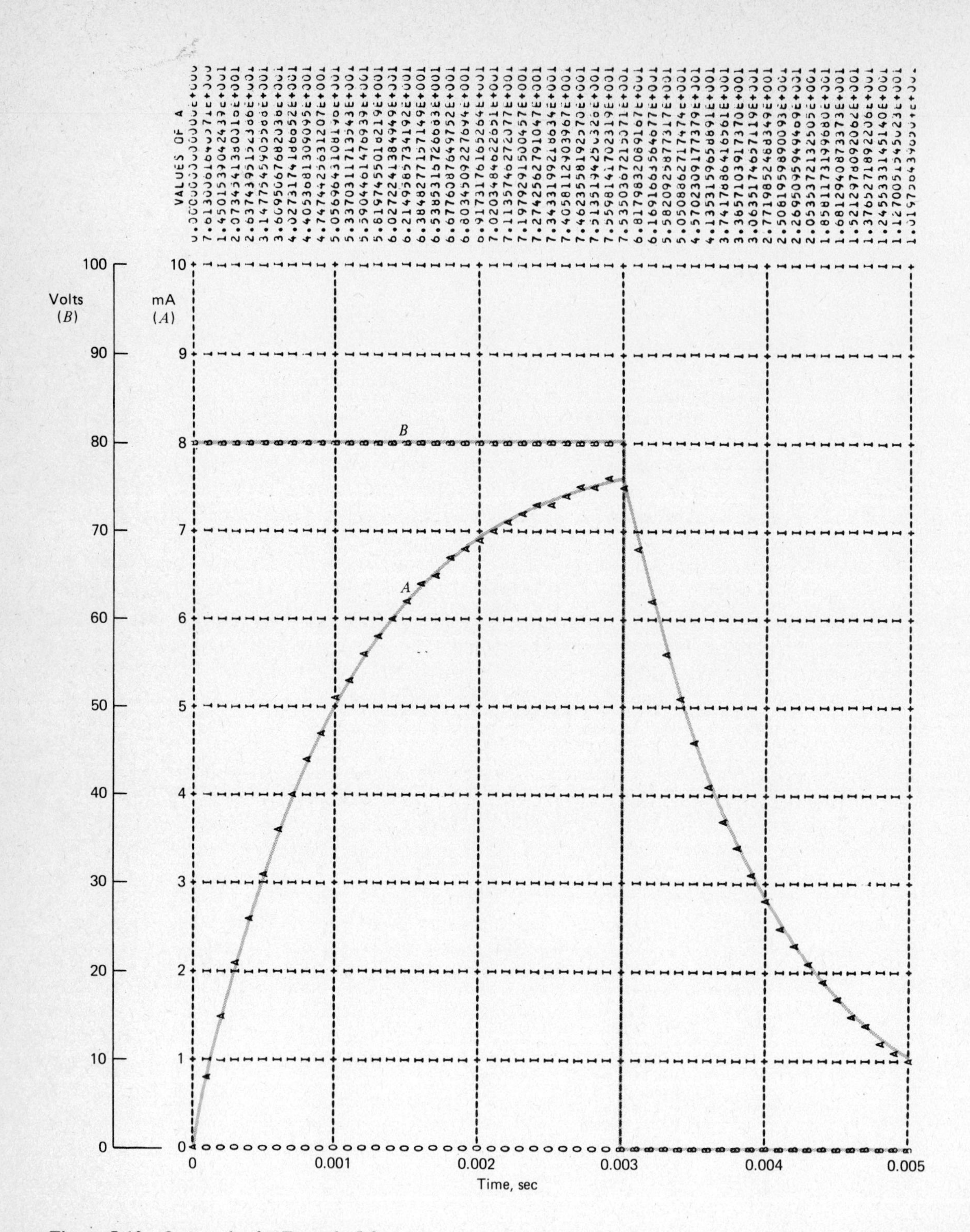

Figure 5.12 Output plot for Example 5.3.

expansion of the function $f(t)$. Here we introduce a quite different type of numerical technique called a *predictor-corrector* method. Such methods use several previous values of the function $y(t)$ and $y'(t)$ to calculate each new value of $y(t)$. As a result, they must be "started" by using a one-step method to obtain the required previous values. Predictor-corrector methods have an advantage over the one-step ones in that they provide a means for controlling the error that occurs at each step of the process. Thus, unnecessary evaluations of the defining relation of (5.4) can be eliminated, and as a result, the predictor-corrector methods tend to be more efficient than the one-step ones.

The operation of a typical predictor-corrector method requires the use of two equations, a *predictor equation* and a *corrector equation,* and is illustrated by the flowchart in Fig. 5.13. In this flowchart, for simplicity, we assume that $t = 0$ is the starting value of time, and we use the notation

$$y_i = y(i\ \Delta t) \quad \text{and} \quad y'_i = y'(i\ \Delta t) = g(i\ \Delta t,\ y_i) \tag{5.21}$$

The flowchart starts with the (known) values of y_0 and y'_0. A one-step method is then used to find y_1, y_2, y_3, y'_1, y'_2, and y'_3. Next, the predictor equation is used to make a first estimate of y_4, which is then used in (5.4) to find y'_4. Then the corrector equation is applied iteratively using the value of y'_4 and the other previously determined values of the derivatives to find a corrected value of y_4. This, in turn, is used to find a corrected value of y'_4. Using these corrected values, the corrector equation is again applied, and the process is repeated until the error, defined as the change in y'_4, is less than some preset value errmax, or until the number of applications of the corrector equation is greater than some constant kmax. The overall predictor-corrector cycle is then repeated to find y_5, and the process continues until all n values of y_i $(i = 1, 2, \ldots, n)$ have been computed.

Many different predictor-corrector methods may be found in the literature. In general, the operation of all of them follows a pattern similar to that given in the flowchart of Fig. 5.13. The main difference between the various methods is in the form of the equations used for the predictor and for the corrector. As an example of the predictor-corrector technique, consider the Adams-Bashforth method. It is a frequently used one which has been shown to have good numerical-stability properties. For it, the predictor equation is

$$y_{n+1} = y_n + (55y'_n - 59y'_{n-1} + 37y'_{n-2} - 9y'_{n-3})\,\frac{t}{24} \tag{5.22}$$

The corrector equation is

$$y_{n+1} = y_n + (9y'_{n+1} + 19y'_n - 5y'_{n-1} + y'_{n-2})\,\frac{t}{24} \tag{5.23}$$

Note that the predictor equation uses only the previous values y_n y'_n, y'_{n-1}, etc., to find y_{n+1}, while the corrector equation also uses y'_{n+1}.

A PASCAL program for implementing a predictor-corrector method for the solution of a differential equation must be supplied with the following inputs.

1. The initial value t_i of the independent variable t at which it is desired to start the solution process.

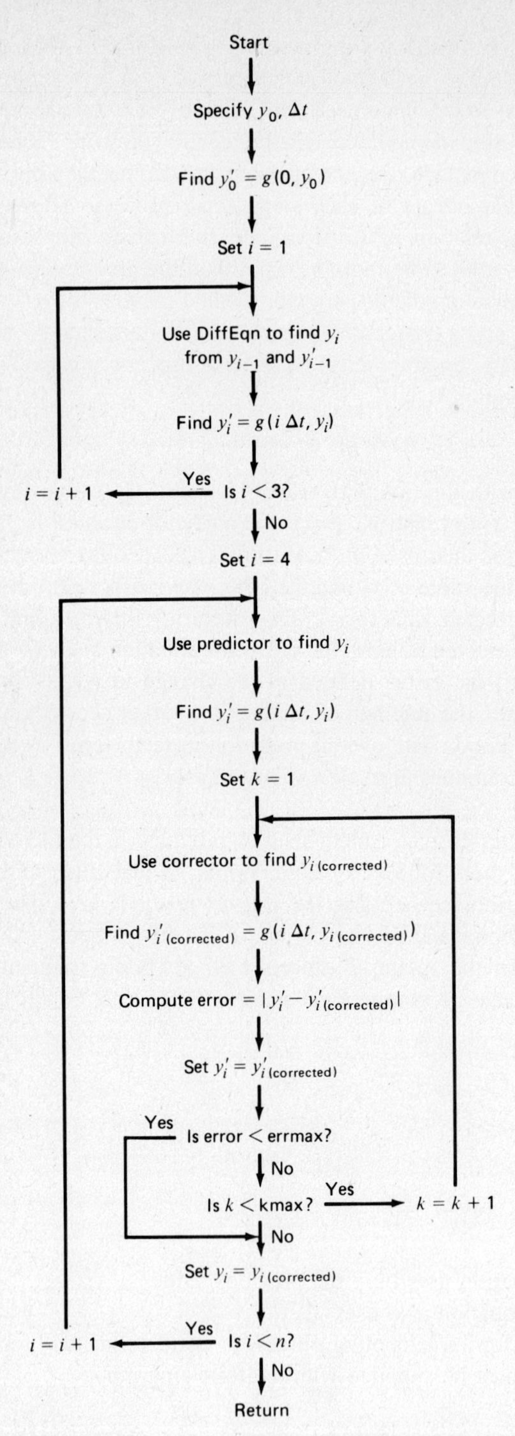

Figure 5.13 Flowchart for predictor-corrector method.

2. The value y_i of the dependent variable $y(t)$ corresponding to the value $t = t_i$.
3. The value t_{stop} of the independent variable t at which it is desired to stop the solution process.
4. The number n of points for which values of $y(t)$ are computed in going from t_i to t_{stop}. (Including the point at $t = t_i$, this makes a total of $n + 1$ points available for plotting.)
5. The name of the PASCAL function G(t, y : real) : real, which specifies the function $g(t, y)$ defined in (5.4).

Because the operation of a predictor-corrector method requires information on previous computations, it is convenient to store the values of $y(t)$ in an array. Anticipating the plotting that is usually done of the values of the function, it is logical to use our standard plotting array for this purpose. In addition, it is convenient to provide an optional "debugging" output. Corresponding to the above, we may specify the following additional parameters for the procedure:

6. A number *scale* that may be used to multiply the data stored in the plotting array by a convenient value to prepare it for plotting.
7. An index $a_{position}$ used to select any of the five positions in the plotting array for the data that is computed by the procedure.
8. The name of the plotting array used for storing and providing the output of the computed data.
9. A test parameter that can be used to output information as to how effectively the corrector portion of the predictor-corrector method is operating. If test = 1, information on the corrector-terminating conditions (the value of the index k and the error) is printed. If test = 0, the output is suppressed.

Using the name AdBash (for *Ad*ams-*Bash*forth predictor-corrector method) for the PASCAL procedure, we may define the following procedure heading:

```
PROCEDURE AdBash (tInitial, yInitial, tstop: real; n: integer;
                  scale: real; aPosition: integer; VAR a: plotArray;
                  FUNCTION G (t, y: real): real; test: integer);
```

where tInitial is the variable name used for the input t_i, etc. A listing of the procedure is given in Fig. 5.14. Its operation follows that indicated by the flowchart shown in Fig. 5.13 except that the plot array a[j, i] is used to store the values y_i (j is set equal to the input aPosition), and only the most recent values of the derivatives y_i' are stored. These are updated in the variables yp[i] (i = 0, 1, 2, 3, 4). The procedure DiffEqn is used to provide the starting values. Example 5.4 illustrates the use of the procedure. Table 5.2 gives a summary of its properties.

```
PROCEDURE AdBash (tInitial, yInitial, tstop : real; n : integer; scale :
                  real; aposition : integer; VAR a : plotArray;
                  FUNCTION G (t, y : real) : real; test : integer);

(* Adams-Bashforth predictor-corrector differential-equation
   solving procedure
     tInitial - Initial value of time
     yInitial - Initial value of dependent variable y(t)
     tstop - Final value of time
     n - Number of intermediate steps used in going to tstop
     scale - Factor for multiplying output data for scaling
     aposition - Index for position for storing data in
          plotting array
     a - Plotting array used to store intermediate values of y(t)
     test - Indicator used to check operation of corrector loop.
          Use test = 1 for output, test = 0 for no output
   Note: A function G(t, y) must be used to define the derivative
   of y(t). This procedure calls the procedure DiffEqn.     *)

   CONST
      iter = 4;           (* Number of iterations for DiffEqn *)
      kmax = 10;          (* Maximum cycles for corrector *)
      errmax = 1.0e-8;    (* Maximum error for corrector *)

   VAR
      t, dt, acor, ypcor, error : real;
      i, j, k : integer;
      yp        : ARRAY[0..4] OF real;

   BEGIN

      (* Calculate variables, store initial values *)

      t := tInitial;
      dt := (tstop - tInitial) / n;
      j := aposition;
      a[j,0] := yInitial;
      yp[0] := G(t, a[j,0]);

      (* Calculate three more values to start predictor *)

      FOR i := 1 TO 3 DO
         BEGIN
            DiffEqn(t, a[j,i-1], t+dt, iter, a[j,i], G);
            t := t + dt;
            yp[i] := G(t, a[j,i])
         END;

      (* Enter the predictor-corrector section of program *)

      FOR i := 4 TO n DO
         BEGIN

            (* Apply the predictor equation *)

            t := t + dt;
            a[j,i] := a[j,i-1] + dt * (55.0*yp[3] - 59.0*yp[2]
                      + 37.0*yp[1] - 9.0*yp[0]) / 24.0;
            yp[4] := G(t, a[j,i]);
            k := 0;
            REPEAT

               (* Start the corrector loop *)

               k := k + 1;
               acor := a[j,i-1] + dt * (9.0*yp[4] + 19.0*yp[3]
                        - 5.0*yp[2] + yp[1]) / 24.0;
```

Figure 5.14 Listing of the procedure AdBash.

```
            ypcor := G(t, acor);
            error := abs(yp[4] - ypcor);
            yp[4] := ypcor
         UNTIL (k >= kmax) OR (error < errmax);
         a[j,i] := acor;
         IF test > 0 THEN
            writeln(' AdBash Corrector, k =', k:3,
                    '  Error =', error);

         (* Update the derivative information *)

         FOR k := 1 TO 4 DO
            yp[k-1] := yp[k]
      END;  (* End of predictor-corrector section *)

   IF scale <> 1.0 THEN
      FOR i := 0 TO n DO
         a[j,i] := a[j,i] * scale
END;
```

Figure 5.14 (*Continued*)

TABLE 5.2 SUMMARY OF THE CHARACTERISTICS OF THE PROCEDURE AdBash

Identifying Statement PROCEDURE AdBash (tInitial, yInitial, tstop : real; n : integer; scale : real; aPosition : integer; VAR a : plotArray; FUNCTION G (t, y : real) : real; test : integer);

Purpose To solve the differential equation

$$y' = g(t, y)$$

and thus to find the value of $y(t)$ at the value of t specified as t_{stop}, starting from t_i, a known value of t, and y_i, where $y_i = y(t_i)$, using the Adams Bashforth predictor-corrector method.

Additional Subprograms Required This procedure calls the function identified by the statement

FUNCTION G (t, y : real) : real;

The function must be used to define the differential equation, that is, to specify $g(t, y)$. This procedure also calls the procedure DiffEqn.

Input Arguments

tInitial	The initial value t_i of the independent variable t
yInitial	The initial value y_i of $y(t)$, that is, $y(t_i)$
tstop	The final value t_{stop} of t for which the value of $y(t)$ is to be found
n	The number of points of $y(t)$ which are calculated in going from t_i to t_{stop}
scale	The factor used to multiply the data stored in the plotting array to prepare it for plotting
aPosition	The index used to select the position in the plotting array
test	A parameter used to control the output from the procedure. If test = 1, information on the terminating conditions of the corrector (the number of iterations and the error) is printed. If test = 0, this output is suppressed.

Output Argument

a	The plotting array used to store the output data

User-Defined Type

plotArray = ARRAY[1..5,0..100] OF real;

EXAMPLE 5.4

As an example of the use of the procedure AdBash to solve a differential equation, consider the problem used in Example 5.1. A solution for this using AdBash is given by the listing shown in Fig. 5.15. The output is shown in Fig. 5.16. Note that the values are stored in the a[j,i] array starting with i = 0, whereas those in the output of Example 5.1 are stored starting with i = 1.

```
PROGRAM Main (output);    (* Example 5.4 *)

(* Main program for linear time-varying Adams-Bashforth example
     a - Plotting array                                   *)

TYPE
   plotArray = ARRAY[1..5,0..100] OF real;

VAR
   a         : plotArray;

FUNCTION G (t, amps : real) : real;
   BEGIN
      G := (4.0 - amps) / t
   END;

BEGIN
   AdBash(0.05, 64.0, 2.5, 49, 1.0, 1, a, G, 0);
   Print5(a, 1, 49)
END.
```

Figure 5.15 Main program for Example 5.4.

J	A(1,J)
0	8.4000000000000E+001
1	4.4000000000000E+001
2	3.0666666666667E+001
3	2.4000000000000E+001
4	1.9869939707226E+001
5	1.7203269929276E+001
6	1.5308301465009E+001
7	1.3891901600234E+001
8	1.2791567278904E+001
9	1.1911831573915E+001
10	1.1192275221260E+001
11	1.0592753917126E+001
12	1.0085522734418E+001
13	9.6507837902200E+000
14	9.2740276265960E+000
15	8.9443765082044E+000
16	8.6535144008144E+000
17	8.3949744797686E+000
18	8.1636520306531E+000
19	7.9554636731493E+000
20	7.7671040481958E+000
21	7.5956689156985E+000
22	7.4395244294077E+000
23	7.2962091102925E+000
24	7.1643593547210E+000
25	7.0426521396907E+000
26	6.9299604635347E+000
27	6.8253183372736E+000
28	6.7278930205574E+000
29	6.6369628113427E+000
30	6.5518991349541E+000
31	6.4721519918571E+000
32	6.3972380515696E+000
33	6.3267308478414E+000
34	6.2602526547944E+000
35	6.1974677171348E+000
36	6.1380765782364E+000
37	6.0816113040011E+000
38	6.0284314408865E+000
39	5.9777205812669E+000
40	5.9294834307484E+000
41	5.8835432946370E+000
42	5.8397399151403E+000
43	5.7979276035056E+000
44	5.7579736212118E+000
45	5.7197567723153E+000
46	5.6831661755006E+000
47	5.6481001896264E+000
48	5.6144654708411E+000
49	5.5821761428451E+000

Figure 5.16 Output for Example 5.4

5.7 CONCLUSION

In this chapter we have illustrated the power and effectiveness of numerical methods for the solution of first-order ordinary differential equations. We have seen that a procedure for performing this solution may be successfully applied not only to the case of linear differential equations with constant coefficients but also to the cases where the coefficients are not constant and where the differential equation is nonlinear. We thus have developed a capability for performing analyses for some very real engineering problems that, in general, cannot be solved by any methods other than numerical ones. We shall see many other applications of our differential-equation-solving capabilities in the following chapters.

PROBLEMS

5.1. Over a period of time from 0 to 2 sec, a time-varying inductor has the value $L(t) = 2 + \sin 4\pi t$ H. If a voltage $v(t) = 150 \sin 2\pi t$ V is applied to the inductor over this time range, use DiffEqn to find and plot 41 points for $i(t)$, the current through the inductor, in amperes. Also plot $v(t)$ and $L(t)$. Assume that $i(0) = 0$.

5.2. **(a)** A current $i(t) = 75t^2 + 50t$ A is applied to the terminals of a time-varying capacitor defined by the relation $C(t) = t + 1$ F. Use DiffEqn to find the voltage $v(t)$ appearing across the terminals of the capacitor over a range of time from 0 to 2 sec. Assume that $v(0) = 0$. Plot the applied current and the resulting voltage, using 51 points to cover the specified time range.

(b) From the resulting plot of $v(t)$, make an estimate of a simple, explicit mathematical relationship defining the voltage waveshape. Substitute your estimate into the basic differential equation to determine its validity.

5.3. **(a)** A simple shunt *RC* circuit excited by a current source is shown in Fig. P5.3*a*. If the current source supplies a square-wave current $i(t)$ defined by the waveshape shown in Fig. P5.3*b*, use DiffEqn to plot 51 points of the voltage $v(t)$ appearing across the

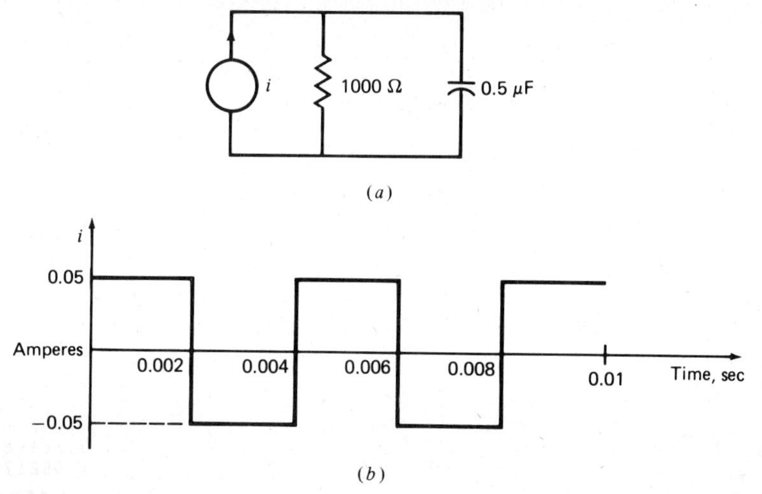

Figure P5.3

terminals of the resistor and the capacitor over a range of time from 0 to 0.005 sec. Assume $v(0) = 0$.

(b) If the current $i(t)$ remains constant for $t > 0.004$ sec, what final value of voltage $v(t)$ will eventually be reached?

5.4. If the circuit described in Prob. 5.3 is excited by a sinusoidal current $i(t) = 0.05 \sin 2000t$ A, use DiffEqn to find and plot 51 points of the voltage $v(t)$ appearing across the terminals of the capacitor for a range of time from 0 to 0.007 sec. Also plot $i(t)$. Assume $v(0) = 0$.

5.5. If the circuit described in Prob. 5.3 is excited by an exponentially decaying sinusoidal current $i(t) = 0.07e^{-t/0.002} \sin 2000t$ A, use DiffEqn to find and plot the resulting voltage $v(t)$ appearing across the terminals of the capacitor for a range of time from 0 to 0.007 sec. Use 51 points on the plot. Also plot $i(t)$. Assume $v(0) = 0$.

5.6. The circuit shown in Fig. P5.6 contains a nonlinear resistance whose value depends on the value of the current passing through it, that is, $R = R(i)$, where $i(t)$ is the current in amperes and $i(0) = 0$. Use DiffEqn to find and plot the current $i(t)$ for the range of time from 0 to 0.0025 sec. Use 51 points on the plot.

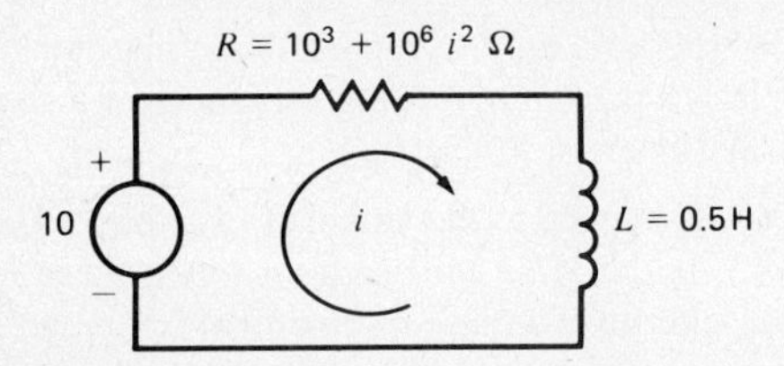

Figure P5.6

5.7. Write a PASCAL procedure for the improved Euler method described in Sec. 5.2. Let the defining statement for the procedure be

```
PROCEDURE Dfieu (tInitial, yInitial, tstop : real; iter : integer;
                 VAR y : real; FUNCTION G (t, y : real) : real);
```

The procedure should solve any equation defined by a function $g(t, y)$. Test the procedure by applying it to any of the problems which use the procedure DiffEqn given in this chapter.

5.8. (a) Another second-order method for solving first-order differential equations is defined by the equations

$$g_1 = g(t_i, y_i)$$

$$g_2 = g\left(t_1 + \frac{\Delta t}{2}, y_i + \frac{g_1\,\Delta t}{2}\right)$$

$$y_{i+1} = y_i + g_2\,\Delta t$$

This method of solution is sometimes referred to as the modified Euler method. Write a PASCAL procedure for the implementation of the above relations. Let the procedure call the function $g(t, y)$ defining the differential equation. Let the defining statement for the procedure be

```
PROCEDURE Dfmeu (tInitial, yInitial, tstop : real; iter : integer;
                 VAR y : real; FUNCTION G (t, y : real) : real);
```

Test the procedure by applying it to any of the problems in this chapter which require the use of the procedure DiffEqn.

(b) Explain the operation of this method by means of a graphical plot similar to the one shown in Fig. 5.2.

5.9. Use the procedure AdBash to solve Prob. 5.1.

5.10. Use the procedure AdBash to solve Prob. 5.2.

5.11. Use the procedure AdBash to solve Prob. 5.4.

5.12. In the vicinity of a discontinuity in the input function, predictor-corrector methods tend to lose accuracy, while one-step methods do not. As an illustration of this, repeat Example 5.3 using the procedure AdBash.

Chapter 6

Solution of Matrix Differential Equations: The General *RLC* Circuit

In Chap. 5 we presented a general method for solving a first-order differential equation. As an example of the application of this method, we showed that it could be used to solve for the variables in any network that consists of resistors and either a single inductor or a single capacitor. In this chapter we show how to extend the technique to higher-order differential equations.

6.1 STATE VARIABLES: MATRIX FIRST-ORDER DIFFERENTIAL EQUATIONS

A very powerful method for the analysis of systems is the *state-variable* method. Basically, the method requires that the variables used to formulate the system equations be chosen in such a way that they may be written as a single matrix equation of the form[1]

$$\mathbf{y}'(t) = \mathbf{A}\mathbf{y}(t) + \mathbf{u}(t) \tag{6.1}$$

where $\mathbf{y}(t)$ is a column matrix of the state variables $y^{(k)}(t)$, t is the independent variable, the prime indicates differentiation with respect to t, and $\mathbf{A}$ is a square matrix called the *system matrix,* which, in the general case, may be a function of t and/or of the state variables (in these cases, if the equation describes a network, we say that the network is time-varying and/or nonlinear). The column matrix $\mathbf{u}(t)$, with elements $u^{(k)}(t)$, is used to represent the input (or inputs) to the system. A discussion of the general methods by which the equations describing a given network are put in a state-variable formulation would take us too far from the purpose of this text. Here we merely content ourselves with pointing out that, for simple networks, inductor currents and capacitor voltages will

[1]Throughout this text we use boldface type to indicate matrices.

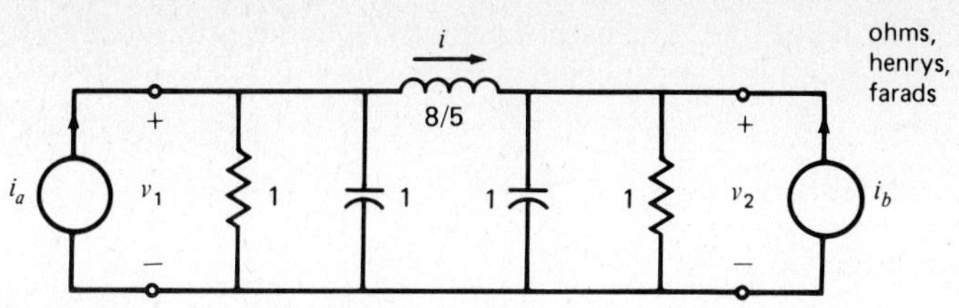

Figure 6.1 An *RLC* circuit.

usually prove to be a satisfactory set of state variables. For example, consider the network shown in Fig. 6.1. By application of Kirchhoff's current law (KCL) at the two nodes and by writing the branch equation for the inductor, we obtain the equations

$$\begin{aligned} i_a(t) - i(t) &= v_1(t) + v_1'(t) \\ i_b(t) + i(t) &= v_2(t) + v_2'(t) \\ \tfrac{8}{5}i'(t) &= v_1(t) - v_2(t) \end{aligned} \tag{6.2}$$

We may easily rearrange the elements of these equations to put them in the form given by (6.1). Thus we obtain

$$\begin{bmatrix} v_1'(t) \\ v_2'(t) \\ i'(t) \end{bmatrix} = \begin{bmatrix} -1 & 0 & -1 \\ 0 & -1 & 1 \\ \frac{5}{8} & -\frac{5}{8} & 0 \end{bmatrix} \begin{bmatrix} v_1(t) \\ v_2(t) \\ i(t) \end{bmatrix} + \begin{bmatrix} i_a(t) \\ i_b(t) \\ 0 \end{bmatrix} \tag{6.3}$$

In general, although there are some exceptions, the number of state variables for a given network is the number of inductors and capacitors that the network contains.

We may rewrite (6.1) in a more concise form, similar to the form of (5.4). Thus we have

$$\mathbf{y}'(t) = \mathbf{g}(t, \mathbf{y}) \tag{6.4}$$

where $\mathbf{g}$ is a matrix function of t and of the matrix $\mathbf{y}$. In view of the above, we may pose the following problem:

Problem. Given a set of n simultaneous first-order ordinary differential equations as represented by the matrix relation of (6.4), and a set of n initial conditions as defined by the initial-condition matrix $\mathbf{y}(t_i)$, find $\mathbf{y}(t)$, for $t > t_i$.

In the following section we discuss the application of numerical techniques to the solution of this problem.

6.2 NUMERICAL TECHNIQUES FOR THE SOLUTION OF SIMULTANEOUS FIRST-ORDER DIFFERENTIAL EQUATIONS

Almost any of the techniques that are used to solve a single first-order differential equation may be used to solve a set of simultaneous first-order differential equations, that is, a matrix first-order differential equation. As an example of this we use the Runge-Kutta method described in Sec. 5.3. The basic equations required by this method for the solution of a *single* first-order differential equation of the form

$$y'(t) = g(t, y) \tag{6.5}$$

were shown in Sec. 5.3 to be

$$\begin{aligned}
g_1 &= g(t_i, y_i) \\
g_2 &= g\left(t_i + \frac{\Delta t}{2}, y_i + \frac{\Delta t\, g_1}{2}\right) \\
g_3 &= g\left(t_i + \frac{\Delta t}{2}, y_i + \frac{\Delta t\, g_2}{2}\right) \\
g_4 &= g(t_i + \Delta t, y_i + \Delta t\, g_3) \\
y_{i+1} &= y_i + (g_1 + 2g_2 + 2g_3 + g_4)\frac{\Delta t}{6}
\end{aligned} \tag{6.6}$$

where y_i is a known value of the dependent variable $y(t)$ at some value t_i of the independent variable, y_{i+1} is the new value of $y(t)$ at t_{i+1} found by application of the numerical-solution technique, and $\Delta t = t_{i+1} - t_i$.

In order to extend the Runge-Kutta method to the solution of a set of simultaneous first-order differential equations, as defined by (6.4), we need merely define the relations given in (6.6) as matrix operations. Thus we obtain

$$\begin{aligned}
\mathbf{g}_1 &= \mathbf{g}(t_i, \mathbf{y}_i) \\
\mathbf{g}_2 &= \mathbf{g}\left(t_i + \frac{\Delta t}{2}, \mathbf{y}_i + \frac{\Delta t\, \mathbf{g}_1}{2}\right) \\
\mathbf{g}_3 &= \mathbf{g}\left(t_i + \frac{\Delta t}{2}, \mathbf{y}_i + \frac{\Delta t\, \mathbf{g}_2}{2}\right) \\
\mathbf{g}_4 &= \mathbf{g}(t_i + \Delta t, \mathbf{y}_i + \Delta t\, \mathbf{g}_3) \\
\mathbf{y}_{i+1} &= \mathbf{y}_i + (\mathbf{g}_1 + 2\mathbf{g}_2 + 2\mathbf{g}_3 + \mathbf{g}_4)\frac{\Delta t}{6}
\end{aligned} \tag{6.7}$$

where $\mathbf{y}_i$ is an n-element column matrix or vector defining the value of the dependent variables $y^{(k)}(t)$ ($k = 1, 2, \ldots, n$) at some value t_i of the independent variable t, and $\mathbf{y}_{i+1}$ is the column matrix defining the new values of the n dependent variables $y^{(k)}(t)$ at t_{i+1}, as found by application of the numerical technique. Similarly, $\mathbf{g}_1$ is a column matrix with n elements $g_1^{(k)}$ determined by evaluating the matrix function $\mathbf{g}$ of (6.4) at the values of its argument $(t_i, \mathbf{y}_i)$, and $\mathbf{g}_2$ is a column matrix with elements $g_2^{(k)}$ determined by evaluating $\mathbf{g}$ at the values of its arguments $(t_i + \Delta t/2, \mathbf{y}_i + \Delta t\, \mathbf{g}_1/2)$, etc. The new values of the dependent variables as specified by the column matrix $\mathbf{y}_{i+1}$ may, of course, be used as the starting point for a new determination of additional values of the elements of $\mathbf{y}(t)$, and the process can be continued iteratively to the desired final value of t, in a manner exactly paralleling that which was used in Chap. 5.

6.3 THE PROCEDURE MxDiffEqn

In this section we describe a procedure that is designed to implement the Runge-Kutta method for the solution of the set of first-order matrix differential equations given in Sec. 6.2. First of all, we require a subprogram to implement (6.4), similar to the function $g(t, y)$ which was used in connection with the procedure DiffEqn in Chap. 5. In this case, however, since (6.4) is a matrix equation, in general, we require more than one output; thus, it is more appropriate to use a PASCAL *procedure* rather than a PASCAL *function* to define the relations of (6.4). This procedure must have provision for receiving (as inputs) the value of the independent variable t_i at which the function is to be evaluated. It must also be supplied with the known values of the dependent variables $y_i^{(k)}$ $(k = 1, 2, \ldots, n)$ at $t = t_i$. As output, it must provide the n values of the function defined by (6.4). From a consideration of the above we see that a suitable argument list for the procedure is

```
(t:real; VAR y, g:array10)
```

where t is the value of the independent variable t, y is a one-dimensional array used to store the values of the $y_i^{(k)}$, and g is a one-dimensional array of the output values of the function defined by (6.4). The user-defined type array10 is assumed to have been declared by a main program global declaration having the form

```
TYPE array10 = ARRAY[1..10] OF real;
```

If we label this procedure Gn, the defining statement for it is

```
PROCEDURE Gn (t:real; VAR y, g:array10);
```

In addition to having available a procedure Gn defined as above, a procedure designed to compute a solution to a first-order matrix differential equation must be provided with the following information:

1. The value t_i of the independent variable t at which we desire to start the solution process.
2. The values $y_i^{(k)}$ $(k = 1, 2, \ldots, n)$ of the dependent variables $y^{(k)}(t)$ corresponding to the value $t = t_i$.
3. The number n of dependent variables. This is also the size (or order) of the matrix $\mathbf{A}$ of (6.1).
4. The value of the independent variable t at which it is desired to stop the solution process.
5. The number of iterations, that is, the number of intermediate calculations that it is desired to make in going from the starting value of t to its final value.

```
PROCEDURE MxDiffEqn (tInitial : real; VAR yInitial : array10;
                     nVars : integer; tstop : real; iter : integer;
                     VAR y : array10; PROCEDURE Gn (t : real;
                     VAR y, g : array10));

(* Matrix Runge-Kutta differential equation solving procedure
      tInitial - Initial value of time
      yInitial - Array of initial values of independent variables
      nVars - Number of independent variables
      tstop - Final value of time
      iter - Number of iterations from tInitial to tstop
      y - Array of values of independent variables at tstop
   A procedure Gn (t, y, g) (where y and g are arrays) must
   be used to define the derivatives of the independent
   variables                                                *)

   VAR
      t, dt, dt2, dt6        : real;
      ytemp, g1, g2, g3, g4  : array10;
      i, j                   : integer;

   BEGIN

      (*   Initialize the values of t and the y array and
           compute the value of dt, dt/2, and dt/6      *)

      t := tInitial;
      dt := (tstop - tInitial) / iter;
      dt2 := dt / 2.0;
      dt6 := dt / 6.0;
      FOR i := 1 TO nVars DO
         y[i] := yInitial[i];

      (*   Begin outer loop to compute the nVars values of
           the independent variables at tstop         *)

      FOR i := 1 TO iter DO
         BEGIN

            (*   Compute the arrays g1, g2, g3, and g4   *)

            Gn(t, y, g1);
            FOR j := 1 TO nVars DO
               ytemp[j] := y[j] + g1[j]*dt2;
            t := t + dt2;
            Gn(t, ytemp, g2);
            FOR j := 1 TO nVars DO
               ytemp[j] := y[j] + g2[j]*dt2;
            Gn(t, ytemp, g3);
            FOR j := 1 TO nVars DO
               ytemp[j] := y[j] + g3[j]*dt2;
            t := t + dt2;
            Gn(t, ytemp, g4);

            (*   Combine the arrays g1, g2, g3, and g4
                 to find the intermediate value of the
                 independent variables                *)

            FOR j := 1 TO nVars DO
               y[j] := y[j] + (g1[j] + 2.0*g2[j] + 2.0*g3[j]
                       + g4[j]) * dt6;
         END   (* End of outer loop with i index *)

   END;   (* End of MxDiffEqn procedure *)
```

(*a*)

Figure 6.2 The procedure MxDiffEqn.

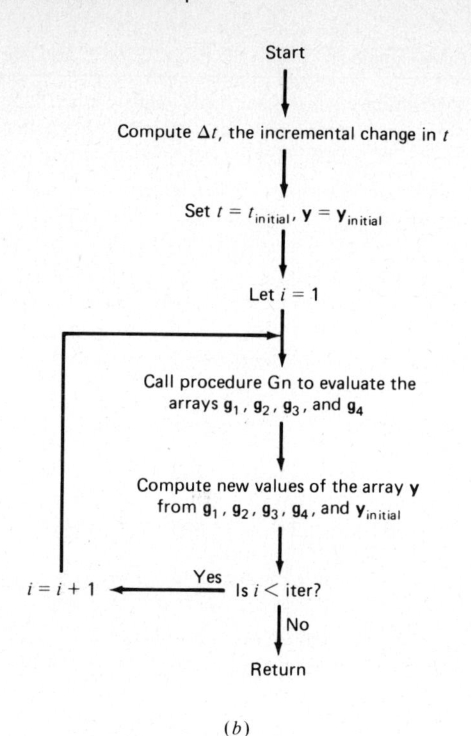

(*b*) **Figure 6.2** (*Continued*)

As an output, the procedure must provide us with the values of dependent variables $y_{i+1}^{(k)}$ ($k = 1, 2, \ldots, n$), that is, the values of the dependent variables $y^{(k)}(t)$ evaluated at the final value of t. Consideration of the above inputs and outputs leads us to a parameter list for the procedure that has the form

```
(tInitial : real; VAR yInitial : array10; ny : integer; tstop : real;
 iter : integer; VAR y : array10; PROCEDURE Gn (t : real;
 VAR y, g : array10);
```

where ny is the variable name for n, etc. The arrays yInitial and y are dimensioned the same as the arrays y and g in the procedure Gn. If we name the procedure MxDiffEqn (for *m*atri*x* Runge-Kutta, fourth-order *diff*erential *eq*uatio*n* solution), it will be identified by the heading

```
PROCEDURE MxDiffEqn (tInitial : real; VAR yInitial : array10; ny : integer;
                     tstop : real; iter : integer; VAR y : array10;
                     PROCEDURE Gn (t : real; VAR y, g : array10));
```

A listing of a set of PASCAL statements and a flowchart defining a procedure to perform the operations listed in (6.7) and requiring a procedure Gn is given in Fig. 6.2. The operation of the procedure can readily be understood by consulting the referenced equations. A summary of the characteristics of the procedure MxDiffEqn is given in Table 6.1.

TABLE 6.1 SUMMARY OF THE CHARACTERISTICS OF THE PROCEDURE MxDiffEqn

Identifying Statement PROCEDURE MxDiffEqn (tInitial : real; VAR yInitial : array10; ny : integer; tstop : real; iter : integer; VAR y : array10; PROCEDURE Gn (t : real; y, g : array10));

Purpose To solve the first-order matrix differential equation $\mathbf{y}' = \mathbf{g}(t, \mathbf{y})$ and thus to find the elements of the column matrix $\mathbf{y}(t)$ at the value of t specified as t_{stop}, starting from t_i, a known value of t, and $\mathbf{y}_i$, where $\mathbf{y}_i = \mathbf{y}(t_i)$

Additional Subprograms Required This subroutine calls the procedure identified by the statement

PROCEDURE Gn (t : real; y, g : array10);

which must be used to define the differential equation, that is, to specify $\mathbf{g}(t, \mathbf{y})$. The arguments y and g in the procedure are both one-dimensional arrays.

Input Arguments

tInitial The initial value t_i of the independent variable t

yInitial The one-dimensional array of variables yInitial[i] in which are stored the initial values of the variables in the first-order differential equation, that is, the elements of the column matrix $\mathbf{y}_i$, where $\mathbf{y}_i = \mathbf{y}(t_i)$

ny The number of elements in the column matrix $\mathbf{y}(t)$, that is, the number of variables

tstop The final value t_{stop} of t for which the value of $y(t)$ is desired

iter The number of iterations used by the subroutine in going from t_i to t_{stop}

Output Argument

y The one-dimensional array of variables y[i] in which are stored the values of the variables in the first-order matrix differential equation at t_{stop}, the final value of the independent variable t, that is, the elements of the column matrix $\mathbf{y}(t_{stop})$

The problem in Example 6.1 illustrates the use of the procedure MxDiffEqn.

EXAMPLE 6.1

In the circuit shown in Fig. 6.3 the currents $i_1(t)$ and $i_2(t)$ are zero at $t = 0$. If the circuit is excited by a step of voltage $v(t) = 120$ V applied at $t = 0$, we may use the procedure MxDiffEqn to find the currents over a range of time from 0 to 5 sec. For the circuit, the loop equations are

$$v(t) = i_1'(t) + 2[i_1(t) - i_2(t)]$$
$$0 = i_2'(t) + 5i_2(t) - 2i_1(t)$$

These are easily rearranged so that they have the form specified in (6.4). Thus, we obtain

$$i_1'(t) = v(t) - 2i_1(t) + 2i_2(t)$$
$$i_2'(t) = 2i_1(t) - 5i_2(t)$$

A procedure Gn and a main program to provide a solution to the above are easily written. A listing of such a set of PASCAL statements and a flowchart are given in Fig. 6.4. A

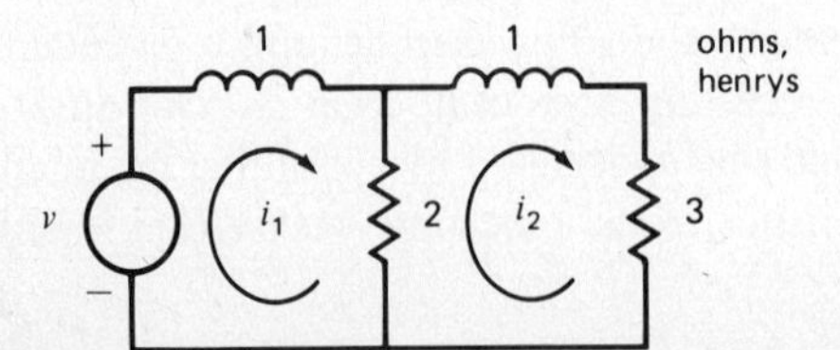

Figure 6.3 Circuit for Example 6.1

```
PROGRAM Main (output);    (* Example 6.1 *)

(* Main program for MxDiffEqn two-mesh resistor-inductor
   network example                                      *)

TYPE
   plotArray = ARRAY[1..5,0..100] OF real;
   array10   = ARRAY[1..10] OF real;

VAR
   t, dt           : real;
   i, iter         : integer;
   oldamps, amps   : array10;
   a               : plotArray;

PROCEDURE Gn (t : real; VAR amps, g : array10);
   BEGIN
      g[1] := 120.0 - 2.0*amps[1] + 2.0*amps[2];
      g[2] := 2.0*amps[1] - 5.0*amps[2]
   END;

BEGIN

   (* Initialize arrays and other variables *)

   t := 0.0;
   dt := 0.1;
   oldamps[1] := 0.0;
   oldamps[2] := 0.0;
   a[1,0] := 0.0;
   a[2,0] := 0.0;
   iter := 2;

   (* Compute and plot the values of the mesh currents *)

   FOR i := 1 TO 50 DO
      BEGIN
         MxDiffEqn(t, oldamps, 2, t+dt, iter, amps, Gn);
         a[1,i] := amps[1];
         a[2,i] := amps[2];
         oldamps[1] := amps[1];
         oldamps[2] := amps[2];
         t := t + dt
      END;
   Plot5(a, 2, 50, 100)
END.
```

(*a*)

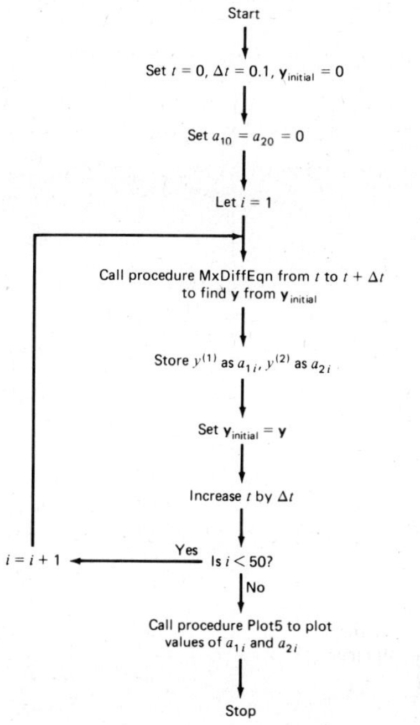

Start

Set $t = 0$, $\Delta t = 0.1$, $\mathbf{y}_{\text{initial}} = 0$

Set $a_{10} = a_{20} = 0$

Let $i = 1$

Call procedure MxDiffEqn from t to $t + \Delta t$ to find $\mathbf{y}$ from $\mathbf{y}_{\text{initial}}$

Store $y^{(1)}$ as a_{1i}, $y^{(2)}$ as a_{2i}

Set $\mathbf{y}_{\text{initial}} = \mathbf{y}$

Increase t by Δt

Is $i < 50$? — Yes → $i = i + 1$ (return to call of MxDiffEqn)

No

Call procedure Plot5 to plot values of a_{1i} and a_{2i}

Stop

(*b*)

Figure 6.4 Listing and flowchart for Example 6.1.

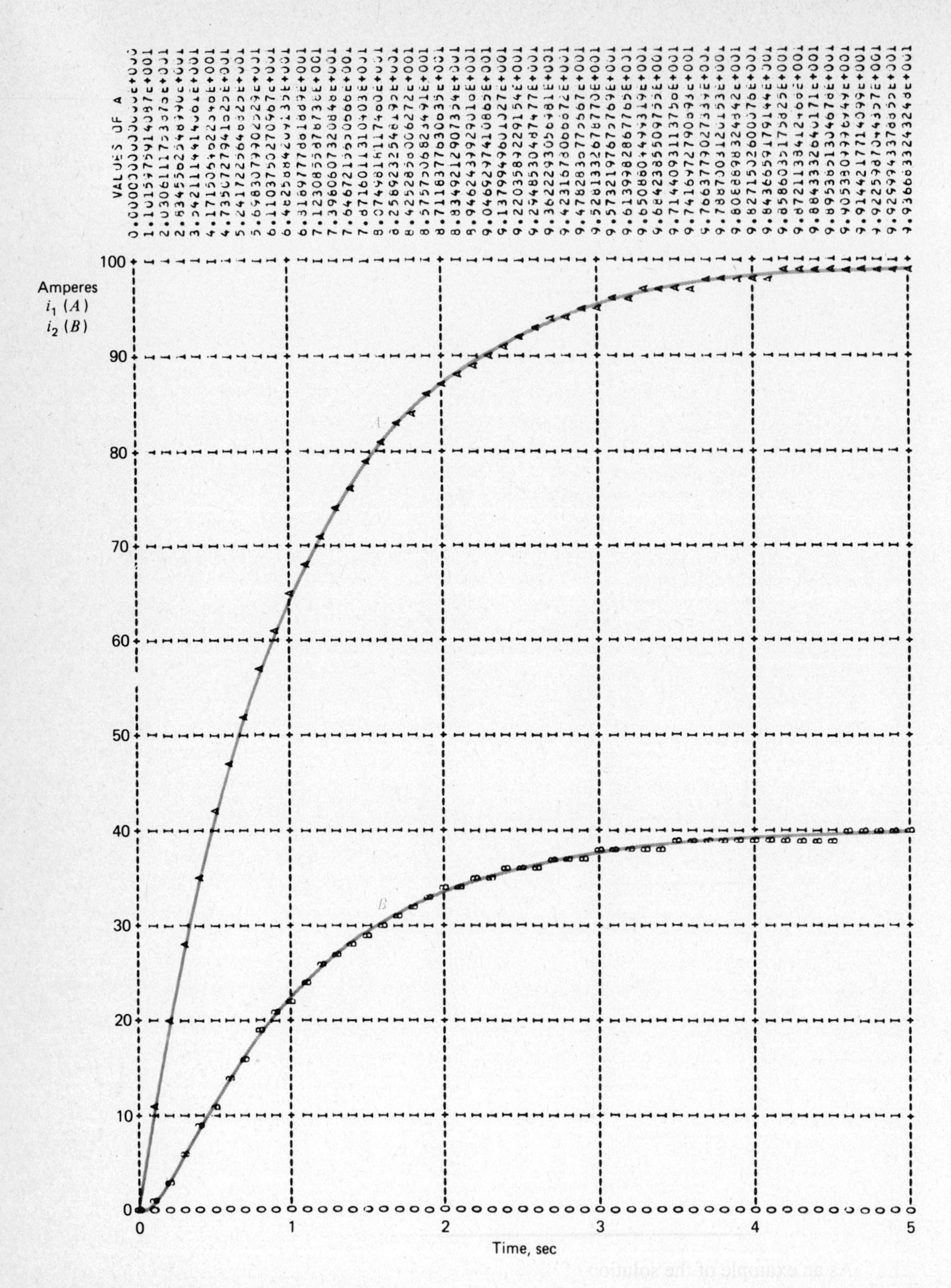

Figure 6.5 Output plot for Example 6.1.

copy of the output plot for the program is shown in Fig. 6.5. Direct circuit analysis readily shows that the solutions for $i_1(t)$ and $i_2(t)$ have the form

$$i_1(t) = 100 - 96e^{-t} - 4e^{-6t}$$
$$i_2(t) = 40 - 48e^{-t} + 8e^{-6t}$$

The agreement of these values with the numerical results is easily verified.

6.4 THE SOLUTION OF HIGHER-ORDER DIFFERENTIAL EQUATIONS

The state-variable approach described in Sec. 6.1 can also be applied to the solution of single differential equations that are higher than first-order. We illustrate this for a third-order differential equation; the extension to the nth-order case will be readily apparent. To see this, consider the third-order differential equation

$$z'''(t) + a_2 z''(t) + a_1 z'(t) + a_0 z(t) = f(t) \tag{6.8}$$

where the primes indicate differentiation with respect to the independent variable t, and $f(t)$ is some excitation or forcing function.

Let us now define a set of variables $y^{(k)}(t)(k = 1, 2, 3)$ by the relations

$$\begin{aligned} y^{(1)}(t) &= z(t) \\ y^{(2)}(t) &= z'(t) = y^{(1)\prime}(t) \\ y^{(3)}(t) &= z''(t) = y^{(2)\prime}(t) \end{aligned} \tag{6.9}$$

We may now write (6.8) in the form

$$y^{(3)\prime}(t) = -a_2 y^{(3)}(t) - a_1 y^{(2)}(t) - a_0 y^{(1)}(t) + f(t) \tag{6.10}$$

From (6.10) and the last two relations of (6.9), we see that we may write the following first-order matrix differential equation:

$$\begin{bmatrix} y^{(1)\prime}(t) \\ y^{(2)\prime}(t) \\ y^{(3)\prime}(t) \end{bmatrix} = \begin{bmatrix} 0 & 1 & 0 \\ 0 & 0 & 1 \\ -a_0 & -a_1 & -a_2 \end{bmatrix} \begin{bmatrix} y^{(1)}(t) \\ y^{(2)}(t) \\ y^{(3)}(t) \end{bmatrix} + \begin{bmatrix} 0 \\ 0 \\ f(t) \end{bmatrix} \tag{6.11}$$

This has exactly the form given in (6.1); therefore, we may use the procedure MxDiffEqn to solve this equation. The solution for the variable $y^{(1)}(t)$ gives us the desired solution for $z(t)$ that satisfies (6.8). Suitable initial conditions for $z(t)$ and its derivatives are required for the initial values of the state variables $y^{(k)}(t)$. Example 6.2 illustrates this method of solution.

EXAMPLE 6.2

As an example of the solution of higher-order differential equations, consider the network shown in Fig. 6.6. If we consider the current $i(t)$ as the excitation function, the voltage

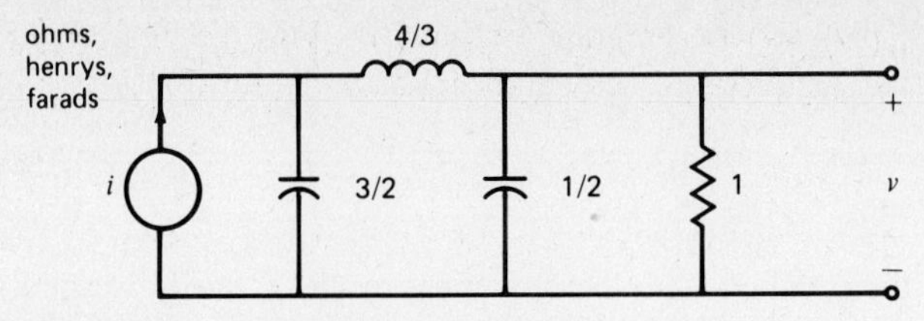

Figure 6.6 Circuit for Example 6.2

```
PROGRAM Main (output);    (* Example 6.2 *)

(* Main program for MxDiffEqn third-order differential
   equation example                                  *)

TYPE
   plotArray = ARRAY[1..5,0..100] OF real;
   array10   = ARRAY[1..10] OF real;

VAR
   t, dt            : real;
   i, j, iter       : integer;
   oldvolts, volts  : array10;
   a                : plotArray;

PROCEDURE Gn (t : real; VAR volts, g : array10);
   BEGIN
      g[1] := volts[2];
      g[2] := volts[3];
      g[3] := -volts[1] - 2.0*volts[2] - 2.0*volts[3]
               + 80.0
   END;

BEGIN

   (* Initialize arrays and other variables *)

   t := 0.0;
   dt := 0.2;
   oldvolts[1] := 0.0;
   oldvolts[2] := 0.0;
   oldvolts[3] := 0.0;
   a[1,0] := 0.0;
   iter := 2;

   (* Compute and plot the value of the voltage v(t) *)

   FOR i := 1 TO 50 DO
      BEGIN
         MxDiffEqn(t, oldvolts, 3, t+dt, iter, volts, Gn);
         a[1,i] := volts[1];
         FOR j := 1 TO 3 DO
            oldvolts[j] := volts[j];
         t := t + dt
      END;
   Plot5(a, 1, 50, 100)
END.
```

Figure 6.7 Main program and procedure for Example 6.2.

$v(t)$ as the dependent variable, and time as the independent variable, then the differential equation relating $v(t)$ and $i(t)$ may be shown to be

$$v'''(t) + 2v''(t) + 2v'(t) + v(t) = i(t)$$

By following the method described in this section, the third-order differential equation given above may be written in the form

$$\begin{bmatrix} y^{(1)\prime}(t) \\ y^{(2)\prime}(t) \\ y^{(3)\prime}(t) \end{bmatrix} = \begin{bmatrix} 0 & 1 & 0 \\ 0 & 0 & 1 \\ -1 & -2 & -2 \end{bmatrix} \begin{bmatrix} y^{(1)}(t) \\ y^{(2)}(t) \\ y^{(3)}(t) \end{bmatrix} + \begin{bmatrix} 0 \\ 0 \\ i(t) \end{bmatrix}$$

The solution for $y^{(1)}(t)$ is also the solution for $v(t)$. A listing of a procedure Gn and a main program for finding the solution over a range of time from 0 to 10 sec and for an input current step $i(t) = 80$ A is given in Fig. 6.7. All initial conditions are assumed to be zero. A plot of the output is shown in Fig. 6.8. The resulting plot of $v(t)$ is easily verified by direct analysis.

In the application of numerical differential-equation-solving techniques of the type we have presented in this and the preceding chapter, care must be taken to achieve a match between the numerical technique that is being used, the problem that is being solved, and the particular computer on which the problem is being run. This matching basically consists in determining the correct step size that must be used. If the step size is chosen too large, the results may be incorrect because of inaccuracies in the numerical method. This is called *truncation error*. If the step size is chosen too small, the results may be incorrect because of inaccuracies caused by accumulation of *round-off error* due to the finite word length (the number of significant decimal places) used in the particular computer installation. A detailed discussion of this problem is beyond the scope of this text; however, its importance is readily evident from an inspection of the data presented in Fig. 6.9.[2] This figure tabulates the maximum error found in applying a fourth-order Runge-Kutta method to the solution of the equations

$$\begin{aligned} y^{(1)\prime}(t) &= y^{(2)}(t) \\ y^{(2)\prime}(t) &= -y^{(1)}(t) \end{aligned} \tag{6.12}$$

These equations represent the state-variable formulation of the differential equation

$$z''(t) + z(t) = 0 \tag{6.13}$$

where $y^{(1)}(t) = z(t)$. The solution of such an equation, for initial conditions $z(0) = 100$ and $z'(0) = 0$, is easily found to be

$$z(t) = 100 \cos t \tag{6.14}$$

This is simply a sinusoidal oscillation with a period of 2π sec. Numerical integration of the relations of (6.12) for a time range from 0 to 13 sec (approximately two cycles) for the indicated number of iterations and the corresponding step size gave the maximum error shown in the figure. Note that if the number of iterations is made too large (too

[2]The author is indebted to Dr. J. V. Wait of the Department of Electrical and Computer Engineering of the University of Arizona for furnishing this example.

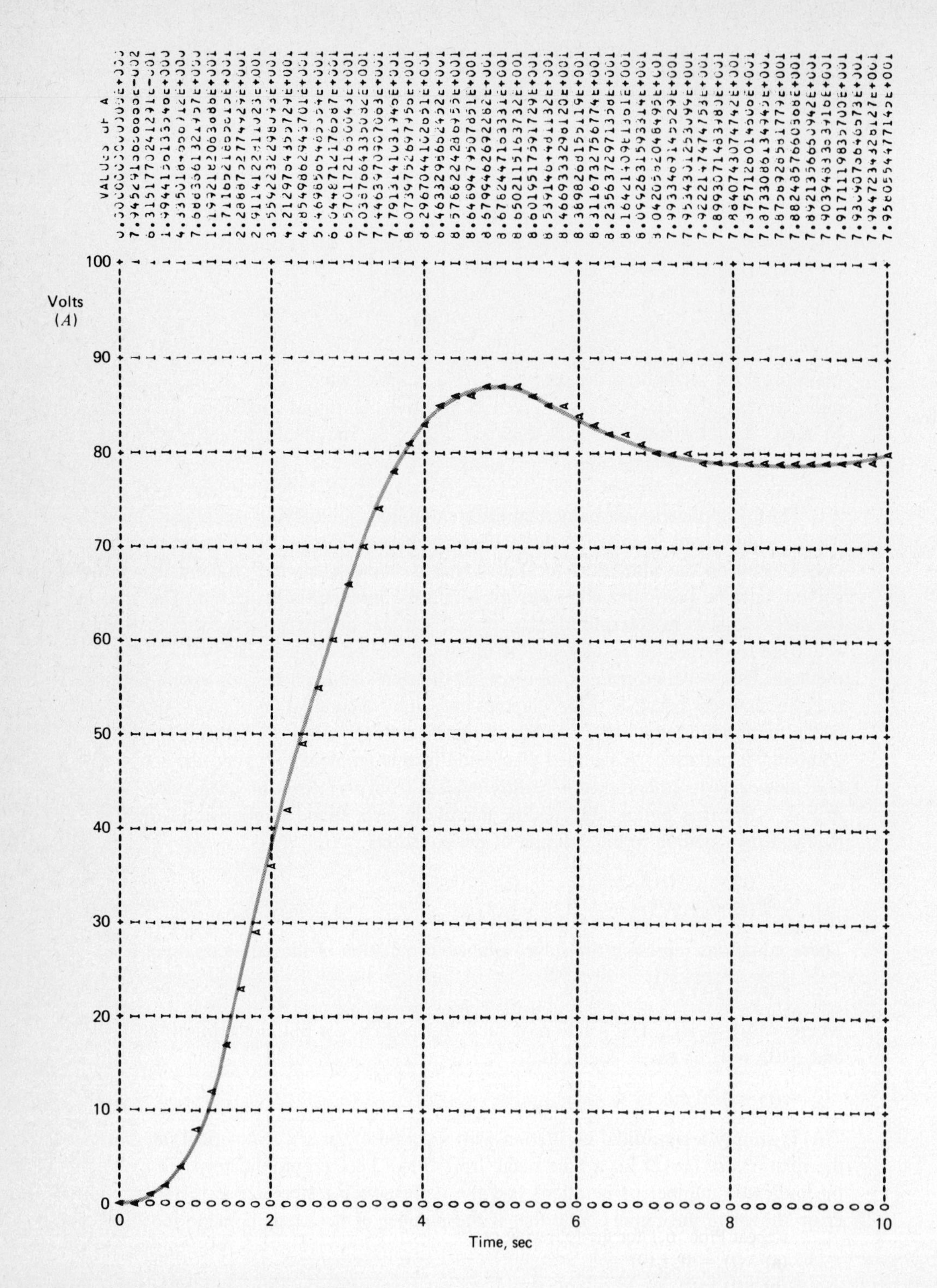

Figure 6.8 Output plot for Example 6.2.

Number of iterations	Step size, sec	Maximum error
13,000	0.001	0.00792
1,300	0.01	0.00078
650	0.02	0.00038
260	0.05	0.00021
130	0.1	0.00104
65	0.2	0.01501
26	0.5	0.59470

Figure 6.9 Truncation and round-off error.

small a step size), the maximum error increases. This increase is due to the effects of round-off error in the computer (an eight-decimal-digit machine was used for this study). Similarly, for too small a number of iterations (too large a step size), the maximum error again increases. This increase, however, is due to the effects of inaccuracies in the numerical method that is being used. For this particular problem, it should be apparent that for a correct match between the numerical method, the problem, and the computer, a step size of about 0.05 sec gives the most accurate results.

6.5 CONCLUSION

In this chapter we have introduced the concept of a simultaneous set of first-order differential equations. We have shown that such a set of equations can be used to represent many network situations. The numerical techniques used to solve such a set of equations are directly applicable to the cases where the elements of the network are nonlinear and/or time-varying. Thus, they may be used to solve problems that cannot be solved by direct analytic means. Indeed, the methods of solution described in this chapter form the basis for some of the most sophisticated digital-computer network-analysis programs developed to date.

PROBLEMS

6.1. An *RLC* circuit is shown in Fig. P6.1. If a step function of voltage $v(t) = 50$ V is applied to the circuit at $t = 0$ and if $i_L(0)$ and $v_C(0)$ are zero, that is, if there are no initial conditions, find 51 values of the variables $i_L(t)$ and $v_C(t)$ covering a range of time from 0 to 2.5 sec, and plot the results.

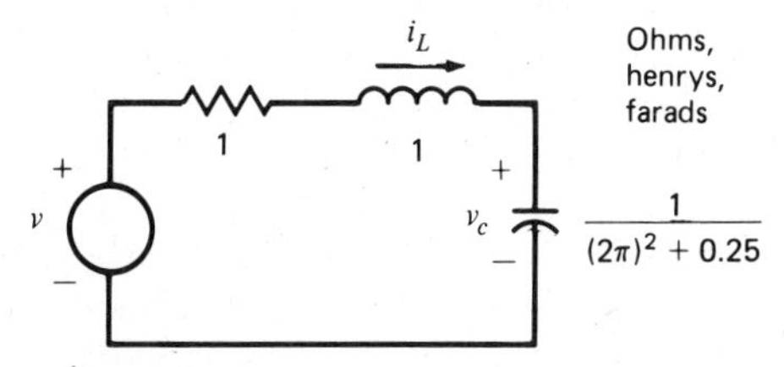

Figure P6.1

6.2. Repeat Prob. 6.1 for the following cases:

(a) $v(t) = 0,\ i_L(0) = 8,\ v_C(0) = 0$

(b) $v(t) = 0,\ i_L(0) = 0,\ v_C(0) = 50$

(c) $v(t) = 0,\ i_L(0) = 5.6,\ v_C(0) = 35$

The last case may be considered as the superposition of the first and the second cases (scaled by a factor of 0.7). Show that the output responses for this case are proportional to the sum of the outputs for the two preceding cases.

6.3. In the circuit shown in Fig. P6.3, $i(0) = 25$ A and $v(0) = 0$. Over a range of time from 0 to 1 sec, plot 51 points of the current $i(t)$ (in amperes), the energy $w_C(t)$ stored in the capacitor, and the energy $w_L(t)$ stored in the inductor (in joules). (*Note:* $w_C(t) = Cv^2(t)/2$ and $w_L(t) = Li^2(t)/2$.)

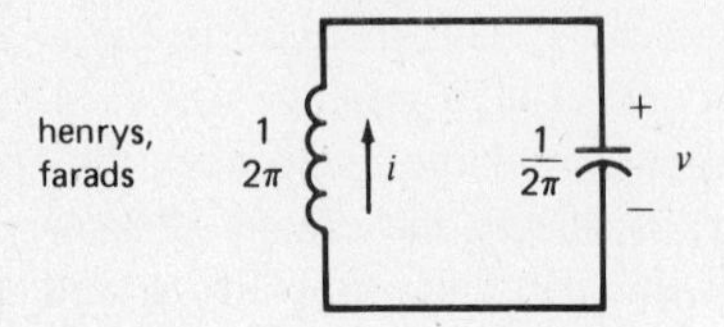

Figure P6.3

6.4. In Fig. P6.4 an *RLC* circuit is shown. If this circuit has the initial conditions $i(0) = 0$, $v(0) = 20/\sqrt{C}$, construct a plot of the total energy $w(t)$ in joules stored in the circuit over the period of time from 0 to 2 sec. Use 51 points in the plot. Also plot the values of $i(t)$ and the voltage across the capacitor, $v(t)$ (in volts).

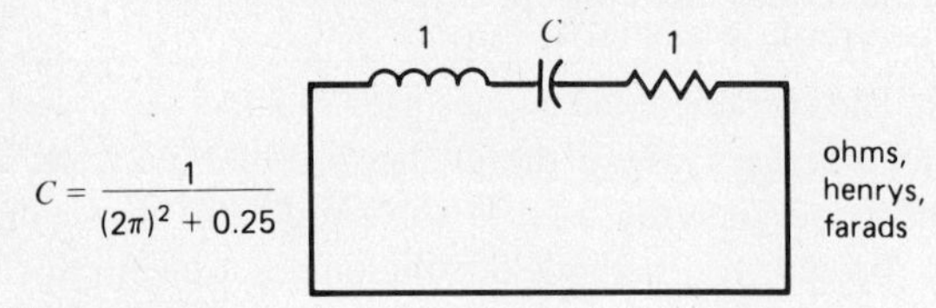

Figure P6.4

6.5. A system has the following relation between its input and its output:

$$o''(t) + 0.2o'(t) + o(t) = i'(t)$$

where the prime indicates differentiation with respect to time. If an input step $i(t) = 50$ V is applied over a time range from 0 to 10 sec, find and plot 51 values of the output $o(t)$ over this time range. Assume that all initial conditions are zero. Verify the numerical answer by solving the differential equation by using standard mathematical techniques.

6.6. The excitation of a system by an impulse function is easily simulated by numerical techniques. To illustrate this, consider a system with the relation between its input and output given by the equation

$$o''(t) + 0.2o'(t) + o(t) = i(t)$$

where the prime indicates differentiation with respect to time. If $i(t) = 60\delta(t)$ V, that is, if the input is an impulse of strength 60, such an impulse may be simulated by allowing i to have a value of 150 for a range of time from 0 to 0.4 sec, and to have a value of zero for all time greater than 0.4 sec. For such an input, find and plot the values of $o(t)$ over a time range from 0 to 10 sec. Use 51 points. Assume that all initial conditions are zero.

6.7. A distributed *RC* network (as used in integrated circuits) may be modeled by a cascade of *RC* sections, as shown in Fig. P6.7. A typical distributed network is found to have a total resistance of 70,000 Ω and a total capacitance of 2400 pF. If we frequency-normalize such a network by a factor of 10^{-5} (that is, 100 kHz → 1 Hz) and impedance-normalize it by a

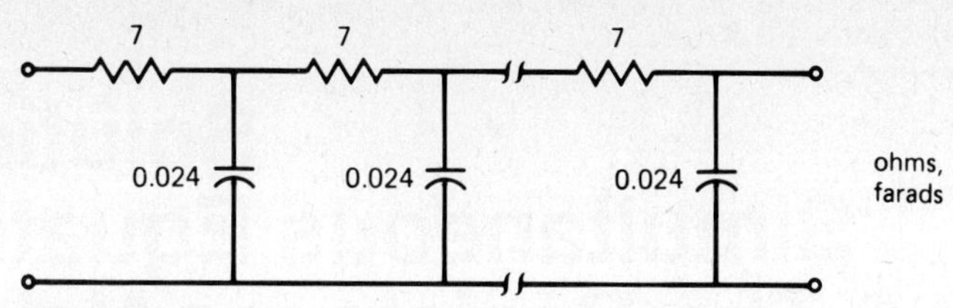

Figure P6.7

factor of 10^{-3} (that is, 1 kΩ → 1 Ω), we obtain a normalized network with a total resistance of 70 Ω and a total capacitance of 0.24 F. If such a network is represented by 10 lumped *RC* sections, each with a resistance of 7 Ω and a capacitance of 0.024 F, find the output voltage across the last capacitor if a 1-V step function of voltage is applied at the input terminals of the network at $t = 0$. Plot 51 values of this output voltage over a (normalized) time range from 0 to 50 sec.

6.8. A system has the following relation between its input and its output:

$$o'''(t) + 2o''(t) + 2o'(t) + o(t) = i(t)$$

where the prime indicates differentiation with respect to time. If an input step $i(t)u(t)$ is applied over a time range of 10 sec, plot 51 values of the output $o(t)$ over the same range. Assume that all the initial conditions are zero.

6.9. Repeat Prob. 6.6 for the differential equation given in Prob. 6.8.

Chapter 7

Solution of Simultaneous Equations: The Resistance Network

In the preceding chapters we have been concerned with the application of numerical techniques to investigate the behavior of physical quantities such as voltage and current under conditions where these quantities vary with time. In this chapter we investigate a somewhat different situation—the dc case, in which the physical variables are constant. An important example of such a case is the resistance network. We develop digital-computational techniques for setting up the simultaneous equations describing such a network and also for solving the equations.

7.1 THE RESISTANCE NETWORK: MESH EQUATIONS

A multiloop resistance network is shown in Fig. 7.1. From basic circuit theory we know that applying Kirchhoff's voltage law (KVL) to each of the meshes defined in the figure produces a set of simultaneous equations. These may be written as the matrix equation

$$\mathbf{RI} = \mathbf{V} \tag{7.1}$$

where $\mathbf{V}$ is an $n \times 1$ column matrix with elements v_i, $\mathbf{R}$ is an $n \times n$ square matrix with elements r_{ij}, and $\mathbf{I}$ is an $n \times 1$ column matrix with elements i_i. The elements of these matrices have the following significance:

v_i The algebraic summation of the voltage sources encountered in traversing the ith mesh. The sources are treated as positive if they represent a voltage rise while traversing the mesh in the direction of the reference mesh current. Otherwise they are treated as negative.

r_{ij} If $i = j$, this element represents the total resistance encountered in traversing the ith mesh. If $i \neq j$, this element is the negative of the resistance that is mutual to

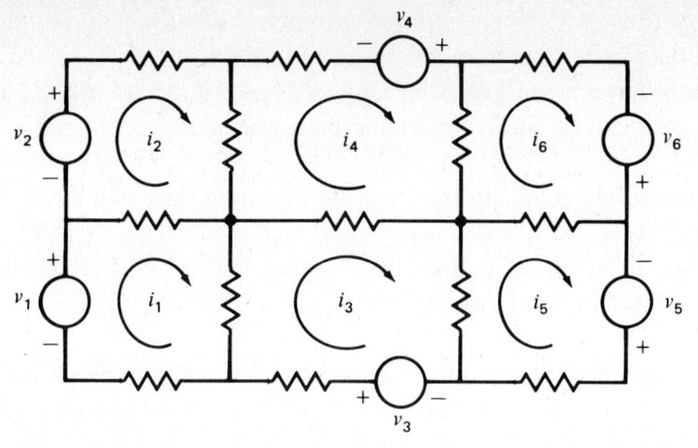

Figure 7.1 A multiloop resistance network.

the ith and jth meshes (assuming that the reference currents of the ith and jth meshes traverse the mutual resistance in opposite directions).

i_i This element represents the current in the ith mesh. As such it is an unknown whose value is to be determined.

The first problem that we shall treat in this chapter is that of formulating a set of equations of the form given in (7.1) from a specified circuit of the type shown in Fig. 7.1. Thus we may define the following problem:

Problem. Given a circuit diagram for a resistance network, find the matrix **R** of (7.1) relating the mesh voltages and the mesh currents of the network.

In the following section we discuss a solution to this problem which is easily implemented for the digital computer.

7.2 FORMULATING THE EQUATIONS FOR A RESISTANCE NETWORK: THE PROCEDURE ResisMesh

In this section we describe a procedure which performs the function of determining the **R** matrix of (7.1) from input information giving the values of the resistors and the identifying numbers of the meshes in which they are located. We make the following assumptions with respect to the topological structure of the network:[1]

1. All the meshes are chosen as shown in Fig. 7.1; that is, the paths followed by them do not enclose any branches not included in the mesh. There are also many other possible configurations of paths that might be chosen to set up a

[1]These restrictions exclude some classes of resistance networks. For example, they do not include the nonplanar-network class.

group of linearly independent loop currents. A determination of these, however, would take us too far from the focus of this text.

2. All mesh currents are given the same reference direction; that is, currents flowing in a clockwise direction around the meshes are considered as positive currents.
3. All resistive elements are traversed by not more than two mesh currents.

For the class of networks defined above, each resistor will affect the elements of the matrix **R** of (7.1) in one of two ways:

1. If a resistive element of value R_i Ω occurs only in the jth mesh, then the quantity R_i must be added to the element r_{jj} of the matrix **R** to take account of the effect of the resistor.
2. If a resistive element of value R_i Ω occurs in the jth and the kth meshes, then the quantity R_i must be added to the elements r_{jj} and r_{kk} of the matrix **R** and subtracted from the elements r_{jk} and r_{kj} of that matrix.

The above two possibilities are easily programmed into a procedure. As an input such a procedure requires the following information:

1. The number of resistive elements n_r that are to be found in the resistance network which is being analyzed.
2. The values R_i of the resistors of the network.
3. The numbers of the first mesh $m_i^{(1)}$ in which the corresponding resistors R_i are located.
4. The numbers of the second mesh (if any) $m_i^{(2)}$ in which the corresponding resistors R_i are located.

As outputs, the procedure must provide the values of the elements r_{ij} of the matrix **R** for the given network. Considerations of the above inputs and outputs for the procedure lead us to define a parameter list for it that has the form

```
(nMesh, nRes: integer; VAR res: array20;
 VAR mesh1, mesh2: iarray20; VAR r: matrix10)
```

where nMesh is the number of meshes, nRes is the variable name for n_r, the number of resistors, res is a one-dimensional array [1..20] of reals, in which are stored the values of the resistors R_i, mesh1 and mesh2 are one-dimensional arrays [1..20] of integers, in which are stored the values $m_i^{(1)}$ and $m_i^{(2)}$ for the resistors R_i, and r is a two-dimensional [1..10,1..10] matrix of reals in which are stored the elements of the matrix **R**. The various arrays are declared as global user-defined types by the following main program statements:

```
TYPE  array20 = ARRAY[1..20] OF real;
      iarray20 = ARRAY[1..20] OF integer;
      matrix10 = ARRAY[1..10,1..10] OF real;
```

Let us use the name ResisMesh for the procedure. Its heading will be

PROCEDURE ResisMesh (nMesh, nRes : integer;
VAR res : array20; VAR mesh1, mesh2 : iarray20;
VAR r : matrix10)

A listing of a set of PASCAL statements defining the procedure ResisMesh is given in Fig. 7.2. The operation of the procedure is readily understood in terms of the discussion given above. A summary of the important features of this procedure is given in Table 7.1.

```
PROCEDURE ResisMesh (nMesh, nRes : integer; VAR res : array20; VAR

                mesh1, mesh2 : iarray20; VAR r : matrix10);

(* Subprogram to compute elements of resistance matrix r
   from the values of the resistors and the numbers of
   the meshes in which these resistors are located.
   Each of the meshes must have the same reference
   direction, and no resistor can be present in more
   than two meshes.
     nMesh - Number of meshes
     nRes - Number of resistors
     res - Array of resistor values
     mesh1 - First array of mesh numbers
     mesh2 - Second array of mesh numbers (if resistor
             occurs only in a single mesh, the
             corresponding element in mesh2 is zero)
     r - Output resistance matrix                    *)

  VAR
     i, j, k : integer;

  BEGIN

     (*   Set the elements of the resistance matrix r to zero    *)

     FOR i := 1 TO nMesh DO
        BEGIN
           FOR j := 1 TO nMesh DO
              r[i,j] := 0.0;
        END;

     (*   Modify the elements of the resistance matrix r for
          each of the resistors                              *)

     FOR i := 1 TO nRes DO
        BEGIN
           j := mesh1[i];
           k := mesh2[i];

           (*   Test to determine whether the resistor occurs
                only in a single mesh                        *)

           IF k <> 0 THEN
              BEGIN
                 r[k,k] := r[k,k] + res[i];
                 r[j,k] := r[j,k] - res[i];
                 r[k,j] := r[j,k]
              END;
           r[j,j] := r[j,j] + res[i]
        END
  END;
```

Figure 7.2 Listing of the procedure ResisMesh.

TABLE 7.1 SUMMARY OF THE CHARACTERISTICS OF THE PROCEDURE ResisMesh

Identifying Statement PROCEDURE ResisMesh (nMesh, nRes : integer; VAR res : array20; VAR mesh1, mesh2 : iarray20, VAR r : matrix10);

Purpose To determine the resistance matrix **R** for a given resistance network

Additional Subprograms Required None

Input Arguments

nMesh	The number of meshes in the network
nRes	The number of resistors contained in the resistance network
res	The one-dimensional array of variables res[i] in which are stored the values of the resistors R_i (in ohms)
mesh1	The one-dimensional array of variables mesh1[i] in which are stored the numbers of the first mesh in which the resistors R_i are located
mesh2	The one-dimensional array of variables mesh2[i] in which are stored the numbers of the second mesh (if any) in which the resistors R_i are located

Output Argument

r	The two-dimensional array of variables r[i,j] in which are stored the values of the elements r_{ij} of the resistance matrix **R** for the specified network

User-Defined Types

```
array20 = ARRAY[1..20] OF real;
iarray20 = ARRAY[1..20] OF integer;
matrix10 = ARRAY[1..10,1..10] OF real
```

Notes: **1.** All meshes are assumed to have their reference currents taken in the same direction, for example, clockwise.
2. It is assumed that the meshes are drawn so that no resistor will be present in more than two meshes.
3. If the ith resistor is present in only one mesh, then mesh2[i] = 0.

7.3 THE SOLUTION OF A SET OF SIMULTANEOUS EQUATIONS: THE PROCEDURE GjSimEqn

In Secs. 7.1 and 7.2 we formulated the problem of forming a matrix equation of the type

$$\mathbf{RI} = \mathbf{V} \tag{7.2}$$

where **V** is an n-element column matrix with elements v_i, **R** is an $n \times n$ square matrix with elements r_{ij}, and **I** is an n-element column matrix with elements i_i. We showed how, if this was the matrix equation for a resistance network, the square matrix could be found directly from a knowledge of the values of the resistors and the identifying numbers for the meshes in which such resistors were located. In this section we concern ourselves with the solution of such a matrix equation. The problem may be formulated as follows:

Problem. Given the matrices **V** and **R** of the matrix equation (7.2), find the matrix **I**.

Such a problem is frequently spoken of as finding the solution of a set of simultaneous equations.

For a solution to the problem formulated above to exist, the determinant of the square matrix **R** must be nonzero. In the following developments, we assume that this is so. We also assume that the elements on the main diagonal of the matrix are nonzero.

Although this is not a necessary condition for a solution, it is a condition that usually exists in physical situations, and it will simplify the resulting computations. To see one way in which the solution of such a set of simultaneous equations may be accomplished, let us write the simultaneous equations represented by the matrix equation (7.2). We obtain

$$\begin{aligned} r_{11}i_1 + r_{12}i_2 + \cdots + r_{1n}i_n &= v_1 \\ r_{21}i_1 + r_{22}i_2 + \cdots + r_{2n}i_n &= v_2 \\ \cdots\cdots\cdots\cdots\cdots\cdots \\ r_{n1}i_1 + r_{n2}i_2 + \cdots + r_{nn}i_n &= v_n \end{aligned} \tag{7.3}$$

We now discuss two operations that may be made on the separate equations given in (7.3) without changing the solutions to these equations, that is, the values of the elements i_i that result for given values of the r_{ij} and the v_i. The first operation is the multiplication of any equation by a nonzero constant. For example, if we multiply the first equation of (7.3) by the constant k, we obtain

$$kr_{11}i_1 + kr_{12}i_2 + \cdots + kr_{1n}i_n = kv_1 \tag{7.4}$$

Obviously, any values of the elements i_i satisfying the equations given in (7.3) will also satisfy (7.4). Thus, we conclude that any equation of the set of equations given in (7.3) may be multiplied by a constant without changing the solutions to the set. The second operation that we discuss is the subtraction of a multiple of any of the equations of (7.3) from another equation of (7.3). For example, if we multiply the first equation of (7.3) by the constant k and subtract it from the second equation of (7.3), we obtain the equation

$$(r_{21} - kr_{11})i_1 + (r_{22} - kr_{12})i_2 + \cdots + (r_{2n} - kr_{1n})i_n = v_2 - kv_1 \tag{7.5}$$

It may be shown that the values of the elements i_i satisfying the first and second equations of (7.3) will also satisfy (7.5). Thus we conclude that we may replace any equation in (7.3) with the difference between that equation and some multiple of one of the other equations of (7.3) without changing the solutions to the set.

The two operations described above may be applied to "reduce" the set of equations given in (7.3) to the form

$$\begin{aligned} 1i_1 + 0i_2 + \cdots + 0i_n &= v_1' \\ 0i_1 + 1i_2 + \cdots + 0i_n &= v_2' \\ \cdots\cdots\cdots\cdots\cdots \\ 0i_1 + 0i_2 + \cdots + 1i_n &= v_n' \end{aligned} \tag{7.6}$$

From this reduced form, the solution for the i_i is obviously given by the relations

$$i_i = v_i' \qquad i = 1, 2, \ldots, n \tag{7.7}$$

Such a method of solving for the unknown quantities in a set of simultaneous equations is known as the *Gauss-Jordan reduction method*.

In considering the steps necessary to put the equations of (7.3) in the form shown in (7.6), we note that all the mathematical operations will be made on the elements r_{ij} and v_i. Therefore, as a first step let us form the *augmented* matrix $\mathbf{R}'$, an $n \times (n + 1)$

rectangular matrix formed by using the elements r_{ij} of the matrix **R** and adding the elements v_i of the matrix **V** as the $(n + 1)$st column. Thus we see that

$$\mathbf{R}' = \begin{bmatrix} r_{11} \cdots r_{1n} & v_1 \\ \cdots\cdots\cdots & \\ r_{n1} \cdots r_{nn} & v_n \end{bmatrix} \tag{7.8}$$

We may also write **R**′ in the form

$$\mathbf{R}' = \begin{bmatrix} r_{11} \cdots r_{1n} & r_{1,n+1} \\ \cdots\cdots\cdots & \\ r_{n1} \cdots r_{nn} & r_{n,n+1} \end{bmatrix} \tag{7.9}$$

where $r_{i,n+1} = v_i$ $(i = 1, 2, \ldots, n)$.

We may now apply the first operation described above by multiplying the first equation of (7.3) by $1/r_{11}$ to set the element in the first row and column of **R**′ to 1. Let us now apply the second operation described above and subtract the new first row multiplied by r_{i1} from the ith row of **R**′ for all values of i except $i = 1$. If we call the resulting matrix **R**′, we see that

$$\mathbf{R}'' = \begin{bmatrix} 1 & \dfrac{r_{12}}{r_{11}} & \cdots & \dfrac{r_{1n}}{r_{11}} & \dfrac{v_1}{r_{11}} \\ 0 & r_{22} - \dfrac{r_{21}r_{12}}{r_{11}} & \cdots & r_{2n} - \dfrac{r_{21}r_{1n}}{r_{11}} & v_2 - \dfrac{r_{21}v_1}{r_{11}} \\ \cdots & \cdots & \cdots & \cdots & \cdots \\ 0 & r_{n2} - \dfrac{r_{n1}r_{12}}{r_{11}} & \cdots & r_{nn} - \dfrac{r_{n1}r_{1n}}{r_{11}} & v_n - \dfrac{r_{n1}v_1}{r_{11}} \end{bmatrix} \tag{7.10}$$

We see that this cycle of operations has had the effect of reducing the first column of **R**′ to zero except for the element in the first row which has been set equal to unity. Additional cycles similar to that described above can be used to set the elements in succeeding columns to zero, except for the elements on the main diagonal, which are set to unity. As a result, we obtain a *reduced* augmented matrix which has the form

$$\begin{bmatrix} 1 & 0 \cdots 0 & r'_{1,n+1} \\ 0 & 1 \cdots 0 & r'_{2,n+1} \\ \cdots & \cdots & \cdots \\ 0 & 0 \cdots 1 & r'_{n,n+1} \end{bmatrix} \tag{7.11}$$

The first n columns of the matrix given in (7.11) form an identity matrix. When **R**′ has been reduced to this form, the solutions to the set of simultaneous equations, that is, the values of the i_i, are given by the elements $r'_{i,n+1}$, by using the relation

$$r'_{i,n+1} = i_i \qquad i = 1, 2, \ldots, n \tag{7.12}$$

In other words, the last column of the **R**′ matrix after reduction to the form given in (7.11) is the matrix **I**, whose elements i_i yield the solution to the set of simultaneous equations.

The method described above may be implemented by some explicit relations. To normalize the diagonal element of the ith row, we need merely replace the value of r_{ij} by the quantity r_{ij}/r_{ii} for all values of $j \geq i$. This may be written

$$\frac{r_{ij}}{r_{ii}} \rightarrow r_{ij} \qquad j = n + 1, n, n - 1, \ldots, i \tag{7.13}$$

where the arrow may be read "is used to replace." As indicated, the replacement elements must be computed in descending order to avoid destroying the value of r_{ii} before the last step. The operation stops at the element in the ith column, since all previous elements in the ith row have been set to zero by earlier operations. To set the other elements in the ith column to zero, we may apply the substitution relation

$$r_{kj} - r_{ki}r_{ij} \rightarrow r_{kj} \qquad j = n + 1, n, n - 1, \ldots, i \tag{7.14}$$

to each row except the ith row. In other words, we let k in (7.14) take on all values from 1 to n except the value i. If the relations of (7.13) and (7.14) are repeated for all values of i from 1 to n, we obtain the reduced form of the augmented matrix **R**′ given in (7.11).

A procedure for performing the operations described above is easily written. Such a procedure must be provided with the following information:

1. The elements r_{ij} of the matrix **R** of (7.2), representing the coefficients of the set of simultaneous equations.
2. The elements of the matrix **V** of (7.2), representing the known (right-hand-side) quantities in the set of simultaneous equations.
3. The number n of simultaneous equations, that is, the size of the square matrix **R**.

As an output such a procedure should supply the values of the i_i, that is, the elements of the matrix **I**, giving the values of the unknown variables which satisfy the simultaneous equations. In consideration of the above, we see that the parameter list for the procedure should have the form

```
(VAR v:array10; VAR r:matrix10;
 VAR amps:array10; n:integer)
```

where v is a one-dimensional array [1..10] of reals in which are stored the (known) values of the mesh voltages, etc. If we use the name GjSimEqn (for *G*auss-*J*ordan *sim*ultaneous *eq*uatio*n* solution) for the procedure, the heading for it will be of the form

```
PROCEDURE GjSimEqn (VAR v:array10; VAR r:matrix10;
                    VAR amps:array10; n:integer);
```

```
PROCEDURE GjSimEqn (VAR v : array10; VAR r : matrix10;
                    VAR amps : array10; n : integer);

(* Subprogram to solve a set of simultaneous equations
   v=r*amps using Gauss-Jordan reduction.  The r matrix
   is preserved.
     v - Column matrix of known variables
     r - Square non-singular matrix
     amps - Output column matrix
     n - Number of equations                  *)

   TYPE
     matrix11 = ARRAY[1..10,1..11] of real;

   VAR
     ra          : matrix11;
     i, j, k, np : integer;
     beta, alfa  : real;

   BEGIN

     (*   Enter r matrix in ra array and enter v as
          n+1th column of ra array                  *)

     FOR i := 1 TO n DO
       BEGIN
         FOR j := 1 TO n DO
           ra[i,j] := r[i,j];
         ra[i,n+1] := v[i]
       END;

     (*   Portion of program to reduce augmented
          matrix ra to solve a set of n simultaneous equations.
          np is no. of columns in augmented matrix,
          alfa and beta are temporary storage                  *)

     np := n + 1;
     FOR i := 1 TO n DO
       BEGIN

         (*   Set main diagonal elements to unity   *)

         alfa := ra[i,i];
         FOR j := i TO np DO
           ra[i,j] := ra[i,j] / alfa;

         (*   Set elements of ith column to zero   *)

         FOR k := 1 TO n DO
           BEGIN
             IF (k - i) <> 0 THEN
               BEGIN
                 beta := ra[k,i];
                 FOR j := i TO np DO
                   ra[k,j] := ra[k,j] - beta * ra[i,j];
               END
           END   (*   End of k index loop   *)
       END;   (* End of i index loop   *)

     (*   Set output matrix amps equal to last column of
          augmented matrix ra   *)

     FOR i := 1 TO n DO
       amps[i] := ra[i,np];
   END;
```

(*a*)

Figure 7.3 The procedure GjSimEqn.

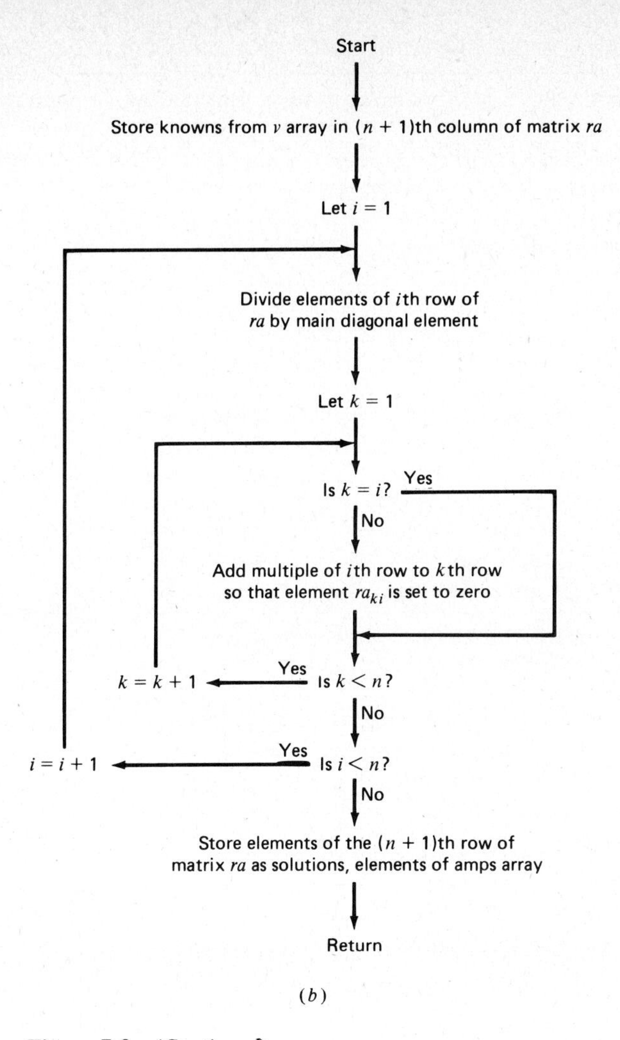

Figure 7.3 *(Continued)*

A listing of a set of PASCAL statements and a flowchart for a procedure that will provide the operations described above are given in Fig. 7.3. The operation of the procedure is readily verified by comparing the statements in Fig. 7.3 with the expressions given in (7.13) and (7.14). A summary of the characteristics of the procedure is given in Table 7.2.

The procedure GjSimEqn may be used in conjunction with the procedure ResisMesh to solve for the mesh currents of a resistance network from the known values of mesh voltages and resistors. Example 7.1 is an illustration of such a use.

EXAMPLE 7.1

In Fig. 7.4 a four-mesh resistance network is shown. We may use the procedures ResisMesh and GjSimEqn to solve for the four mesh currents i_1, i_2, i_3, and i_4. A listing of a main

TABLE 7.2 SUMMARY OF THE CHARACTERISTICS OF THE PROCEDURE GjSimEqn

Identifying Statement PROCEDURE GjSimEqn (VAR v : array10; VAR r : matrix10; VAR amps : array10; n : integer);

Purpose To provide a solution to a set of simultaneous equations of the form **V** = **RI**, where **R** is a given square matrix with elements r_{ij}, **V** is a given column matrix with elements v_i, and **I** is a column matrix whose elements i_i are to be found

Additional Subprograms Required None

Input Arguments

v The one-dimensional array of variables v[i] in which are stored the elements v_i of the column matrix **V** giving the known quantities for the set of simultaneous equations

r The two-dimensional array of variables r[i,j] in which are stored the elements r_{ij} of the square matrix **R** defining the set of equations

n The order of the matrix **R**, that is, the number of simultaneous equations which are being solved

Output Arguments

amps The one-dimensional array of variables amps[i] in which are stored the elements i_i of the column matrix **I** giving the unknown quantities of the set of simultaneous equations

User-Defined Types

array10 = ARRAY[1..10] OF real;
matrix10 = ARRAY[1..10,1..10] OF real;

Notes: **1.** The elements on the main diagonal of the matrix **R** must be nonzero.
2. The matrix **R** must be nonsingular.

program to accomplish this is shown in Fig. 7.5. A listing of the input data and the resulting answers for the currents (in amperes) is also given in the figure.

7.4 TREATMENT OF CONTROLLED SOURCES

When resistance networks are used to model the dc characteristics of active elements, *controlled sources* must be used. There are four types of such sources. They are identified as follows:

Abbreviation	Name
VCVS	Voltage-controlled voltage source
VCIS	Voltage-controlled current source
ICVS	Current-controlled voltage source
ICIS	Current-controlled current source

The symbols and gain constants for these sources are identified in Fig. 7.6.

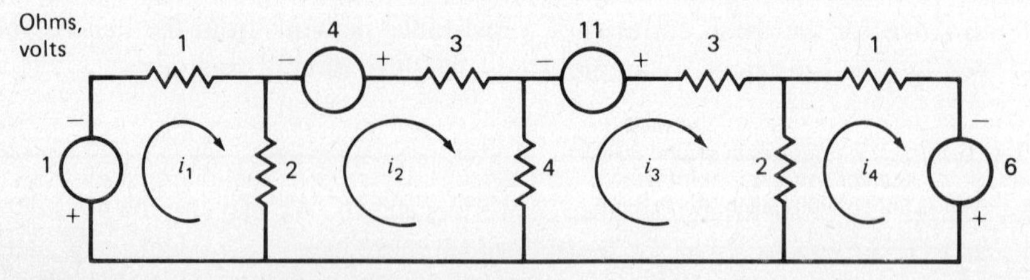

Figure 7.4 A four-loop resistance network.

```
PROGRAM Main (input, output);    (* Example 7.1 *)
(* Example of the use of the procedures ResisMesh and GjSimEqn
   to solve for the currents in a resistance network
     nMesh - Number of meshes
     nRes - Number of resistors
     res - Array of values of resistors
     mesh1 - Array of first mesh in which resistors
          are located
     mesh2 - Array of second mesh in which resistors
          are located
     r - Resistance matrix
     v - Array of excitation voltages
     amps - Array of mesh currents                      *)
TYPE
   array10 = ARRAY[1..10] OF real;
   array20 = ARRAY[1..20] OF real;
   iarray20 = ARRAY[1..20] OF integer;
   matrix10 = ARRAY[1..10,1..10] OF real;
VAR
   mesh1, mesh2       : iarray20;
   res                : array20;
   volts, amps        : array10;
   r                  : matrix10;
   i, j, nMesh, nRes  : integer;
BEGIN
   (*   Read the data defining the circuit   *)
   readln(nMesh, nRes);
   FOR i := 1 TO nRes DO
      readln(mesh1[i], mesh2[i], res[i]);
   FOR i := 1 TO nMesh DO
      readln(volts[i]);
   (* Use ResisMesh to compute the r matrix   *)
   ResisMesh(nMesh, nRes, res, mesh1, mesh2, r);
   writeln(' R MATRIX');
   FOR i := 1 TO nMesh DO
      BEGIN
         writeln;
         FOR j := 1 TO nMesh DO
            write(r[i,j]:6:1)
      END;
   (*   Use GjSimEqn to solve the equations   *)
   GjSimEqn(volts, r, amps, nMesh);
   writeln; writeln;
   writeln(' MESH CURRENTS'); writeln;
   FOR i := 1 TO nMesh DO
      write(amps[i]:8:4)
END.
```

(*a*)

Input Data

```
4 7   <----  Number of meshes, number of resistors
1 0     1.0  <----  First mesh, second mesh, value of resistance
1 2     2.0
2 0     3.0
2 3     4.0
3 0     3.0
3 4     2.0
4 0     1.0
-1.0  <----  Sum of voltages in first mesh
4.0   <----  Sum of voltages in second mesh
11.0  <----  Sum of voltages in third mesh
6.0   <----  Sum of voltages in fourth mesh
```

(*b*)

Output

```
R MATRIX

  3.0  -2.0   0.0   0.0
 -2.0   9.0  -4.0   0.0
  0.0  -4.0   9.0  -2.0
  0.0   0.0  -2.0   3.0

MESH CURRENTS

  1.0000  2.0000  3.0000  4.0000
```

(*c*)

Figure 7.5 Program, input, and output for Example 7.1.

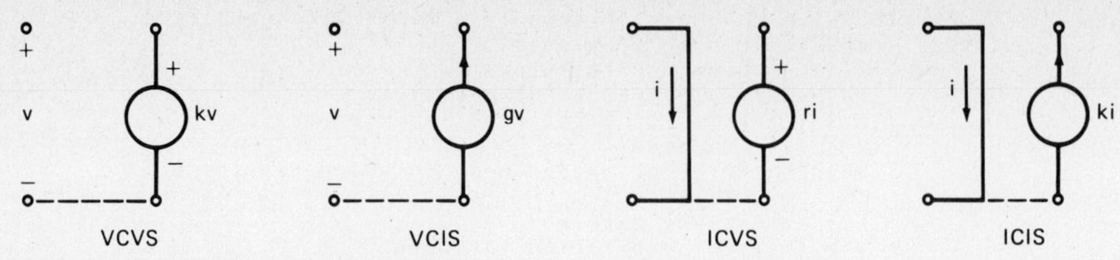

Figure 7.6 Types of controlled sources.

The procedures for the solution of a resistance network described in Secs. 7.2 and 7.3 are easily modified for the case where one or more controlled sources are present in the network. In such a situation all that is required is to use the procedure ResisMesh to find the resistance matrix of the network *without* considering the controlled sources, and then to modify this matrix to take account of the effects of the sources. An example follows.

EXAMPLE 7.2

In Fig. 7.7*a*, a simple resistance network containing an ICVS (current-controlled voltage source) is shown. This source affects the summation of the voltage drops in both mesh 1 and mesh 2. In mesh 1, applying KVL we obtain

$$10 - 5i_1 = 1i_1 + 2(i_1 - i_2)$$
$$10 = 3i_1 - 2i_2 + 5i_1 = 8i_1 - 2i_2$$

Thus, the effect of the controlled source is to add 5 to the coefficient multiplying i_1. In mesh 2, applying KVL, we obtain

$$5i_1 = 2(i_2 - i_1) + 3i_2 + 1i_2$$
$$0 = 6i_2 - 2i_1 - 5i_1 = -7i_1 + 6i_2$$

Thus, the effect of the controlled source is to subtract 5 from the coefficient multiplying i_1. In Fig. 7.7*b*, the network of Fig. 7.7*a* has been redrawn without the controlled source. For this network the resistance matrix is

$$R_{\text{no source}} = \begin{bmatrix} 3 & -2 \\ -2 & 6 \end{bmatrix}$$

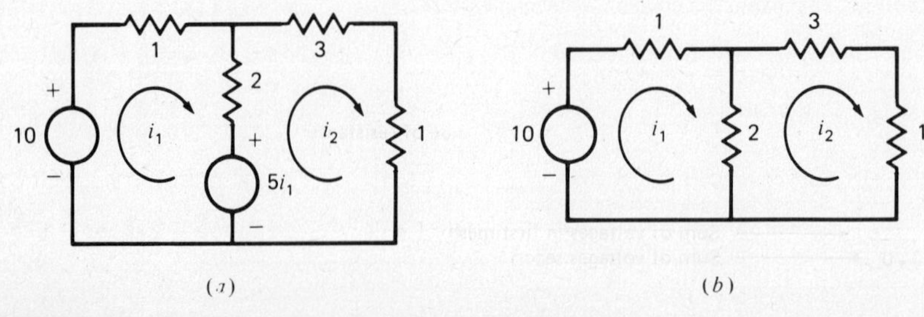

Figure 7.7 Resistance network with controlled source.

```
PROGRAM Main (input, output);    (* Example 7.2 *)

(* Main program for solving for the mesh currents of a
   a resistance network containing a controlled source
     nMesh - Number of meshes
     nRes - Number of resistors
     res - Array of values of resistors
     mesh1 - Array of first mesh in which resistors
          are located
     mesh2 - Array of second mesh in which resistors
          are located
     r - Resistance matrix
     volts - Array of excitation voltages
     amps - Array of mesh currents                  *)

TYPE
   array10 = ARRAY[1..10] OF real;
   array20 = ARRAY[1..20] OF real;
   iarray20 = ARRAY[1..20] OF integer;
   matrix10 = ARRAY[1..10,1..10] OF real;

VAR
   mesh1, mesh2        : iarray20;
   res                 : array20;
   volts, amps         : array10;
   r                   : matrix10;
   i, j, nMesh, nRes   : integer;

BEGIN

   (*   Read the data defining the circuit and compute
        the resistance matrix                            *)

   readln(nMesh, nRes);
   FOR i := 1 TO nRes DO
      readln(mesh1[i], mesh2[i], res[i]);
   FOR i := 1 TO nMesh DO
      readln(volts[i]);
   ResisMesh(nMesh, nRes, res, mesh1, mesh2, r);

   (* Modify the r matrix to take account of the
      controlled source                             *)

   r[1,1] := r[1,1] + 5.0;
   r[2,1] := r[2,1] - 5.0;
   writeln(' R MATRIX');
   FOR i := 1 TO nMesh DO
      BEGIN
         writeln;
         FOR j := 1 TO nMesh DO
            write(r[i,j]:6:1)
      END;

   (*   Use GjSimEqn to solve the equations    *)

   GjSimEqn(volts, r, amps, nMesh);
   writeln; writeln;
   writeln(' MESH CURRENTS'); writeln;
   FOR i := 1 TO nMesh DO
      write(amps[i]:8:4)
END.
```

(*a*)

Input Data

```
2 4 ←———— Number of meshes, number of resistors
1 0 1.0 ←———— First mesh, second mesh, value of resistance
1 2 2.0
2 0 3.0
2 0 1.0
10.0 ←———— Sum of voltages in first mesh
0.0 ←———— Sum of voltages second mesh
```

(*b*)

Output

```
R MATRIX

  8.0  -2.0
 -7.0   6.0

MESH CURRENTS

 1.7647  2.0588
```

(*c*)

Figure 7.8 Program, input, and output for Example 7.2.

From the KVL equations given above, the effect of the controlled source is to modify the matrix to the form

$$R_{\text{with source}} = \begin{bmatrix} 8 & -2 \\ -7 & 6 \end{bmatrix}$$

Thus, the set of equations to be solved to find the values of i_1 and i_2 is

$$\begin{bmatrix} 10 \\ 0 \end{bmatrix} = \begin{bmatrix} 8 & -2 \\ -7 & 6 \end{bmatrix} \begin{bmatrix} i_1 \\ i_2 \end{bmatrix}$$

The solution process is implemented in the program and input shown in Fig. 7.8. From this program we readily find that the currents are $i_1 = 1.7647$ A and $i_2 = 2.0588$ A.

As illustrated in the preceding example, the effects of controlled sources of the ICVS type are easily accounted for in mesh analysis. Similarly, in node analysis (see Conclusion and Prob. 7.4) the effects of controlled sources of the VCIS type are readily treated. The effects of sources of the VCVS and ICIS types may be treated by using source-conversion techniques. A description of such techniques may be found in any of the books on basic circuit theory listed in the bibliography.

7.5 CONCLUSION

The techniques described in this chapter are readily extended to many other network situations. For example, the formulation given for the procedure ResisMesh applies directly to the analysis of resistance networks on a node basis (the element values must be given in mhos, that is, reciprocal ohms, in this case). As another example, these techniques may be used to find the operating points of networks containing nonlinear elements, if the latter are described by piecewise-linear characteristics. The details of these and other applications are left to the reader as exercises.

PROBLEMS

7.1. The resistive network shown in Fig. P7.1*a* contains a nonlinear element R_0 with the *v*-*i* characteristic shown in Fig. P7.1*b*. Find the loop currents i_1, i_2, and i_3, and determine on which section of its characteristic the nonlinear element operates.

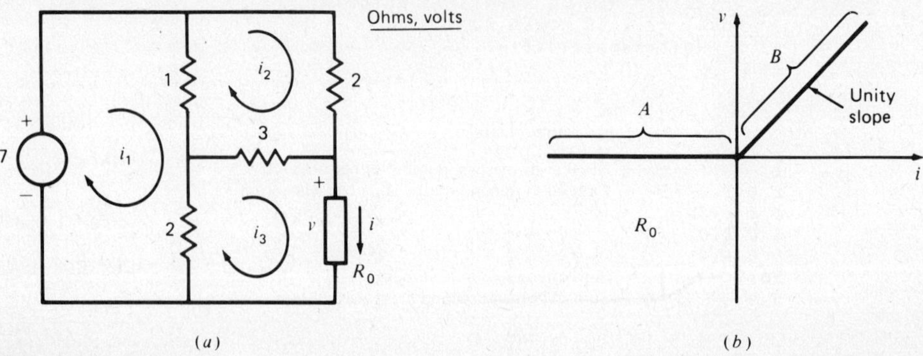

Figure P7.1

7.2. Find a Thevenin equivalent circuit for the resistive network shown in Fig. P7.2, as it appears at terminals a and b.

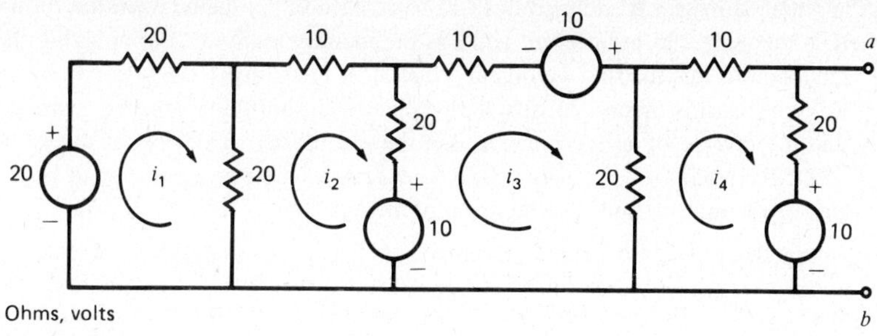

Figure P7.2

7.3. Find the input resistance at terminals a and b of the resistive network shown in Fig. P7.3.

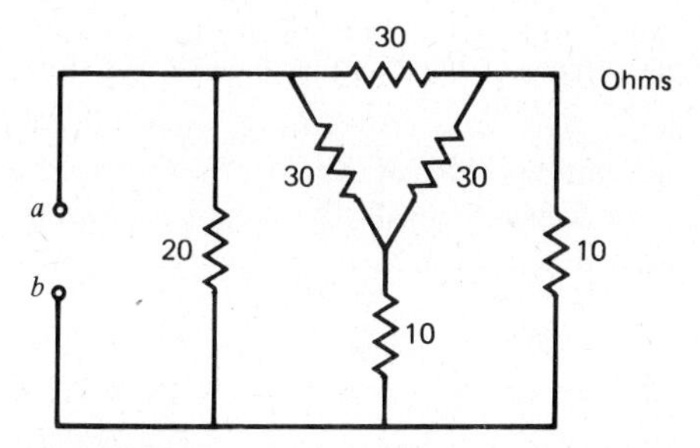

Figure P7.3

7.4. Since the matrix operation by means of which a conductance matrix is formed for resistive elements connected between various nodes (with element values given in mhos, that is, reciprocal ohms) is exactly the same as the operation of forming the resistance matrix for resistors located in various loops, the procedure ResisMesh may be used directly to find the values of nodal voltages if a network is analyzed on a node basis rather than a loop basis. Illustrate this fact by using the procedures ResisMesh and GjSimEqn to solve for the nodal voltages of the network shown in Fig. P7.4.

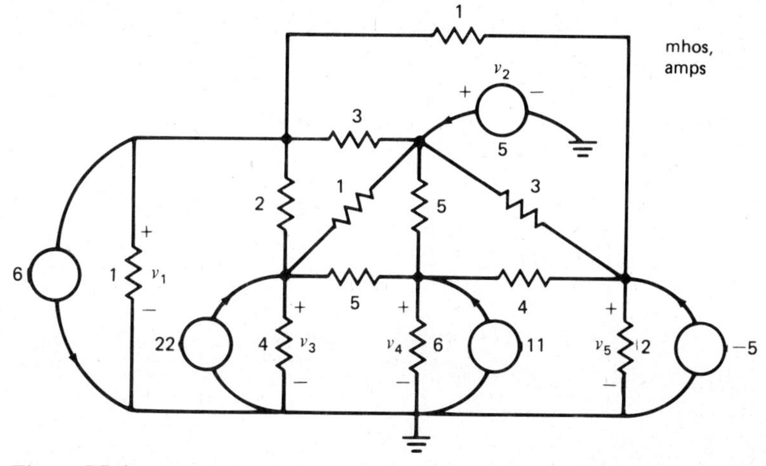

Figure P7.4

7.5. The algorithm given in Sec. 7.3 for solving a set of simultaneous equations can easily be extended to cover the problem of obtaining the inverse of a given matrix. To do this, let the augmented matrix $\mathbf{R}'$ defined in (7.8) be replaced by a matrix with the form $[\mathbf{R} : \mathbf{I}]$, where $\mathbf{R}$ is the matrix to be inverted and $\mathbf{I}$ is the identity matrix. After applying the Gauss-Jordan reduction to this matrix, we obtain a matrix of the form $[\mathbf{I} : \mathbf{E}]$. It may be shown that $\mathbf{E} = \mathbf{R}^{-1}$. Using this theory, modify the procedure GjSimEqn to produce a procedure which will find the inverse of a given matrix. Let the identifying statement for the procedure be PROCEDURE MxInv (r, e : matrix10, n : integer), where r is the matrix to be inverted, n is the size of the matrix, and e is the resulting inverse.

7.6. Apply the procedure MxInv developed in Prob. 7.5 to find the inverse of the following matrix:

$$\begin{bmatrix} 4 & 1 & 1 & 0 \\ 1 & 7 & 1 & 1 \\ 2 & 6 & 4 & 1 \\ -5 & 2 & -1 & 5 \end{bmatrix}$$

Verify the result by (a) taking the inverse of the resulting inverse and (b) multiplying the inverse by the original matrix.

7.7. Add negative-value resistors in each of the meshes of the network used in Example 7.1 such that the main diagonal elements of the $\mathbf{R}$ matrix are all zero. Repeat the example and observe the effects of trying to run the program.

Chapter 8

The Laplace Transformation: Manipulation of Polynomials

In this chapter we look at the problem of applying digital-computing techniques to the general concepts of system analysis from a viewpoint entirely different from that used in the preceding chapters. Our previous approach was to consider physical variables such as voltage, current, and power, which are functions of time. Starting with this chapter, however, we consider variables that are *transformed* quantities, that is, those which are functions of the complex frequency variable.

8.1 THE LAPLACE TRANSFORMATION: COMPLEX FREQUENCY

One of the most important ideas of basic network and system theory is the concept of *complex frequency,* using the *complex frequency variable* s. A function of the complex frequency variable is related to a function of time by a transformation process called the *Laplace transformation*. If we let $f(t)$ be an arbitrary function of time and let the transformed function derived from $f(t)$ be $F(s)$, the Laplace transformation is defined by the relation[1]

$$F(s) = \int_0^\infty f(t)e^{-st}\,dt \tag{8.1}$$

Specific expressions for $f(t)$ and $F(s)$ which are related by (8.1) are known as *transform pairs*. Several examples of transform pairs for some common functions are given in Fig. 8.1. All the functions $f(t)$ listed in the figure are considered to be zero for $t < 0$.

[1]More formally, the lower limit of the integral should be specified as $0-$, where

$$0- = \lim_{\substack{t \to 0 \\ t > 0}} t$$

See L. P. Huelsman, *Basic Circuit Theory,* 2d ed., Chap. 8, Prentice-Hall, Inc., Englewood Cliffs, N.J., 1984.

Figure 8.1 Some common Laplace-transform pairs.

$f(t)$	$F(s)$
$\delta(t)$*	1
1	$1/s$
t	$1/s^2$
e^{-at}	$1/(s+a)$
$\sin bt$	$b/(s^2+b^2)$
$\cos bt$	$s/(s^2+b^2)$
$e^{-at}\sin bt$	$b/(s^2+2as+a^2+b^2)$
$e^{-at}\cos bt$	$(s+a)/(s^2+2as+a^2+b^2)$

*$\delta(t)$ is the symbol for an impulse function that has the property that

$$\int_{-\infty}^{t} \delta(t)\,dt = u(t)$$

where $u(t)$ is a unit step occurring at $t = 0$.

In the study of networks, the complex frequency variable is usually used to define such terms as impedance and admittance, or, more generally, *network functions*. As an example of a network function, consider the circuit shown in Fig. 8.2. For this circuit we may define a network function relating the transformed input voltage $V_1(s)$ and the transformed output voltage $V_2(s)$. Routine circuit analysis shows that the network function $N(s)$ relating these variables is

$$N(s) = \frac{V_2(s)}{V_1(s)} = \frac{1/LC}{s^2 + (R/L)s + 1/LC} \tag{8.2}$$

More generally, a network function is defined as a function of the complex frequency variable that gives the ratio of some transformed output variable to some transformed input variable. If we specify the expression for the transformed input variable, then this expression, multiplied by the network function, gives us an expression that is the transform of the output variable. An easily remembered link between a variable in the time domain, that is, a function of time, and its transformed counterpart in the frequency domain, that is, a function of the complex frequency variable s, is found by considering s to be a symbol representing the operation of differentiation (and $1/s$ to be a symbol representing the operation of integration). This is one of the basic ideas of ''operational'' calculus. Thus, for the example network shown in Fig. 8.2, with the network function given in (8.2), we see that we may write (in the time domain)

$$v_2''(t) + \frac{R}{L}v_2'(t) + \frac{1}{LC}v_2(t) = \frac{1}{LC}v_1(t) \tag{8.3}$$

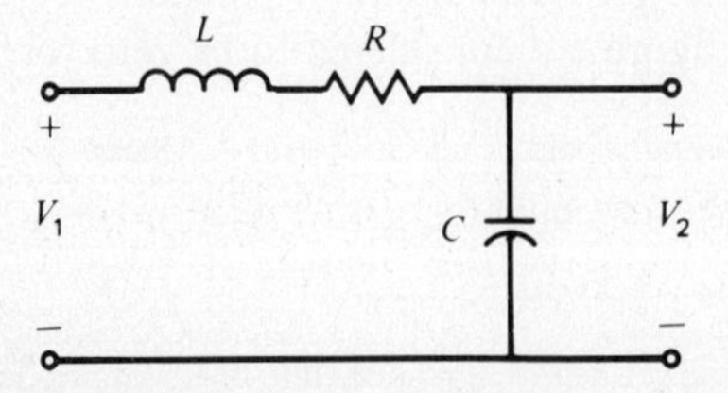

Figure 8.2 An *RLC* network.

From a consideration of (8.2) and (8.3), we note that the network function does not take into account any initial conditions that may be associated with the network. Such initial conditions may, however, be introduced by directly using (8.1) to transform any differential equations of the type illustrated in (8.3).

It is a property of the transform pairs given in Fig. 8.1 that, given a function $F(s)$, a process referred to as the inverse Laplace transformation can be applied to this function to generate the original function $f(t)$. A discussion of the mechanism for this would carry us too far afield from the goals of this text. Here we merely content ourselves with assuming the validity of such an inverse transformation and confine our attention to seeking some means of implementing it through the use of the digital computer. If the function $F(s)$ is a proper rational function, that is, one that can be expressed as a ratio of polynomials in which the degree of the numerator is less than the degree of the denominator, then one approach to the inversion problem consists in making a partial-fraction expansion of the function into a sum of terms of the type shown in the right column of Fig. 8.1 and finding the inverse transformation of each such term. Because of the linear nature of the transform, the resulting inverse transformation is simply the sum of the terms realized by the inversion process. For example, consider the following function and its partial-fraction expansion:

$$F(s) = \frac{6}{s^3 + 7s^2 + 6s} = \frac{1}{s} - \frac{6/5}{s+1} + \frac{1/5}{s+6}$$

The function $f(t)$ found by using the relations of Fig. 8.1 is

$$f(t) = 1 - 6/5e^{-t} + 1/5e^{-6t} \qquad t > 0$$

The above may be summarized by stating the following problem:

Problem. Given a rational function $F(s)$, expand it in the form of terms given in Fig. 8.1, and find the inverse transformation.

In the following sections we discuss means for doing this by digital-computational techniques.

8.2 COMPLEX VARIABLES: THE PROCEDURES ValPoly AND DifPoly

In implementing the operations involved in finding the inverse Laplace transformation of network functions, we will need to make use of PASCAL variables that are complex, that is, that have separate real and imaginary parts. In general, if z is such a variable, it can be written

$$z = x + jy \tag{8.4}$$

where x and y are real variables representing the real and imaginary parts, respectively, of the complex variable z. Thus, x and y may be defined as

$$x = \text{Re}(z) \qquad y = \text{Im}(z) \tag{8.5}$$

Complex variables may be implemented in PASCAL by making a (main program) declaration of a user-defined type which is a *record* (see Appendix A.10). The fields of the record are used to define separately the real and imaginary parts of a complex variable. The type declaration has the form

```
TYPE complex = RECORD
               re, im: real
               END;
```
(8.6)

Using such a declaration, we may define a variable z as complex by using the VAR declaration

```
VAR z: complex;
```
(8.7)

The real and imaginary components of z may now be separately referred to as z.re and z.im, respectively. For example, to assign the complex number $3 + j4$ to the complex variable z would require the use of two assignment statements as follows:

```
z.re := 3.0;    z.im := 4.0;
```
(8.8)

To implement the usual arithmetic operations of addition, subtraction, multiplication, and division for complex numbers, separate procedures must be defined for each of these operations. For these we use the names Add, Sub, Mul, and Dvd, respectively (note that we cannot use Div, since this is a PASCAL reserved word). These procedures are shown in Fig. 8.3. Two other operations involving complex numbers are frequently useful.

```
(* These procedures require a type definition as follows:
TYPE complex = RECORD
               re, im : real
               END;                                        *)

PROCEDURE Add (x, y : complex; VAR z : complex);
   BEGIN
      z.re := x.re + y.re;
      z.im := x.im + y.im
   END;

PROCEDURE Sub (x, y : complex; VAR z : complex);
   BEGIN
      z.re := x.re - y.re;
      z.im := x.im - y.im
   END;

PROCEDURE Mul (x, y : complex; VAR z : complex);
   VAR q : complex;
   BEGIN
      q.re := x.re*y.re - x.im*y.im;
      q.im := x.re*y.im + x.im*y.re;
      z := q
   END;

PROCEDURE Dvd (x, y : complex; VAR z : complex);
   VAR q : complex;
       d : real;
   BEGIN
      d := sqr(y.re) + sqr(y.im);
      q.re := (x.re*y.re + x.im*y.im) / d;
      q.im := (x.im*y.re - x.re*y.im) / d;
      z := q
   END;
```

Figure 8.3 Procedures for complex arithmetic operations.

```
FUNCTION Mag (x : complex) : real;
   BEGIN
      Mag := sqrt(sqr(x.re) + sqr(x.im))
   END;

FUNCTION Arg (x : complex) : real;
   CONST pi = 3.141592653;
   VAR a : real;
   BEGIN
      WITH x DO
         IF re = 0.0 THEN
            IF im = 0.0 THEN
               a := 0.0
            ELSE BEGIN
               a := 0.5 * pi;
               IF im < 0.0 THEN
                  a := -a
               ELSE
            END
         ELSE BEGIN
            a := arctan(im/re);
            IF re < 0.0 THEN
               IF im = 0.0 THEN
                  a := pi
               ELSE
                  IF im < 0.0 THEN
                     a := a - pi
                  ELSE
                     a := a + pi
            ELSE
         END;
         Arg := a
   END;
```

Figure 8.4 Functions for evaluating complex numbers.

These are determining the magnitude of a complex number and determining its argument (or phase). For the complex variable z defined in (8.4), these operations are defined as

$$|z| = \sqrt{x^2 + y^2} \qquad \text{Arg } z = \tan^{-1}\left(\frac{y}{x}\right) \tag{8.9}$$

To implement these operations in PASCAL, we use the functions Mag and Arg given in Fig. 8.4. Note that the function Arg provides an output which is defined over four quadrants; that is, it has the range

$$-\pi \leq \text{Arg } z \leq \pi$$

Also note that if the magnitude of z is zero, the argument is arbitrarily set to zero. This feature avoids the possibility that program execution might be terminated because an undefined output was being produced by the function. Example 8.1 illustrates the use of these procedures.

EXAMPLE 8.1

As an example of the use of the subprograms defined in Figs. 8.3 and 8.4, consider the expression

$$z = \frac{a + b}{c}$$

```
PROGRAM TestComplex (output);
TYPE complex = RECORD
                  re, im : real
               END;
VAR a, b, c, z, temp : complex;

BEGIN
   a.re := 1.0;  a.im := 2.0;
   b.re := -3.0; b.im := 4.0;
   c.re := -5.0; c.im := -6.0;
   Add(a, b, temp);
   Dvd(temp, c, z);
   writeln(' Magnitude = ', Mag(z));
   writeln(' Argument = ', Arg(z), ' radians')
END.
```

Figure 8.5 Program for Example 8.1.

where a, b, c, and z are complex numbers. It is desired to find $|z|$ and argument z for the case where $a = 1 + j2$, $b = -3 + j4$, and $c = -5 - j6$. A program for doing this is given in Fig. 8.5. For brevity, the procedures and functions defined in Figs. 8.3 and 8.4 are not shown in the listing. The output from this program is "Magnitude = 0.80978" and "Argument = -2.1251 radians." These results are readily verified by direct computation.

Now let us apply the procedures for manipulating complex variables defined above to implement the evaluation of a polynomial at some (complex) value of its argument. Let the polynomial be $P(s)$, where

$$P(s) = a_0 + a_1 s + a_2 s^2 + \cdots + a_n s^n \tag{8.10}$$

and where the coefficients a_i ($i = 0, 1, 2, \ldots, n$) are real, and where s is a complex variable. To evaluate this polynomial at some value $s = s_0$ requires that we find

$$P(s_0) = a_0 + a_1 s_0 + a_2 s_0^2 + \cdots + a_n s_0^n \tag{8.11}$$

where s_0 is a complex number. To compute the quantity given in (8.11), it is more efficient to rewrite the equation in a form that does not require that the quantity s_0 be directly raised to the various indicated powers. To avoid this, we may put (8.11) in the form

$$P(s_0) = a_0 + s_0\{a_1 + s_0[a_2 + \cdots + s_0(a_{n-1} + s_0 a_n)]\} \tag{8.12}$$

The operations indicated in (8.12) are readily programmed as an algorithm. In this algorithm we use value as an intermediate (complex) variable. The algorithm is defined by the following steps:

1. Set value equal to a_n.
2. Multiply value by s_0 and add a_{n-1} to the product. Set value equal to the result.
3. Multiply value by s_0 and add a_{n-2} to the product. Set value equal to the result.
4. Continue this operation until value is multiplied by s_0 and added to a_0. Set value equal to the result. At this point value equals $P(s_0)$.

The algorithm defined by the steps given above is readily implemented in a PASCAL procedure called ValPoly (for complex e*val*uation of a *poly*nomial) and identified by the following heading:

PROCEDURE ValPoly (n: integer; VAR a: array11;
s: complex; VAR val: complex);

where n is the degree of the polynomial, a is a one-dimensional 11-element array (array11 is a user-defined type) of real coefficients, s is the complex value at which the polynomial is to be evaluated, and val is the complex output, that is, $P(s_0)$. A listing of the procedure is given in Fig. 8.6. A flowchart for it is given in Fig. 8.7. A comparison of the flowchart with the relation of (8.12) will serve to explain its operation. A summary of the information concerning the procedure is given in Table 8.1.

A second operation that we shall frequently be called upon to make on a given polynomial is the operation of differentiation. If we define a polynomial $P(s)$ by the equation

$$P(s) = a_0 + a_1 s + a_2 s^2 + a_3 s^3 + \cdots + a_n s^n \tag{8.13}$$

then $P'(s)$, where the prime indicates differentiation with respect to s, is defined by the relation

$$P'(s) = a_1 + 2a_2 s + 3a_3 s^2 + \cdots + n a_n s^{n-1} \tag{8.14}$$

```
PROCEDURE ValPoly (n : integer; VAR a : array11;
                   s : complex; VAR val : complex);

(* Procedure for finding the (complex) value of a
   polynomial when evaluated at a (complex) value
   of its argument
     n - Degree of polynomial
     a - Array of coefficients of polynomial
     s - Complex value of argument of polynomial
     value - Value of polynomial
   The polynomial is assumed to have the form
   a[0] + a[1]*s + a[2]*s**2 + a[3]*s**3 + ...
   Note: This procedure calls the procedures Add and Mul *)

   VAR
     i : integer;
     z : complex;

   BEGIN
     z.im := 0.0;
     val.re := a[n];
     val.im := 0.0;
     i := n;
     IF n > 0 THEN
       REPEAT
         BEGIN
           i := i - 1;
           z.re := a[i];
           Mul(val, s, val);
           Add(val, z, val)
         END
       UNTIL i = 0
   END;
```

Figure 8.6 Listing of procedure ValPoly.

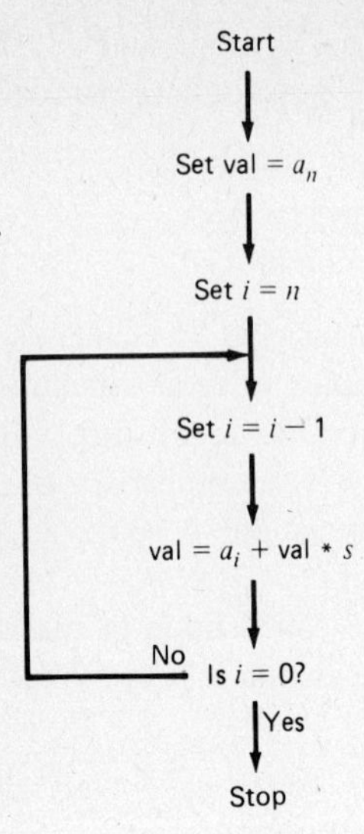

Figure 8.7 Flowchart for procedure ValPoly.

This may also be written in the form

$$P'(s) = b_0 + b_1 s + b_2 s^2 + \cdots + b_{n-1} s^{n-1} \tag{8.15}$$

where $b_0 = a_1$, $b_1 = 2a_2$, etc. The operation of differentiation is easily programmed for the digital computer. A procedure for performing such an operation must be supplied with the following information:

1. The coefficients a_i which define the original polynomial $P(s)$.
2. The degree n of the original polynomial $P(s)$.

TABLE 8.1 SUMMARY OF THE CHARACTERISTICS OF THE PROCEDURE ValPoly

Identifying Statement PROCEDURE ValPoly (n : integer; VAR a : array11; s : complex; VAR val : complex);

Purpose To determine the value of a polynomial $P(s)$ of the form

$$P(s) = a_0 + a_1 s + a_2 s^2 + a_3 s^3 + \cdots + a_n s^n$$

at a given (complex) value of the argument s.

Additional Subprograms Required This procedure calls the procedures Mul and Add.

Input Arguments

n The degree n of the polynomial $P(s)$
a The array of coefficients a_i of the polynomial $P(s)$
s The complex value of the variable s at which it is desired to evaluate the polynomial $P(s)$

Output Argument

val The (complex) value of $P(s)$ evaluated for the specified value of s

User-Defined Types

complex = RECORD re, im : real END;
array11 = ARRAY[0..10] OF real;

```
PROCEDURE DifPoly (n : integer; VAR a : array11;
                   VAR m : integer; VAR b : array11);

(* Procedure for differentiating a polynomial
   n - Degree of input polynomial
   a - Array of coefficients of input polynomial
   m - Degree of differentiated polynomial
   b - Array of coefficients of differentiated polynomial
   Note: The polynomials are assumed to have the form
   a[0] + a[1]*s + a[2]*s**2 + a[3]*s**3 + ...        *)

   VAR
      i : integer;

   BEGIN
      m := n - 1;
      FOR i := 1 TO n DO
         b[i-1] := i * a[i]
   END;
```

Figure 8.8 Listing of procedure DifPoly.

As outputs, the procedure must calculate the coefficients b_i of the differentiated polynomial $P'(s)$ and the degree $m(= n - 1)$ of the new polynomial.

Consideration of the above inputs leads us to define a procedure DifPoly (for *dif*ferentiation of a *poly*nomial) identified by the following heading:

```
PROCEDURE DifPoly (n:integer; VAR a:array11;
                   VAR m:integer; VAR b:array11);
```

where n is the degree of the polynomial $P(s)$, a and b are the one-dimensional arrays of real coefficients, and m is the degree of the polynomial $P'(s)$. A listing of such a procedure is given in Fig. 8.8. Comparison of the statements of this listing with (8.13) to (8.15) readily verifies its operation. A summary of information concerning the procedure is given in Table 8.2.

TABLE 8.2 SUMMARY OF THE CHARACTERISTICS OF THE PROCEDURE DifPoly

Identifying Statement PROCEDURE DifPoly (n : integer; VAR a : array11; VAR m : integer; VAR b : array11);

Purpose To differentiate a polynomial $P(s)$ of the form

$$P(s) = a_0 + a_1 s + a_2 s^2 + \cdots + a_n s^n$$

The resulting polynomial will be $P'(s)$, where

$$P'(s) = b_0 + b_1 s + b_2 s^2 + \cdots + b_m s^m$$

Additional Subprograms Required None

Input Arguments

n The degree n of the polynomial $P(s)$

a The one-dimensional array of variables a[i] in which are stored the values of the coefficients a_i defining the polynomial $P(s)$

Output Arguments

m The degree m of the polynomial $P'(s)$ $(= n - 1)$

b The one-dimensional array of variables b[i] in which are stored the values of the coefficients b_i of the polynomial $P'(s)$

User-Defined Type

array11 = ARRAY[0..10] OF real;

In this section we have introduced two procedures—ValPoly and DifPoly. In the next section we see how these procedures may be used to solve the problem formulated in Sec. 8.1.

8.3 THE DETERMINATION OF RESIDUES AND THE INVERSE TRANSFORM: THE PROCEDURES ParFracExp AND InvLaplace

Let us assume that a given rational function $F(s)$ may be written in the form

$$F(s) = \frac{A(s)}{B(s)} = \frac{a_0 + a_1 s + a_2 s^2 + \cdots + a_m s^m}{b_0 + b_1 s + b_2 s^2 + \cdots + b_n s^n} \tag{8.16}$$

where s is the complex frequency variable, coefficients a_i and b_i are real, and the degree of the numerator polynomial $A(s)$ is less than the degree of the denominator polynomial $B(s)$ $(m < n)$. The *poles* of the function $F(s)$ [the roots of the polynomial $B(s)$] can be found using a root-solving algorithm. Such an algorithm is implemented by the PASCAL procedure Root. This has the heading

```
PROCEDURE Root (n: integer; b: array11;
                VAR p: carray10);
```

where n is the degree of the polynomial $B(s)$, b is the one-dimensional array of real coefficients, and p is the output one-dimensional complex array of roots (array11 and carray10 are user-defined types). A listing and a discussion of the operation of this procedure are given in Appendix B. A summary of its characteristics is given in Table

TABLE 8.3 SUMMARY OF THE CHARACTERISTICS OF THE PROCEDURE Root

Identifying Statement PROCEDURE Root (n : integer; b : array11; VAR p : carray10);

Purpose To find the roots of a polynomial $B(s)$ identified as

$$B(s) = b_0 + b_1 s + b_2 s^2 + b_3 s^3 + \cdots + b_n s^n$$

Additional Subprograms Required None

Input Arguments

n The degree of the polynomial $B(s)$

b The array of coefficients of the polynomial $B(s)$

Output Argument

p The complex array of the roots of $B(s)$

User-Defined Types

```
complex = RECORD re, im: real END;
array11 = ARRAY[0..10] OF real;
carray10 = ARRAY[1..10] OF complex;
```

Note: Root requires that the highest-degree coefficient b_n be set to unity for proper execution. If the value of this coefficient is not 1, the user should divide all the coefficients by b_n before calling Root.

8.3. If the function $F(s)$ has a simple pole located at the value of the complex frequency variable p_i, then, in the limit as s approaches p_i, the function takes on the value

$$\lim_{s \to p_i} F(s) = \frac{k_i}{s - p_i} \tag{8.17}$$

The coefficient k_i in (8.17) is referred to as the *residue* of the simple pole at p_i. If p_i is real, k_i will be real. If p_i is complex, k_i will be complex. In addition, since poles that are complex will always occur in conjugate pairs (this assumes that the b_i are real), it may be shown that the residues for such a conjugate pair will also be conjugate (if the a_i are real). The value of the residue k_i may readily be determined directly from the function $F(s)$. A relation for determining it is

$$k_i = \frac{A(p_i)}{B'(p_i)} \tag{8.18}$$

where p_i is a simple pole of $F(s)$, $A(p_i)$ is the numerator polynomial of $F(s)$ evaluated at $s = p_i$, and $B'(p_i)$ is the derivative of the denominator polynomial of $F(s)$ (taken with respect to s), evaluated at $s = p_i$. The relation of (8.18) is easily programmed for the digital computer. A flowchart describing the operation of a procedure that will perform this computation is given in Fig. 8.9. From an examination of the flowchart and of (8.16) and (8.17), we see that such a procedure must be supplied with the following input quantities:

1. The values of the coefficients a_i of the numerator polynomial $A(s)$ for the rational function $F(s)$ of (8.16).

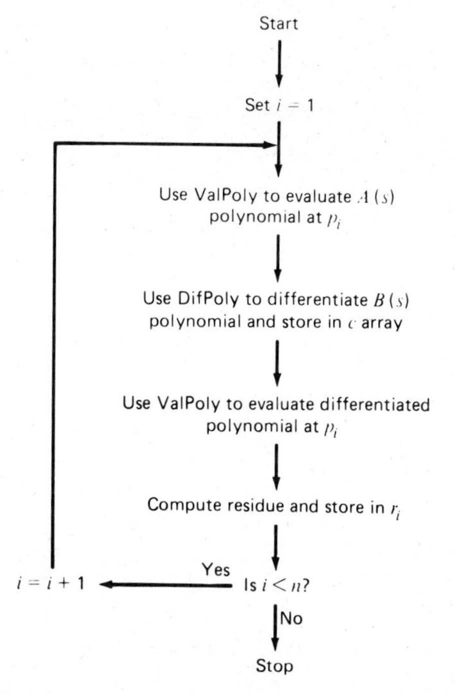

Figure 8.9 Flowchart for procedure ParFracExp.

2. The degree m of the numerator polynomial $A(s)$.
3. The values of the coefficients b_i of the denominator polynomial $B(s)$ for the rational function $F(s)$ of (8.16).
4. The degree n of the denominator polynomial $B(s)$.
5. The value Re p_i of the real part of the pole location p_i.
6. The value Im p_i of the imaginary part of the pole location p_i.

As outputs, the procedure must provide the values of R_{real} and R_{imag}, which are, respectively, the real and imaginary parts of the residue of the pole at p_i.

Consideration of the above leads us to define a procedure ParFracExp (for *par*tial *frac*tion *exp*ansion) identified by the following heading:

PROCEDURE ParFracExp (m: integer; VAR a: array11; n: integer;
VAR b: array11; VAR p, r: carray10);

where m and a specify the numerator of $F(s)$, n and b specify the denominator, and p specifies the (complex) pole locations. The outputs from the procedure are the residues contained in the (complex) array r. Thus, the real parts of the residues are contained in the variables r[i].re and the imaginary parts (these will be zero for real poles) in r[i].im. The types array11 and carray10 are the same as used in the procedure Root. Such a procedure is easily constructed by using the procedures ValPoly and DifPoly, which were presented in Sec. 8.2. A listing of the PASCAL statements for such a procedure is given in Fig. 8.10. An examination of this listing and the flowchart shown in Fig. 8.9 should

```
PROCEDURE ParFracExp (m : integer; VAR a : array11;

                      n : integer; VAR b : array11;

                      VAR p, r : carray10);

(* Procedure for finding the residues of the poles of
   a rational function F(s) = A(s)/B(s) containing
   only simple poles
     m - Degree of numerator polynomial A(s)
     a - Array of coefficients of A(s)
     n - Degree of denominator polynomial B(s)
     b - Array of coefficients of B(s)
     p - Complex array of poles of F(s)
     r - Complex array of residues
   Note: This procedure calls the procedures ValPoly,
   DifPoly, and Dvd                                       *)

   VAR
      i, k : integer;
      w, z : complex;
      c    : array11;

   BEGIN
      FOR i := 1 TO n DO
         BEGIN
            ValPoly(m, a, p[i], w);
            DifPoly(n, b, k, c);
            ValPoly(k, c, p[i], z);
            Dvd(w, z, r[i])
         END
   END;
```

Figure 8.10 Listing of procedure ParFracExp.

TABLE 8.4 SUMMARY OF THE CHARACTERISTICS OF THE PROCEDURE ParFracExp

Identifying Statement PROCEDURE ParFracExp (m : integer; VAR a : array11; n : integer; VAR b : array11; VAR p, r : carray10);

Purpose To determine the residues at the simple poles p_i of a rational function $F(s)$ having the form

$$F(s) = \frac{A(s)}{B(s)} = \frac{a_0 + a_1 s + a_2 s^2 + \cdots + a_m s^m}{b_0 + b_1 s + b_2 s^2 + \cdots + b_n s^n}$$

Additional Subprograms Required This procedure calls the procedures ValPoly, DifPoly, and Dvd.

Input Arguments

m The degree of the numerator polynomial $A(s)$
a The array of coefficients a_i of the polynomial $A(s)$
n The degree of the denominator polynomial $B(s)$
b The array of coefficients b_i of the polynomial $B(s)$
p The complex array of the poles of the function $F(s)$

Output Argument

r The complex array of the residues for the poles

User-Defined Types

```
complex = RECORD re, im : real END;
array11 = ARRAY[0..10] OF real;
carray10 = ARRAY[1..10] OF complex;
```

enable the reader to understand its operation. A summary of the information concerning the procedure is given in Table 8.4.

To see the manner in which the procedure ParFracExp may be used to find the inverse transformation of a given rational function, let us assume that such a function has j simple real poles p_i and h pairs of complex conjugate poles located at $p_i^{(c)}$ and $\bar{p}_i^{(c)}$, where $\bar{p}_i^{(c)}$ is the complex conjugate of $p_i^{(c)}$. The function may then be expanded in the form

$$F(s) = \sum_{i=1}^{j} \frac{k_i}{s - p_i} + \sum_{i=1}^{h} \frac{k_i^{(c)}}{s - p_i^{(c)}} + \sum_{i=1}^{h} \frac{\bar{k}_i^{(c)}}{s - \bar{p}_i^{(c)}} \tag{8.19}$$

where k_i and $k_i^{(c)}$ are residues. We may easily take the inverse transformation of each of the j simple real poles p_i by using the relation given in Fig. 8.1. Thus we obtain

$$\sum_{i=1}^{j} k_i e^{p_i t} \tag{8.20}$$

For the complex simple poles, let the residues have the form $k_i^{(c)} = a_i + jb_i$, where a_i is the real part of the residue and b_i is the imaginary part. If we write $p_i^{(c)} = p_i^{(r)} + jp_i^{(i)}$, where $p_i^{(r)}$ is the real part of the location of $p_i^{(c)}$ and $p_i^{(i)}$ is the imaginary part, then it may be shown that each pair of complex conjugate poles and the residues associated with them will have an inverse transform of the form

$$2e^{p_i^{(r)}t}[a_i \cos(p_i^{(i)}t) - b_i \sin(p_i^{(i)}t)] \tag{8.21}$$

Thus, the complete inverse transform for a function of the type given in (8.19) will be

$$f(t) = \sum_{i=1}^{j} k_i e^{p_i t} + 2\sum_{i=1}^{h} e^{p_i^{(r)}t}[a_i \cos(p_i^{(i)}t) - b_i \sin(p_i^{(i)}t)] \tag{8.22}$$

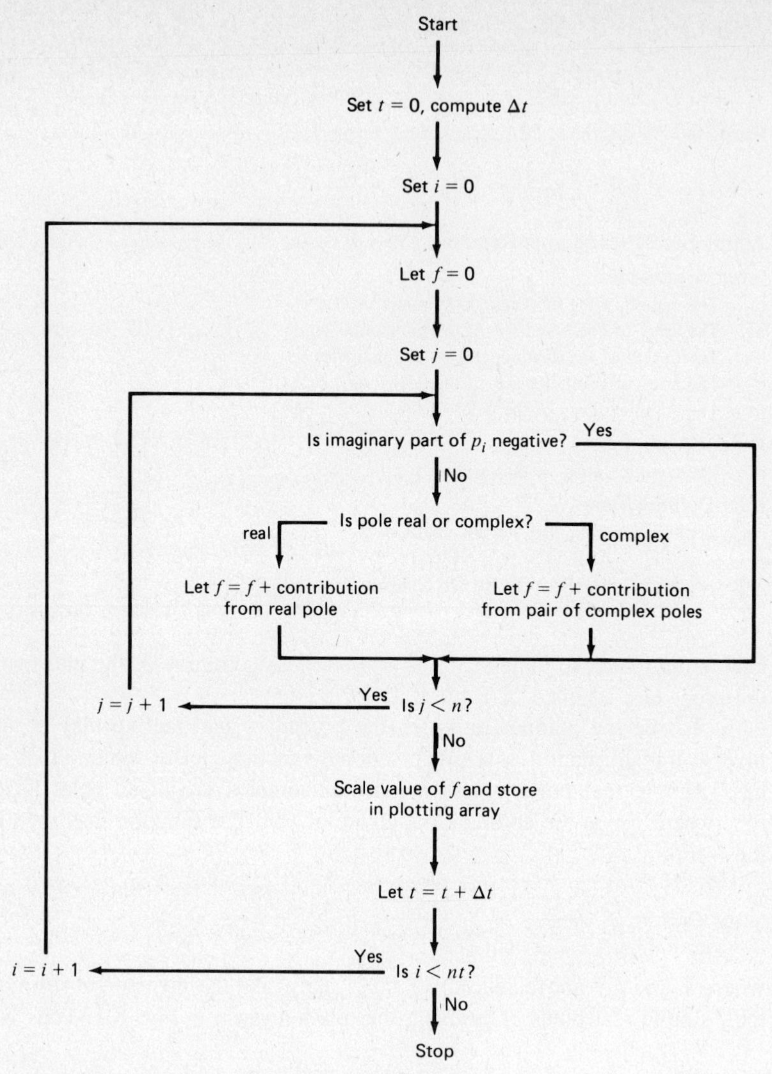

Figure 8.11 Flowchart for procedure InvLaplace.

```
PROCEDURE InvLaPlace (nt : integer; tr : real; n : integer;

                VAR p, r : carray10; na : integer;

                scale : real; VAR a : plotArray);

(* Procedure for computing and plotting the inverse
   Laplace transform of a rational function F(s)
     nt - Number of time increments in the plot
     tr - Range of values of time
     n - Degree of denominator polynomial of F(s)
     p - Complex array of pole locations
     r - Complex array of residues
     na - Position in plotting array where data is stored
     scale - Scale factor for plotted data
     a - Plotting array
   Note: This procedure calls the procedures Expc and Mul *)

   VAR
      t, dt, f : real;
      i, j     : integer;
      pt, ex   : complex;

   BEGIN

      (* Initialize the time variable and compute dt *)

      t := 0.0;
      dt := tr / nt;

      (* Start outer loop to make computations at each
         value of time                                  *)

      FOR i := 0 TO nt DO
         BEGIN
            f := 0.0;

               (* Start inner loop to sum the effects of
                  each of the poles                       *)
               FOR j := 1 TO n DO
                  BEGIN
                     IF p[j].im > -1.0E-5 THEN
                     IF abs(p[j].im) < 1.0E-5
                        THEN

                           (* Contribution from real pole *)

                           f := f + r[j].re*exp(p[j].re*t)
                        ELSE

                           (* Contribution from complex pole *)
                           BEGIN
                              pt.re := p[j].re * t;
                              pt.im := p[j].im * t;
                              Expc(pt, ex);
                              Mul(ex, r[j], ex);
                              f := f + 2.0 * ex.re
                           END
                  END; (* End of inner loop with j index *)
            a[na,i] := f * scale;
            t := t + dt
         END  (* End of outer loop with i index *)
   END;

PROCEDURE Expc (x : complex; VAR z : complex);
   VAR r : real;
   BEGIN
      r := exp(x.re);
      z.re := r * cos(x.im);
      z.im := r * sin(x.im)
   END;
```

Figure 8.12 Listing of procedures InvLaplace and Expc.

This relation is easily programmed into a PASCAL inverse-transformation procedure that can separate the effects of the real and complex poles, and perform the summations indicated in (8.22) for a series of values of t, finally storing the results in an array for subsequent plotting by the procedure Plot5. Such an inverse-transformation procedure must be supplied with the following inputs:

1. The number of time points n_t at which the function $f(t)$ of (8.22) is to be evaluated.
2. The total time range t_r (starting from $t = 0$) over which the function $f(t)$ of (8.22) is to be evaluated.
3. The degree n of the denominator polynomial $B(s)$ of (8.16).
4. The pole locations p_i of the function $F(s)$ of (8.16).
5. The residues r_i of the poles p_i of the function $F(s)$ of (8.16).
6. The position n_a (where $n_a = 1, 2, 3, 4,$ or 5) in the plotting array where the values of $f(t)$ of (8.22) are to be stored.
7. Any scale factor that is to be applied to the values of $f(t)$ of (8.22) for convenience in plotting.

As an output, the procedure will provide a plotting array a in which are stored the values of $f(t)$ of (8.22).

Consideration of the above leads us to define a procedure InvLaplace for (*inv*erse *Laplace* transform) identified by the following heading:

```
PROCEDURE InvLaplace (nt:integer; tr:real; n:integer;
                      VAR p, r:carray10; na:integer;
                      scale:real; VAR a:plotArray);
```

TABLE 8.5 SUMMARY OF THE CHARACTERISTICS OF THE PROCEDURE InvLaplace

Identifying Statement PROCEDURE InvLaplace (nt : integer; tr : real; n : integer; VAR p, r : carray10; na : integer; scale : real; VAR a : plotArray):

Purpose To compute the data representing the inverse Laplace transform of a rational function characterized by its pole locations and its residues

Additional Subprograms Required This procedure calls the procedure Expc

Input Arguments

nt	Number of time increments to be used in preparing the data
tr	Time range (starting from zero) used in computing the data
n	Degree of the denominator polynomial of the rational function
p	Complex array of pole locations
r	Complex array of the residues for the poles
na	Position in the plotting array where the computed data is to be stored
scale	Scale factor used to multiply the data to prepare it for plotting

Output Argument

a	Plotting array used to store the computed data

User-Defined Types

```
complex = RECORD re, im:real END;
carray10 = ARRAY[1..10] OF complex;
plotArray = ARRAY[1..5,0..100] OF real;
```

where the various arguments correspond with the items and notation used in the list of inputs and outputs given above, and where carray10 is a user-defined one-dimensional array [1..10] of complex reals and plotArray is a user-defined two-dimensional array [1..5,0..100] of reals. A flowchart for the procedure is given in Fig. 8.11. A listing of it is given in Fig. 8.12. Note that it calls a procedure Expc which is also shown in the figure. A summary of information concerning it is given in Table 8.5. Examples 8.2 and 8.3 illustrate the use of the procedures described in this chapter.

EXAMPLE 8.2

As an example of the use of the procedures described in this chapter, we find the inverse transform of a rational function giving the output current $i_2(t)$ for the network shown in Fig. 8.13 under the condition that a step function of voltage $v_1(t) = 3$ V is applied to the network at $t = 0$. The network function relating the transformed input voltage and output current for this network is

$$\frac{I_2(s)}{V_1(s)} = \frac{2}{s^2 + 7s + 6}$$

For the specified input voltage, from Fig. 8.1 we find that the transformed input voltage $V_1(s) = 3/s$. Thus, we may rewrite the equation given above to specify $I_2(s)$ as a rational function in the complex frequency variable s. We obtain

$$I_2(s) = \frac{6}{s^3 + 7s^2 + 6s}$$

By using the procedure ParFracExp to find the terms in the partial-fraction expansion, and the procedure InvLaplace to compute the terms in the partial-fraction expansion at various values of the independent variable time and to store them in the plotting array, a plot is readily obtained for the quantity $i_2(t)$. A listing and a flowchart for such a program are shown in Fig. 8.14. A plot of the output over the period of time from 0 to 5 sec is given in Fig. 8.15. The network used in this example and shown in Fig. 8.13 is the same network as that used in Example 6.1. A comparison of the results obtained from the two examples readily establishes the equivalence of the approach used in this chapter (which is based on the frequency domain) and the one presented in Chap. 6 (based on the time domain).

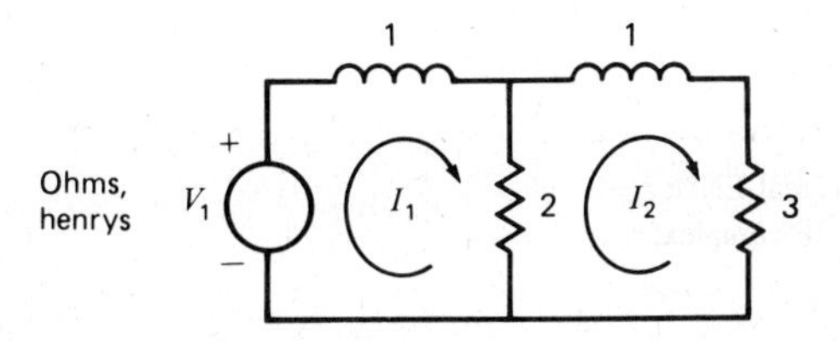

Figure 8.13 Circuit for Example 8.2.

```
PROGRAM Main (input, output);    (* Example 8.2 *)

(* Main program for analyzing two-mesh resistor/
   inductor network
     m, a - Degree and array of coefficients of numerator
     n, b - Degree and array of coefficients of denominator
     poles - Complex array of pole locations
     resid - Complex array of residues for poles
     plt - plotting array
   This program calls the procedures Root, ParFracExp,
   InvLaplace, and Plot5.                                  *)

TYPE
   complex = RECORD
               re, im : real
             END;
   array11 = ARRAY[0..10] OF real;
   carray10 = ARRAY[1..10] OF complex;
   plotArray = ARRAY[1..5,0..100] OF real;

VAR
   i, m, n : integer;
   a, b    : array11;
   poles, resid    : carray10;
   plt      : plotArray;

BEGIN

   (* Read the input data defining the problem *)

   readln(m, n);
   writeln(' Numerator');
   FOR i := 0 TO m DO
      BEGIN
         readln(a[i]);
         writeln(i, a[i])
      END;
   writeln(' Denominator');
   FOR i := 0 TO n DO
      BEGIN
         readln(b[i]);
         writeln(i, b[i])
      END;

   (* Call the Root procedure to find the pole locations
      and the procedure ParFracExp to find the residues    *)

   Root(n, b, poles);
   writeln(' Poles');
   FOR i := 1 TO n DO
      writeln(poles[i].re, poles[i].im);
   ParFracExp(m, a, n, b, poles, resid);
   writeln(' Residues');
   FOR i := 1 TO n DO
      writeln(resid[i].re, resid[i].im);

   (* Call the procedures InvLaPlace and Plot5 to get the output *)

   InvLaPlace(50, 5.0, 3, poles, resid, 1, 100.0, plt);
   Plot5(plt, 1, 50, 100);
END.
```

Input Data

```
0 3  <------  Degree of numerator, degree of denominator
6.0  <------  Zero degree coefficient of numerator
0.0  <------  Zero degree coefficient of denominator
6.0  <------  First degree coefficient of denominator
7.0  <------  Second degree coefficient of denominator
1.0  <------  Third degree coefficient of denominator
```

(*a*)

Figure 8.14 Listing and flowchart for Example 8.2.

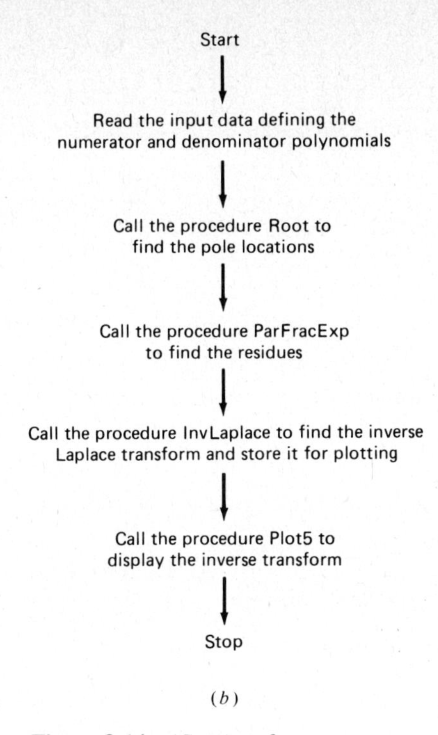

(*b*)

Figure 8.14 (*Continued*)

```
NUMERATOR
         0   6.0000000000000E+000
DENOMINATOR
         0   0.0000000000000E+000
         1   6.0000000000000E+000
         2   7.0000000000000E+000
         3   1.0000000000000E+000
POLES
-6.0000000000000E+000   0.0000000000000E+000
-1.0000000000000E+000   0.0000000000000E+000
 0.0000000000000E+000   0.0000000000000E+000
RESIDUES
 2.0000000000000E-001   0.0000000000000E+000
-1.2000000000000E+000   0.0000000000000E+000
 1.0000000000000E+000   0.0000000000000E+000
```

(*a*)

Figure 8.15 Output for Example 8.2.

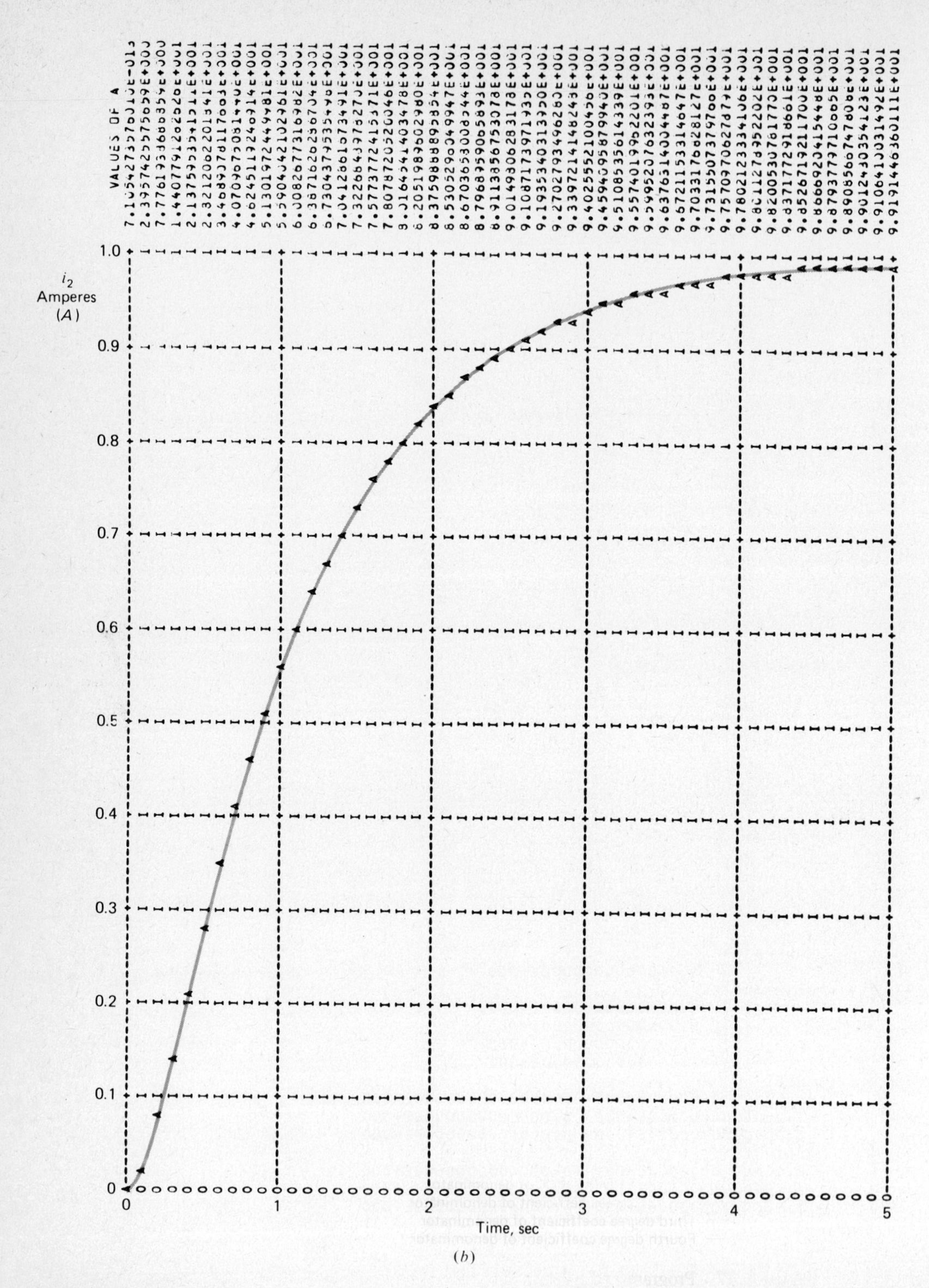

(b)

Figure 8.15 (*Continued*)

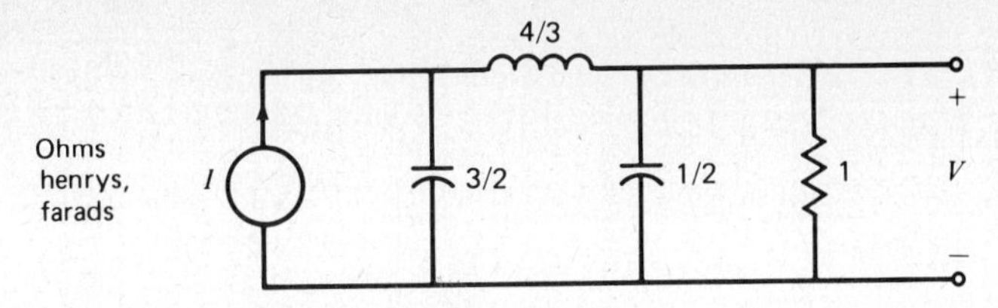

Figure 8.16 Circuit for Example 8.3.

EXAMPLE 8.3

As another example of the use of the procedures ParFracExp and InvLaplace in determining the inverse Laplace transformation of a given function of the complex frequency variable, consider the network shown in Fig. 8.16. The network function relating the input current and the output voltage for this network, in terms of the transformed variables $I(s)$ and $V(s)$, is

$$\frac{V(s)}{I(s)} = \frac{1}{s^3 + 2s^2 + 2s + 1}$$
$$= \frac{1}{(s^2 + s + 1)(s + 1)}$$

If we apply a step current of 80 A to the circuit at $t = 0$, then the transform of the current is $I(s) = 80/s$. We may use the procedures to find an expression for the output voltage $v(t)$. A listing of a PASCAL program for performing this computation and determining 51 values of $v(t)$ over a time range from 0 to 10 sec is shown in Fig. 8.17. A plot of the output is given in Fig. 8.18.

Note that, since the plot array has five positions in which data may be stored, the problem-solving methods illustrated above may readily be extended to situations in which multiple plots are constructed for different values of some of the network elements. Some examples of this technique may be found in the problems. It should also be noted that, although the method described in this chapter applies only to simple poles, the case

```
Program is the same as was used for Example 8.1 except
that the call to InvLaplace is changed to the following:

InvLaplace(50, 10.0, 4, p, r, 1, 1.0, plt);

                     Input Data

0 4   <------  Degree of numerator, degree of denominator
80.0  <------  Zero degree coefficient of numerator
0.0   <------  Zero degree coefficient of denominator
1.0   <------  First degree coefficient of denominator
2.0   <------  Second degree coefficient of denominator
2.0   <------  Third degree coefficient of denominator
1.0   <------  Fourth degree coefficient of denominator
```

Figure 8.17 Program and input data for Example 8.3.

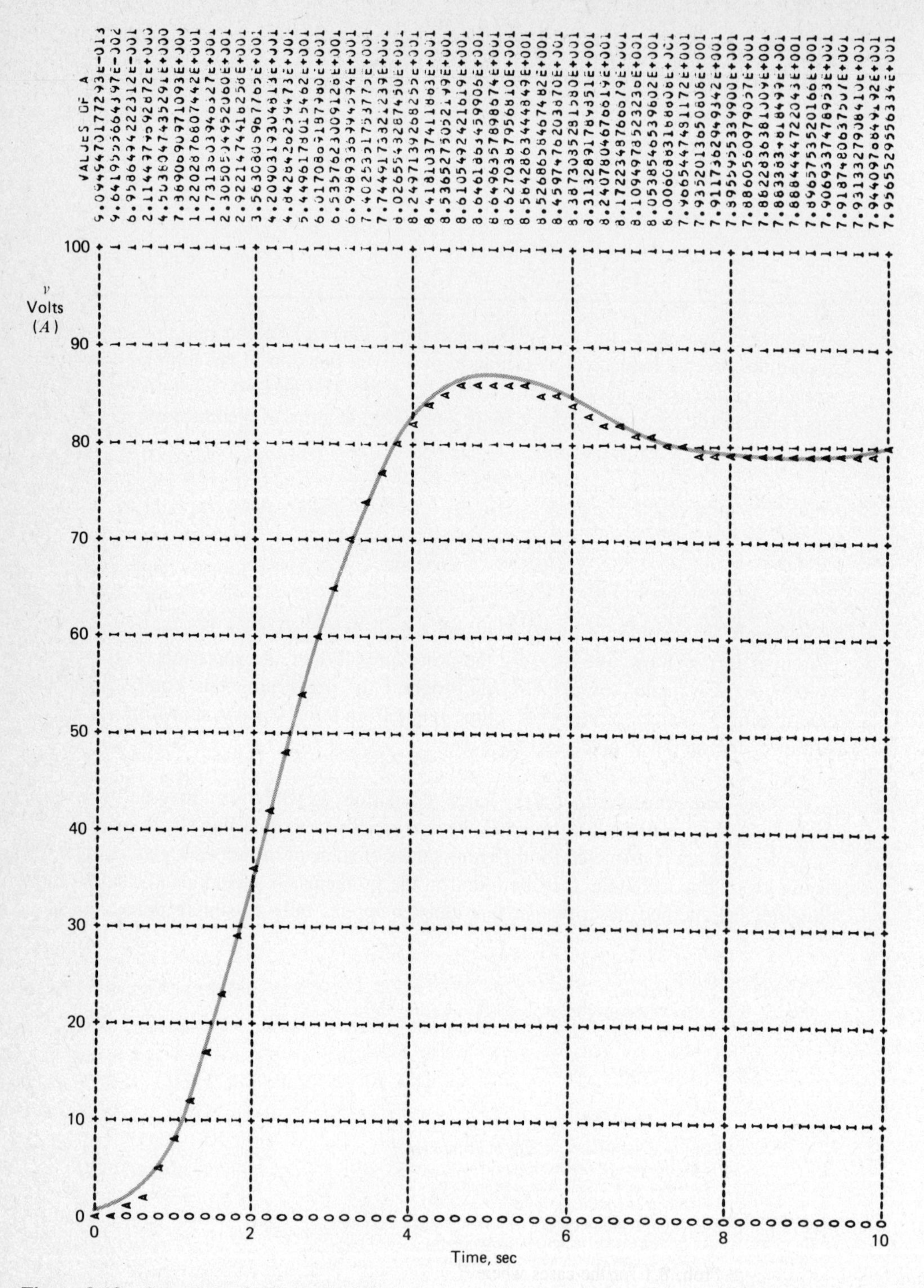

Figure 8.18 Output plot for Example 8.3.

where the poles are of greater order is readily approximated by making a small perturbation of the pole locations so that they may be treated as simple ones. For example, the function $F(s) = 1/(s + 1)^2$ with a second-order pole at $s = -1$ can be approximated as $1/(s + 0.95)(s + 1.05)$. An example of this technique is given in the problems.

8.4 CONCLUSION

In this chapter we have presented a method for finding the time responses of various physical variables by using digital-computational techniques to obtain the inverse Laplace transformation. This procedure is quite different from the ones presented in earlier portions of this book. The earlier techniques relied on operations performed entirely in the time domain, whereas the approach given here is based on the frequency domain. There are advantages and disadvantages to both approaches. For example, the time-domain approach is better adapted to making simultaneous determinations of the values of several variables, since all the variables in the network are calculated at each step of the solution process. The frequency-domain approach, however, is better suited to making determinations of a single variable for different values of the parameters of the network, since the residues for these different values remain the same at all values of time at which the inverse transformation is made.

In general, the digital computer is better suited to performing the operations in the time domain; thus, for large systems, the techniques described in this chapter may not be as useful as those described in earlier chapters. This is especially true for the case of networks containing time-varying and nonlinear elements. There are situations, however, in which the frequency domain provides very desirable information for the network analyst and designer; thus, the techniques of this chapter provide tools that may be useful in many situations.

PROBLEMS

8.1. An *RLC* circuit is shown in Fig. P8.1. If a step function of voltage $v(t) = 50$ V is applied to this circuit at $t = 0$, use the techniques of this chapter to find 51 values of the variable $i_L(t)$ covering a range of time from 0 to 5 sec, and plot the values.

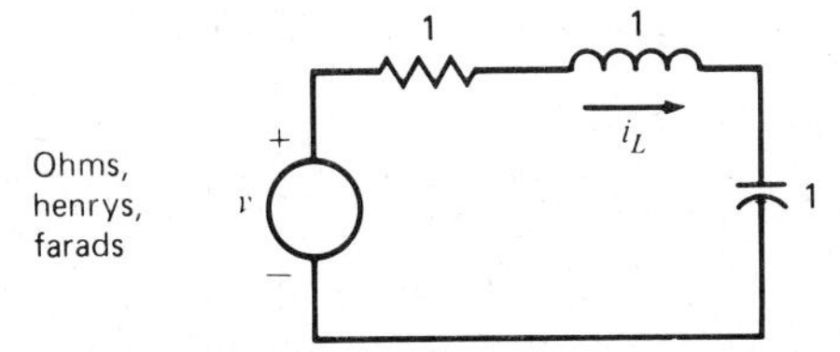

Figure P8.1

8.2. Repeat Prob. 8.1 for the cases where $R = 0.5$ and $R = 1.5\ \Omega$. Plot the results on a single plot and compare them with those obtained in Prob. 8.1.

8.3. Use the techniques described in this chapter to conduct a study of the manner in which the impulse response of a network varies as its pole positions are changed. Let the network function be

$$N(s) = \frac{O(s)}{I(s)}$$

$$= \frac{1}{(s+1)(s^2+as+1)}$$

$$= \frac{1}{s^3+(a+1)s^2+(a+1)s+1}$$

Let the input function $I(s)$ be equal to 1. Plot 51 values of the output function $o(t)$ [the inverse transformation of $O(s)$] for a range of time from 0 to 10 sec for the following values of a: 0, 0.5, 1, and 1.5.

8.4. The functions listed below are second-degree, third-degree, and fourth-degree low-pass maximally flat magnitude functions. Find the output $o(t)$ for the range of time from 0 to 10 sec for each of these functions when excited by an impulse input function of value 140; that is, let

$$i(t) = 140\delta(t)\ \text{A}\ [I(s) = 140]$$

Show all three responses on the same plot, using 51 points to cover the indicated range of time.

Second-degree function:

$$N(s) = \frac{O(s)}{I(s)}$$

$$= \frac{1}{(s+0.70711+j0.70711)(s+0.70711-j0.70711)}$$

$$= \frac{1}{s^2+1.4142s+1}$$

Third-degree function:

$$N(s) = \frac{O(s)}{I(s)}$$

$$= \frac{1}{(s+1)(s+0.5+j0.86603)(s+0.5-j0.86603)}$$

$$= \frac{1}{s^3+2s^2+2s+1}$$

Fourth-degree function:

$$N(s) = \frac{O(s)}{I(s)}$$

$$= \frac{1}{(s+0.92388+j0.38268)(s+0.92388-j0.38268)(s+0.38268+j0.92388)(s+0.38268-j0.92388)}$$

$$= \frac{1}{s^4+2.6131s^3+3.4142s^2+2.6131s+1}$$

8.5. The network functions listed below are second-degree, third-degree, and fourth-degree maximally flat delay functions. Find the output $o(t)$ for each of the functions for a time range from 0 to 5 sec if an input impulse $i(t) = 70\delta(t)$ is applied. Show the results on a single plot, using 51 points.

Second-degree function:

$$N(s) = \frac{O(s)}{I(s)}$$
$$= \frac{3}{(s + 1.5 + j0.86603)(s + 1.5 - j0.86603)}$$
$$= \frac{3}{s^2 + 3s + 3}$$

Third-degree function:

$$N(s) = \frac{O(s)}{I(s)}$$
$$= \frac{15}{(s + 2.3222)(s + 1.8389 + j1.75438)(s + 1.8389 - j1.75438)}$$
$$= \frac{15}{s^3 + 6s^2 + 15s + 15}$$

Fourth-degree function:

$$N(s) = \frac{O(s)}{I(s)}$$
$$= \frac{105}{(s + 2.8962 + j0.86723)(s + 2.8962 - j0.86723)(s + 2.10379 + j2.6574)(s + 2.10379 - j2.6574)}$$
$$= \frac{105}{s^4 + 10s^3 + 45s^2 + 105s + 105}$$

8.6. The network function given below is a non-minimum-phase function, so called because it has right-half-plane zeros. It is also a constant-magnitude or all-pass function, so called because its magnitude for all values of sinusoidal frequency is constant. Use the techniques of this chapter to find the output response $o(t)$ as a function of time over the range of time from 0 to 10 sec if such a function is excited by an input $i(t) = 40\delta(t)$ A, that is, an impulse of strength 40 units. Use 51 points in the plot.

$$N(s) = \frac{O(s)}{I(s)}$$
$$= \frac{(s - 1)(s^2 - s + 1)}{(s + 1)(s^2 + s + 1)}$$

8.7. Repeat Prob. 8.4 for the case where the input is a step of value 70 applied at $t = 0$, that is, $i(t) = 70$, $t \geq 0$.

8.8. Repeat Prob. 8.5 for the case where the input is a step of value 70 applied at $t = 0$, that is, $i(t) = 70$, $t \geq 0$.

8.9. Repeat Prob. 8.6 for the case where the input is a step of value 70 applied at $t = 0$, that is, $i(t) = 70$, $t \geq 0$.

8.10. Repeat the problem given as Example 8.2 for the case where the input $i(t)$ is an impulse of unit strength applied at $t = 0$. Use a time range from 0 to 10 sec.

8.11. To illustrate how second-order poles may be treated using the procedures ParFracExp and InvLaplace, find the inverse Laplace transform of the function

$$F(s) = \frac{1}{(s + 0.95)(s + 1.05)}$$

and compare the resulting time-domain values over a range of time from 0 to 2.5 sec with those obtained from the inverse transform of $1/(s + 1)^2$.

Chapter 9

Sinusoidal Steady-State Analysis: Magnitude and Phase of System Functions

In Chap. 8 we introduced the complex frequency variable s and showed how we could use the inverse Laplace transformation to find output responses (in the time domain) for systems defined by a function of the complex frequency variable. In this chapter we investigate some other uses of the complex frequency variable.

9.1 THE USE OF THE SYSTEM FUNCTION IN SINUSOIDAL STEADY-STATE ANALYSIS

If a system or circuit is described by a function in the complex frequency variable s relating some transformed input variable $I(s)$ and some transformed output variable $O(s)$, there is one type of input excitation (in the time domain) for which it is not necessary to take the inverse Laplace transformation to find the output response. The input excitation for which this is true is a sinusoidal excitation. For example, consider a function $N(s)$ defined as

$$N(s) = \frac{O(s)}{I(s)} = \frac{a_0 + a_1 s + a_2 s^2 + \cdots + a_m s^m}{b_0 + b_1 s + b_2 s^2 + \cdots + b_n s^n} \tag{9.1}$$

If the input waveform $i(t)$ to such a network has the form

$$i(t) = I_0 \cos(\omega t + \alpha) \tag{9.2}$$

then, after any transient terms have disappeared, the response waveform $o(t)$ will have the form

$$o(t) = O_0 \cos(\omega t + \beta) \tag{9.3}$$

We may use the real numbers I_0 and α to define a complex number, usually referred to as a *phasor*, which may be used to represent the input waveform at a specified value of ω, the frequency in radians per second. If we let $\mathcal{I}$ be the symbol for this phasor, then

$$\mathcal{I} = I_0 e^{j\alpha} \tag{9.4}$$

Similarly, we may define an output phasor $\mathcal{O}$ by the relation

$$\mathcal{O} = O_0 e^{j\beta} \tag{9.5}$$

It may be shown that the relation between the input phasor $\mathcal{I}$ and the output phasor $\mathcal{O}$ can be found directly from the function $N(s)$ of (9.1) by substituting $s = j\omega$. Thus we see that

$$\frac{\mathcal{O}}{\mathcal{I}} = \frac{O_0 e^{j\beta}}{I_0 e^{j\alpha}} = N(j\omega) \tag{9.6}$$

In addition, we see that the following relations hold:

$$\frac{O_0}{I_0} = |N(j\omega)| \qquad \beta - \alpha = \text{Arg } N(j\omega) \tag{9.7}$$

In other words, the ratio of the magnitudes of the input and output sinusoids, under conditions of steady-state sinusoidal excitation, is exactly the magnitude of the network function $N(s)$ evaluated for $s = j\omega$, where ω is the value of the sinusoidal variation of the excitation in radians per second. Similarly, the phase or angle of the network function determines the phase difference between the input and output sinusoids. Obviously, since $N(j\omega)$ is a function of ω, the ratio of the magnitudes and the value of the phase difference will be different at different values of input frequency, that is, different values of ω.

The information concerning the magnitude and the phase of a given network function is of considerable interest to the system designer, since it determines the "frequency characteristics" of the system. Determining plots of the magnitude and phase, however, for any but the simplest of system functions can be a tedious chore. In the sections that follow, we present some digital-computer techniques for avoiding this drudgery. Specifically, we present a solution for the following problem:

Problem. Given a function $N(s)$ of the form shown in (9.1), find and plot values of the magnitude and phase of the function $N(j\omega)$ for a range of values of ω.

9.2 DETERMINING THE MAGNITUDE AND PHASE OF A SYSTEM FUNCTION: THE PROCEDURE SinStdySt

In this section we present a method for finding the magnitude and phase of the function $N(j\omega)$ introduced in Sec. 9.1. First, let us examine the function given in (9.1) for the case where $s = j\omega$. We obtain

$$N(j\omega) = \frac{a_0 + a_1 j\omega + a_2(j\omega)^2 + \cdots + a_m(j\omega)^m}{b_0 + b_1 j\omega + b_2(j\omega)^2 + \cdots + b_n(j\omega)^n} \tag{9.8}$$

The numerator and denominator polynomials of (9.8) may be separately evaluated using the procedure ValPoly. The ratio of the results can be found with the procedure Dvd. The magnitude and phase of $N(j\omega)$ can be computed with the procedures Mag and Arg. All these procedures are described in Sec. 8.2.

From a consideration of the above discussion we see that a procedure designed to compute the real part, imaginary part, magnitude, and phase of a given function $N(s)$ as defined in (9.1) for the case where $s = j\omega$ must be supplied with the following information:

1. The degree m of the numerator polynomial of $N(s)$.
2. The values of the coefficients a_i of the numerator polynomial $A(s)$.
3. The degree n of the denominator polynomial of $N(s)$.
4. The values of the coefficients b_i of the denominator polynomial $B(s)$.
5. The value of ω, in radians per second, at which it is desired to determine the real part, the imaginary part, the magnitude, and the phase of $N(j\omega)$.

As outputs, the procedure must supply the values of the real part, the imaginary part, the magnitude, and the phase of the given function at the desired frequency. From a consideration of the above, we see that a procedure designed to make such a compu-

TABLE 9.1 SUMMARY OF THE CHARACTERISTICS OF THE PROCEDURE SinStdySt

Identifying Statement PROCEDURE SinStdySt (m : integer; VAR a : array11; n : integer; VAR b : array11; omega : real; VAR vreal, vimag, vmag, vphase : real);

Purpose To determine the real part, imaginary part, magnitude, and phase of a rational function $N(s)$ of the form

$$N(s) = \frac{a_0 + a_1 s + a_2 s^2 + \cdots + a_m s^m}{b_0 + b_1 s + b_2 s^2 + \cdots + b_n s^n}$$

under sinusoidal steady-state conditions (that is, for $s = j\omega$).

Additional Subprograms Required This procedure calls the procedures ValPoly and Dvd, and the functions Mag and Arg.

Input Arguments

m	The degree m of the numerator polynomial of $N(s)$
a	The array of coefficients a_i of the numerator of $N(s)$
n	The degree n of the denominator polynomial of $N(s)$
b	The array of coefficients b_i of the denominator of $N(s)$
omega	The value of frequency (rad/sec) at which the evaluation is made

Output Arguments

vreal	The value of the real part of $N(j\omega)$
vimag	The value of the imaginary part of $N(j\omega)$
vmag	The value of the magnitude of $N(j\omega)$
vphase	The value of the argument (in radians) of $N(j\omega)$

User-Defined Types

```
array11 = ARRAY[0..10] OF real;
complex = RECORD re, im : real END;
```

```
PROCEDURE SinStdySt (m : integer; VAR a : array11; n : integer;
                     VAR b : array11; omega : real; VAR vreal,
                     vimag, vmag, vphase : real);

(* Procedure for determining the magnitude and phase of a
   network function N(s) = A(s) / B(s) under conditions of
   sinusoidal steady-state excitation (s + jw)
     m - Degree of numerator polynomial A(s)
     a - Array of coefficients of polynomial A(s)
     n - Degree of denominator polynomial B(s)
     b - Array of coefficients of polynomial B(s)
     omega - Frequency w (rad/sec) at which magnitude and
          phase of N(jw) are to be determined
     vreal  - Re N(jw)
     vimag  - Im N(jw)
     vmag   - Magnitude of N(jw)
     vphase - Arg N(jw) (in radians)
   Note: The polynomials are assumed to have the form:
   A(s) = a[0] + a[1]*s + a[2]*s**2 + ...
   Note: This procedure calls the procedures ValPoly,
     Dvd, Mag, and Arg                                  *)

   VAR
      s, numVal, denVal, val : complex;

   BEGIN
      s.re := 0.0;
      s.im := omega;
      ValPoly(m, a, s, numVal);
      ValPoly(n, b, s, denVal);
      Dvd(numVal, denVal, val);
      vreal := val.re;
      vimag := val.im;
      vmag := Mag(val);
      vphase := Arg(val);
   END;
```

Figure 9.1 Listing of procedure SinStdySt.

tation and named SinStdySt (for *sin*usoidal *steady st*ate) may be identified by the following heading:

PROCEDURE SinStdySt (m : integer; VAR a : array11; n : integer;
VAR b : array11; omega : real; VAR vreal, vimag,
vmag, vphase : real);

A listing of a procedure for performing such an evaluation is given in Fig. 9.1. A summary of the characteristics of the procedure is given in Table 9.1.

As an illustration of the use of such a procedure in a typical network problem, consider Example 9.1.

EXAMPLE 9.1

A simple *RLC* network is shown in Fig. 9.2. Routine circuit analysis shows that the open-circuit voltage transfer function for this network for a value of R of 0.1 is

$$\frac{V_2(s)}{V_1(s)} = \frac{2s}{s^2 + 2s + 400}$$

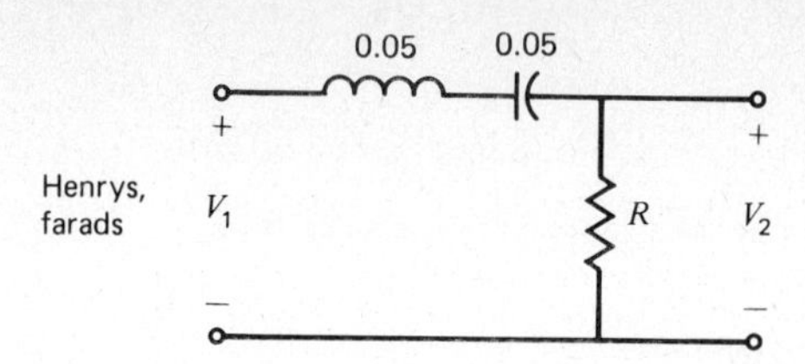

Figure 9.2 An *RLC* network.

We may use the procedure SinStdySt to determine the magnitude and phase of this transfer function under conditions of sinusoidal steady-state excitation, and the procedure Plot5 to display the resulting characteristic. A main program and its input for performing this operation over a frequency range from 0 to 50 rad/sec is shown in Fig. 9.3. The magnitude and the phase are multiplied by 100 and 0.5, respectively, to provide the proper scaling for the plots. The resulting output is shown in Fig. 9.4.

9.3 THE BODE PLOT: THE FUNCTION LogIncrem

Example 9.1 illustrated the use of the procedure SinStdySt to present a plot (using the procedure Plot5) showing how the magnitude of a given network function varies with frequency. Both the magnitude and the frequency were scaled linearly along their respective axes. Another type of plot that is frequently used for displaying the magnitude characteristics of a given network or system function is the Bode plot. Such a plot uses a logarithmic scale for both frequency and magnitude. In addition, the logarithmic magnitude units are sometimes given in decibels (abbreviated dB). The gain in decibels is defined for a voltage transfer function under conditions of sinusoidal steady-state excitation by the relation

$$\text{Decibels}(\omega) = 20 \log_{10}|N(j\omega)| \quad \text{where} \quad N(s) = \frac{V_2(s)}{V_1(s)} \tag{9.9}$$

```
PROGRAM Main (input, output);    (* Example 9.1 *)

(* Main program for computing and plotting the sinusoidal
   steady-state magnitude and phase of a network function
     m - Degree of numerator polynomial
     a - Array of numerator polynomial coefficients
     n - Degree of denominator polynomial
     b - Array of denominator polynomial coefficients
     rad - Frequency (rad/sec)
     pltMag - Plotting array for magnitude
     pltPhase - Plotting array for phase                    *)

CONST
   mscale = 100.0;
   pscale = 0.5;

TYPE
   complex = RECORD
               re, im : real
             END;
   array11 = ARRAY[0..10] OF real;
   plotArray = ARRAY[1..5,0..100] OF real;
```

Figure 9.3 Program and input data for Example 9.1

```
VAR
   rad, vreal, vimag, vmag, vphase : real;
   pltMag, pltPhase                : plotArray;
   a, b                            : array11;
   m, n, i                         : integer;

BEGIN

   (* Initialize the variables and the plot arrays *)

   pltMag[1,0] := 0.0;
   pltPhase[1,0] := 90.0 * pscale;
   readln(m, n);
   writeln(' Numerator');
   FOR i := 0 TO m DO
      BEGIN
         readln(a[i]);
         writeln(i, a[i])
      END;
   writeln(' Denominator');
   FOR i := 0 TO n DO
      BEGIN
         readln(b[i]);
         writeln(i, b[i])
      END;

   (* Use a loop to compute the values of the magnitude and
      phase and store them for plotting                      *)

   FOR i := 1 TO 50 DO
      BEGIN
         rad := i;
         SinStdySt(m, a, n, b, rad, vreal, vimag, vmag, vphase);
         pltMag[1,i] := vmag * mscale;
         pltPhase[1,i] := vphase * pscale * 57.2958
      END;

   (* Plot the magnitude and phase on separate plots *)

   Plot5(pltMag, 1, 50, 100);
   Plot5(pltPhase, 1, 50, 50)
END.
```

(*a*)

Input Data

```
1 2     <------  Degree of numerator, degree of denominator
0.0     <------  Zero degree coefficient of numerator
2.0     <------  First degree coefficient of numerator
400.0   <------  Zero degree coefficient of denominator
2.0     <------  First degree coefficient of denominator
1.0     <------  Second degree coefficient of denominator
```

(*b*)

Figure 9.3 (*Continued*)

```
 NUMERATOR
         0  0.0000000000000E+000
         1  2.0000000000000E+000
DENOMINATOR
         0  4.0000000000000E+002
         1  2.0000000000000E+000
         2  1.0000000000000E+000
```

(*a*)

Figure 9.4 Output for Example 9.1

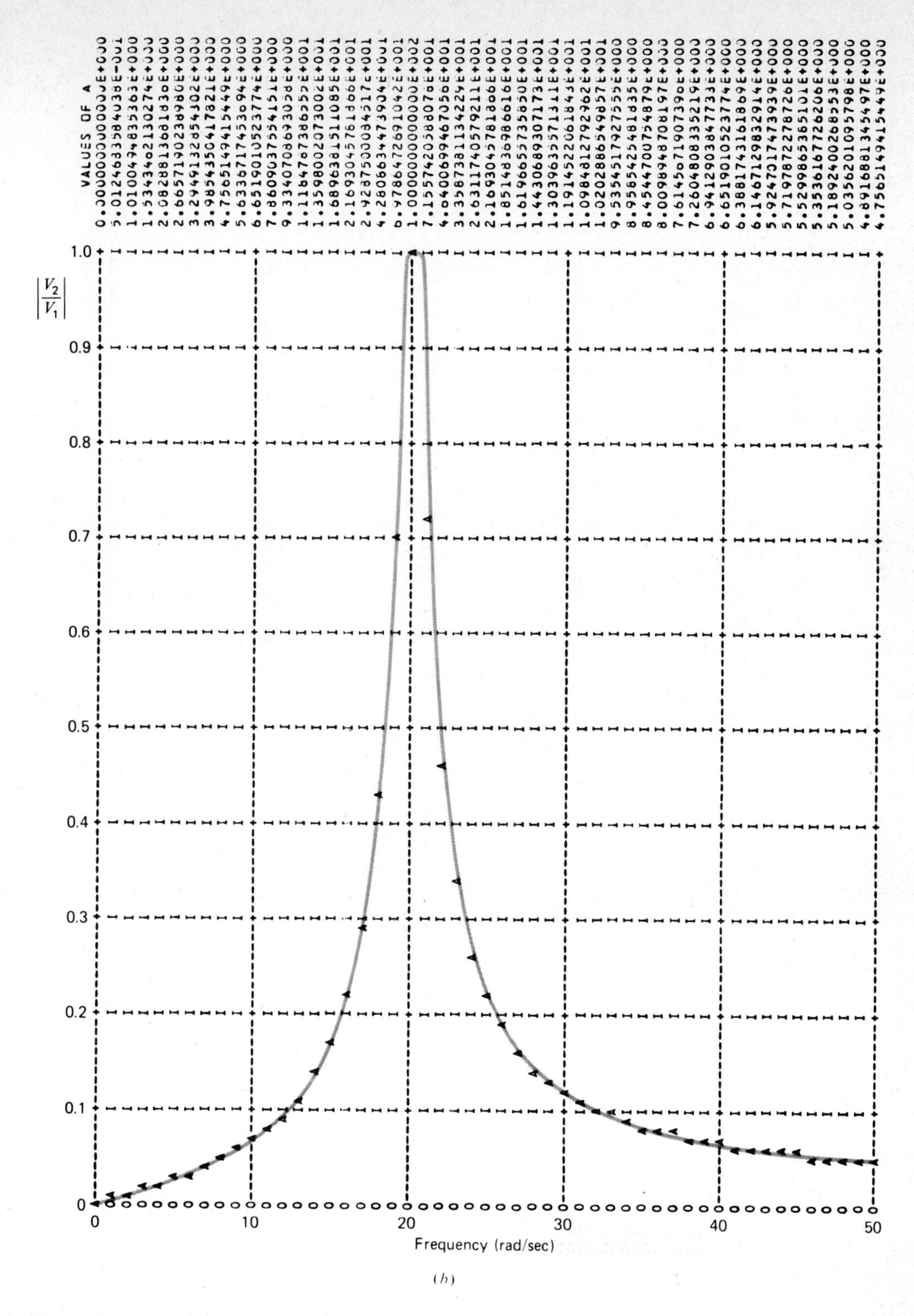

(b)

Figure 9.4 (*Continued*)

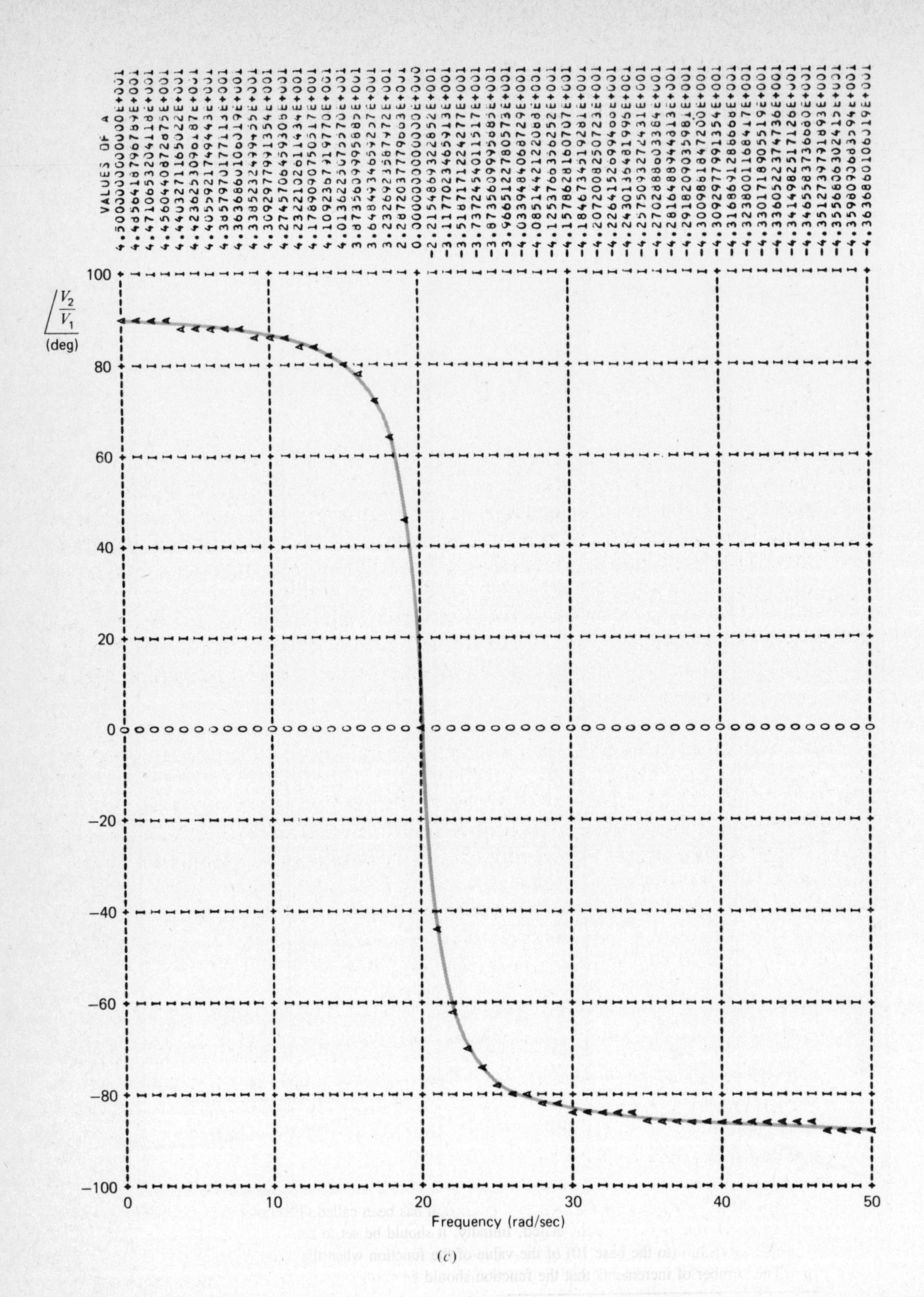

(*c*)

Figure 9.4 (*Continued*)

```
FUNCTION LogIncrem (VAR n : integer; lo, lp : integer) : real;

(* Function for generating a set of logarithmically spaced
   values of a variable
     n  - Index indicating which value of the variable is
          currently being generated
     lo - Logarithm (to the base 10) of the value of the
          variable which is generated when n = 0
     lp - Number of values per decade
   Note: The index n is increased by 1 at each call of
   the function LogIncrem                                    *)

   BEGIN
     LogIncrem := exp(ln(10.0) * (lo + n/lp));
     n := n + 1
   END;
```

Figure 9.5 Listing of function LogIncrem.

The definition of gain in decibels also applies to a current transfer function. It should be noted that if a scale of gain in decibels is provided for a plot, such a scale will be linear (in decibels) although it is, of course, logarithmic in the magnitude of the voltage or current transfer ratio. For immittance transfer functions, a logarithmic scale is still useful; however, it should be simply specified as the logarithm of the magnitude, since the unit of decibel is not applicable in such a case.

In order to make a Bode plot with a logarithmic scale of frequency, we need some method of generating frequency values that are spaced at logarithmic intervals. This may be accomplished by the use of a PASCAL function. Such a function must be supplied with the following information:

1. An index number n indicating that the function is now generating the nth logarithmic value that is required for a given application.
2. A number l_o giving the value of the logarithm (to the base 10) of the frequency at which it is desired to start the plot. This would correspond with a value of 0 for the index number n. For example, if we desired to start the set of values at 0.01 rad/sec, we should set l_o to -2.
3. A number l_p giving the number of logarithmic values that it is desired to generate for each decade of the output. For example, if five values were desired from 0.1 to 1 rad/sec, and five more values from 1 to 10 rad/sec, etc., then we should set l_p to 5.

TABLE 9.2 SUMMARY OF THE CHARACTERISTICS OF THE FUNCTION LogIncrem

Identifying Statement FUNCTION LogIncrem (VAR n : integer; lo, lp : integer) : real;

Purpose To generate a sequence of logarithmically spaced numbers for use in plots having a logarithmic scale for the independent variable.

Additional Subprograms Required None

Input Arguments

n	The index giving the number of times the function has been called. The index is automatically advanced by 1 each time the function is called. Initially, it should be set to zero.
lo	The logarithm (to the base 10) of the value of the function when the index $n = 0$.
lp	The number of increments that the function should generate for each decade of output values.

Note: For most plotting applications, the value of the index variable should be set to zero before the first call of the function in the main program.

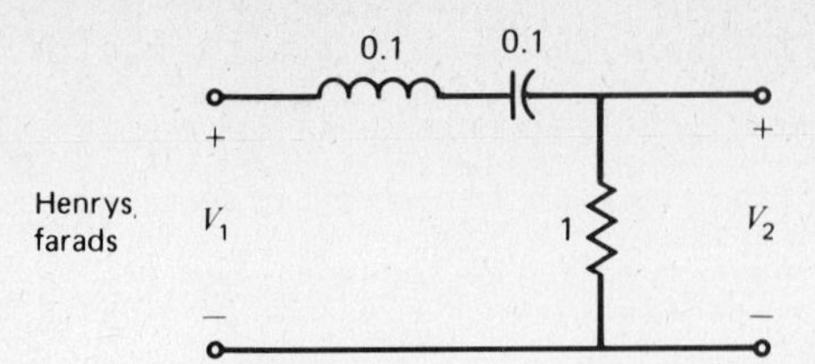

Figure 9.6 An *RLC* network.

Consideration of the above inputs indicates that a function LogIncrem for generating such a set of logarithmically spaced numbers may be identified by the following heading:

FUNCTION LogIncrem (VAR n : integer; lo, lp : integer) : real;

A listing of a function which will provide such a set of logarithmically spaced values is shown in Fig. 9.5. It should be noted that values for lo and lp can be entered directly into the argument listing when the function is called. The variable n, however, is recomputed by the function; therefore, it can only be a variable in the argument listing used in calling the function. For using such a function to generate a set of values to be used with the procedure Plot5 (in which the first set of values is presented at the value of 0 for the index of the independent variable), a value of 0 for n should be established in the program prior to the first time the function is called. As may be seen from the listing, the function will then increase the value of n automatically by the statement n := n + 1 each time the function is called. A summary of the characteristics of the function is given in Table 9.2.

As an illustration of the use of the function LogIncrem in constructing a Bode plot, consider Example 9.2.

EXAMPLE 9.2

A simple *RLC* network is shown in Fig. 9.6. Routine circuit analysis shows that the voltage transfer function for this network is a bandpass function of the form

$$\frac{V_2(s)}{V_1(s)} = \frac{s}{s^2 + s + 100}$$

We may use the procedure SinStdySt and the function LogIncrem together with the procedure Plot5 to generate a Bode plot of this function for a range of frequency from 0.1 to 10,000 rad/sec, using 10 intermediate values for each decade of frequency. A listing of a main program and input for providing such a plot is shown in Fig. 9.7. A plot of the output is given in Fig. 9.8.

```
PROGRAM Main (input, output);     (* Example 9.2 *)

(* Main program for computing and plotting a Bode plot
   for a network function
     m - Degree of numerator polynomial
     a - Array of numerator polynomial coefficients
     n - Degree of denominator polynomial
     b - Array of denominator polynomial coefficients
     rad - Frequency (rad/sec)
     pltMag - Plotting array for magnitude
     pltPhase - Plotting array for phase              *)
```

```
CONST
   pscale = 0.5;

TYPE
   complex = RECORD
             re, im : real
             END;
   array11 = ARRAY[0..10] OF real;
   plotArray = ARRAY[1..5,0..100] OF real;

VAR
   rad, vreal, vimag, vmag, vphase : real;
   pltMag, pltPhase                : plotArray;
   a, b                            : array11;
   m, n, i, indxLogIncrem          : integer;

BEGIN

   (* Initialize the variables and the plot arrays *)

   pltMag[1,0] := -60.0;
   pltPhase[1,0] := 90.0 * pscale;
   readln(m, n);
   writeln(' Numerator');
   FOR i := 0 TO m DO
      BEGIN
         readln(a[i]);
         writeln(i, a[i])
      END;
   writeln(' Denominator');
   FOR i := 0 TO n DO
      BEGIN
         readln(b[i]);
         writeln(i, b[i])
      END;

   (* Use a loop to compute the values of the magnitude and
      phase and store them for plotting                  *)

   indxLogIncrem := 1;
   FOR i := 1 TO 50 DO
      BEGIN
         rad := LogIncrem(indxLogIncrem, -1, 10);
         SinStdySt(m, a, n, b, rad, vreal, vimag, vmag, vphase);
         pltMag[1,i] := 20.0 * ln(vmag) / ln(10.0);
         pltPhase[1,i] := vphase * pscale * 57.2958
      END;

   (* Plot the magnitude and phase on separate plots *)

   Plot5(pltMag, 1, 50, 0);
   Plot5(pltPhase, 1, 50, 50)
END.
```

(*a*)

Input Data

1 2	←	Degree of numerator, degree of denominator
0.0	←	Zero degree coefficient of numerator
1.0	←	First degree coefficient of numerator
100.0	←	Zero degree coefficient of denominator
1.0	←	First degree coefficient of denominator
1.0	←	Second degree coefficient of denominator

(*b*)

Figure 9.7 Program and imput data for Example 9.2.

```
NUMERATOR
         0   0.0000000000000E+000
         1   1.0000000000000E+000
DENOMINATOR
         0   1.0000000000000E+002
         1   1.0000000000000E+000
         2   1.0000000000000E+000
```

(*a*)

Figure 9.8 Output for Example 9.2.

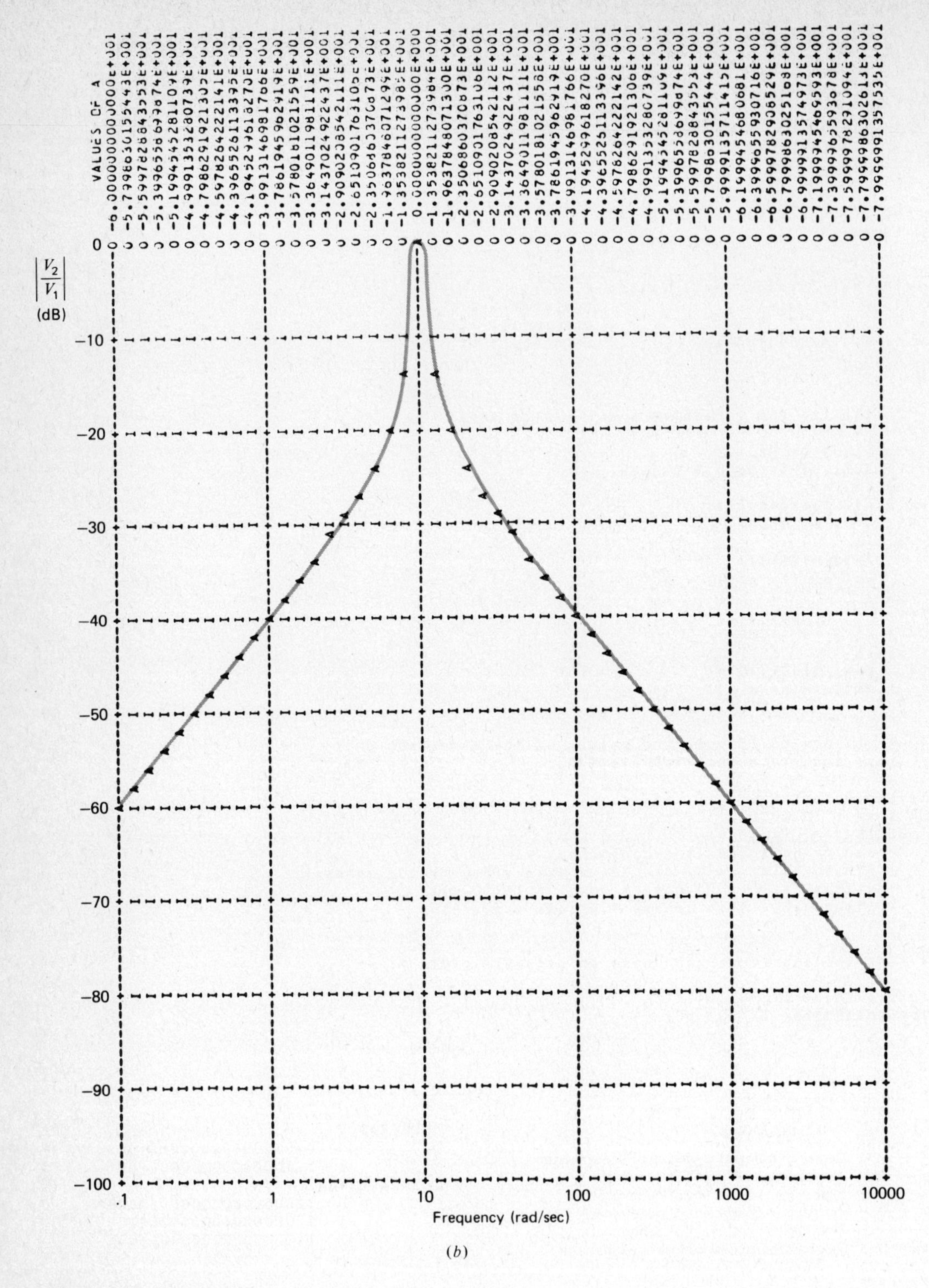

(*b*)

Figure 9.8 (*Continued*)

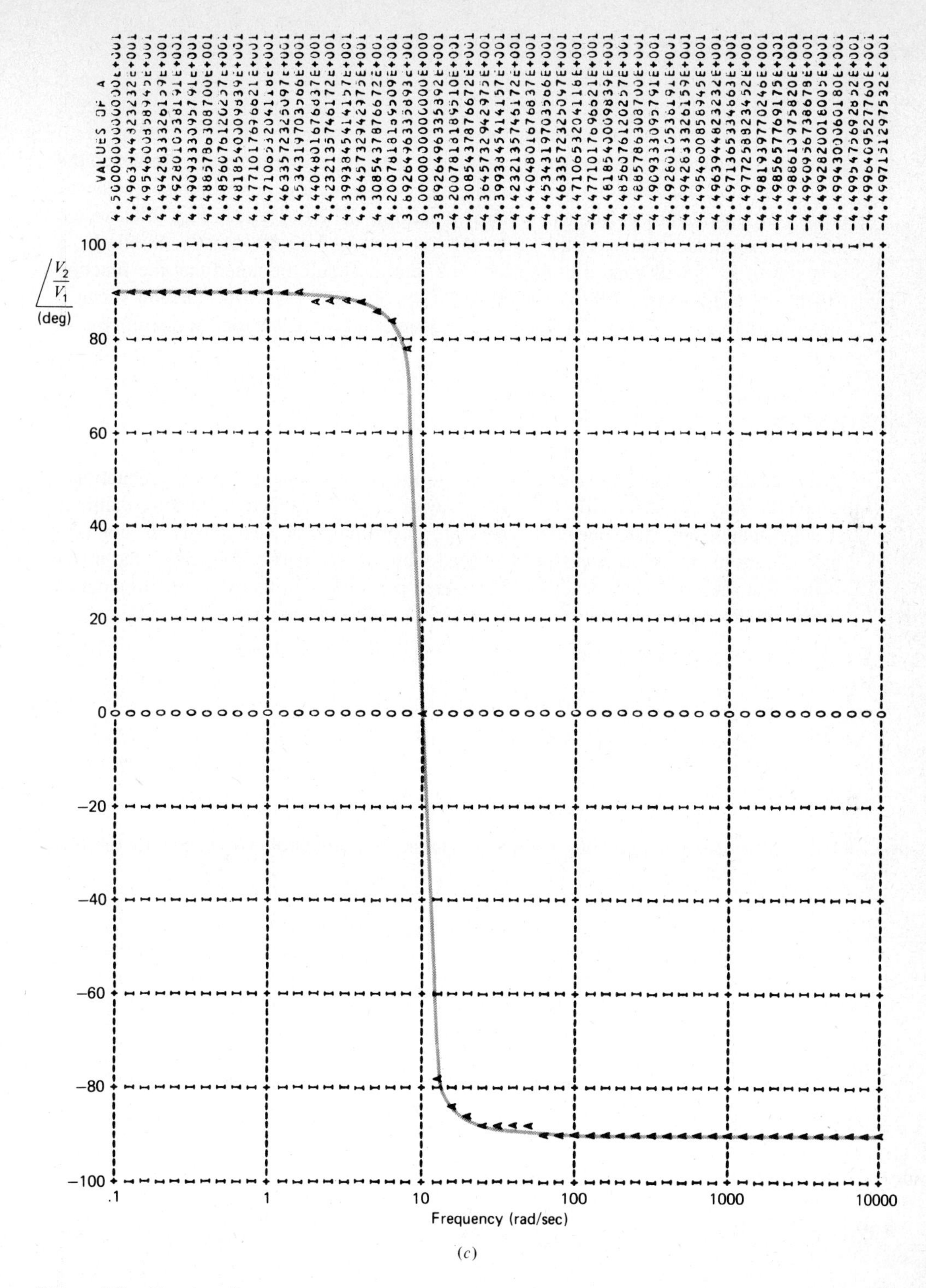

(*c*)

Figure 9.8 (*Continued*)

One of the characteristics of the Bode plot is that, for very large or very small values of frequency, the characteristic approaches a straight line. The slope of such a straight-line portion of the characteristic will always be some multiple (including zero) of 20 dB/decade (6 dB/octave). The multiple is determined by the manner in which the system function behaves at large (or small) values of frequency. For example, a behavior of ks^{-1} at large values of frequency means a multiple of -1. Thus, in Fig. 9.8*a* we see that the magnitude characteristic "falls off" at a rate of 20 dB/decade. Similarly, a behavior of ks^{-2} will yield a multiple of -2, etc. It should be noted that the function LogIncrem can also be used to make a plot of the phase of a network function versus a logarithmic frequency scale. Such a plot is frequently useful in stability studies of control systems.

9.4 CONCLUSION

In this chapter we have presented some basic methods in which digital-computational techniques may be used to illustrate the behavior of system functions under conditions of sinusoidal steady-state behavior. Although the methods of this chapter do not take into account the transient behavior associated with a given network function, the information that the sinusoidal steady-state approach provides is of considerable importance in determining the characteristics of a given network or system.

PROBLEMS

9.1. Plot the magnitude of the voltage transfer function for the network shown in Fig. 9.2 (under conditions of sinusoidal steady-state excitation) for a range of values of frequency from 0 to 50 rad/sec for the following values of the resistor *R*: 0.1, 0.2, 0.5, and 1 Ω. Use 51 values of frequency and put all four characteristics on the same plot.

9.2. For the data given in Prob. 9.1, make a plot of the phase of the voltage transfer function for the same four values of *R*.

9.3. A ladder network will produce zeros of transmission at the poles of series impedances and shunt admittances. Verify this fact by plotting the magnitude of the voltage transfer function (scaled by 80) for the network shown in Fig. P9.3, under conditions of sinusoidal steady-state excitation for a range of frequency from 0 to 2.5 rad/sec. Find the locations where the transmission zeros should appear, and verify from the plot that they are present. Use 51 values of frequency in the plot.

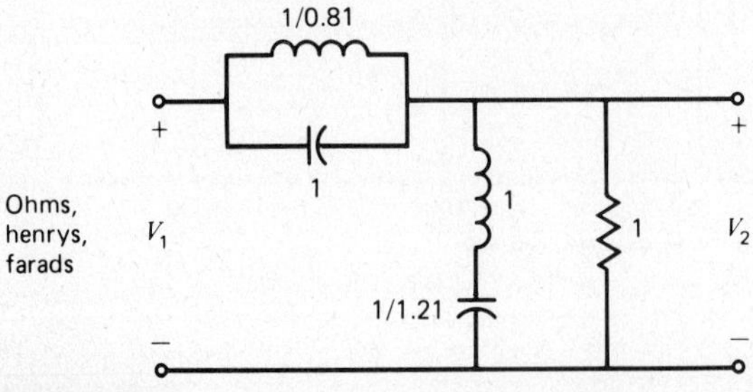

Figure P9.3

9.4. **(a)** Use the techniques given in this chapter to conduct a study of how the magnitude characteristics of a given network function vary as the pole positions are varied. Let the network function be

$$N(s) = \frac{s}{(s+1)(s^2+as+1)}$$

$$= \frac{s}{s^3+(1+a)s^2+(1+a)s+1}$$

On a single plot, show the magnitude of this network function for the following values of a: 0.001, 0.5, 1, and 1.5. Use a range of frequency from 0 to 2.5 rad/sec, and use 51 values of frequency in this range.

(b) Is it possible to make a plot for the case where $a = 0$? Why or why not?

9.5. For the data given in Prob. 9.4, determine the phase characteristics of the network function for the specified values of the constant a.

9.6. Plot the magnitude and phase characteristics for the second-, third-, and fourth-degree maximally flat magnitude low-pass network functions given below.

Second-degree function:

$$N(s) = \frac{100}{s^2+1.4142s+1}$$

Third-degree function:

$$N(s) = \frac{100}{s^3+2s^2+2s+1}$$

Fourth-degree function:

$$N(s) = \frac{100}{s^4+2.6131s^3+3.4142s^2+2.6131s+1}$$

Determine the values of magnitude and phase for 51 points over a range of frequency from 0 to 2.5 rad/sec. Plot all three magnitude characteristics on a single plot. Similarly, put all three phase characteristics on a single plot.

9.7. Repeat Prob. 9.6 for the low-pass maximally flat delay second-, third-, and fourth-degree network functions given below. Use 51 points in the plots, and use a range of frequency from 0 to 5 rad/sec.

Second-degree function:

$$N(s) = \frac{300}{s^2+3s+3}$$

Third-degree function:

$$N(s) = \frac{1500}{s^3+6s^2+15s+15}$$

Fourth-degree function:

$$N(s) = \frac{10{,}500}{s^4+10s^3+45s^2+105s+105}$$

9.8. Determine the magnitude and phase characteristics of the all-pass network function given below.

$$N(s) = \frac{(s-1)(s^2-s+1)}{(s+1)(s^2+s+1)} = \frac{s^3-2s^2+2s-1}{s^3+2s^2+2s+1}$$

Use a range of frequency from 0 to 2.5 rad/sec. Use 51 points in the plot.

9.9. For the three network functions given in Prob. 9.6, make a Bode plot of the magnitude of the function (in decibels) versus the frequency (on a logarithmic scale). Use 51 points for the characteristic of each function. Use 20 points per decade on the plot, starting at 0.1 rad/sec.

9.10. Construct Bode plots of the low-pass network function given below for the following values of the constant a: 0.5, 1.414 (the maximally flat magnitude case), and 2.5. Put all three characteristics on the same plot. Use 51 points on the plot to cover the range of frequency from 0.01 to 1000 rad/sec.

$$N(s) = \frac{1}{s^2+as+1}$$

Chapter 10

Fourier-Series Analysis

In Chap. 9 we introduced sinusoidal steady-state analysis techniques and showed how they could be used to determine the properties of network and system functions. In this chapter we illustrate a further application of such techniques—their use in Fourier-series analysis.

10.1 THE FOURIER SERIES

The techniques developed in Chap. 9 were shown to provide a method for finding the steady-state output response of a system defined by a system function under the condition that the input was a *sinusoidal* periodic function of time. These techniques may be extended to the case where the input is a *nonsinusoidal* periodic function. Such a function can be represented by a *Fourier series* having the form

$$f(t) = a_0 + \sum_{i=1}^{\infty} (a_i \cos i\omega_0 t + b_i \sin i\omega_0 t) \tag{10.1}$$

where ω_0 is the *fundamental frequency* (in radians per second) and each of the quantities $i\omega_0$ $(i \neq 1)$ is referred to as the *ith harmonic frequency*. Note that a periodic function is one satisfying the relation

$$f(t) = f(t + T) \tag{10.2}$$

where T is called the *period*. The fundamental frequency ω_0 is derived from the period by the relation

$$\omega_0 = \frac{2\pi}{T} \tag{10.3}$$

The coefficients a_i and b_i of the Fourier series of (10.1) are defined by the relations

$$a_0 = \frac{1}{T}\int_0^T f(t)\, dt$$
$$a_i = \frac{2}{T}\int_0^T f(t) \cos i\omega_0 t\, dt \qquad (i = 1, 2, 3, \ldots) \tag{10.4}$$
$$b_i = \frac{2}{T}\int_0^T f(t) \sin i\omega_0 t\, dt \qquad (i = 1, 2, 3, \ldots)$$

Note that the coefficient a_0 gives the average (or dc) value of $f(t)$.

The a_i and b_i coefficients, together with (10.1), define what is called the *rectangular* (or *quadrature* or *sine/cosine*) form of the Fourier series. Another form for the series is the *polar* (or *magnitude/phase*) form. It is

$$f(t) = c_0 + \sum_{i=1}^{\infty} c_i \cos (i\omega_0 t + \phi_i) \tag{10.5}$$

where

$$c_0 = a_0 \qquad c_i = \sqrt{a_i^2 + b_i^2} \qquad \phi_i = \tan^{-1} \frac{-b_i}{a_i} \tag{10.6}$$

Since the Fourier series of nonsinusoidal periodic function consists of a sum of sinusoidal terms, when such a function is applied as the input to a linear system (for which superposition applies), the resulting steady-state nonsinusoidal output can be found as the sum of the sinusoidal outputs which result from the individual sinusoidal terms of the input series. In practice, even though the series defined in (10.1) and (10.5) contain an infinite number of terms, the use of only a small number of these (usually less than 10) provides very good accuracy in most situations.

In terms of the above discussion, we may now formulate the following problem for solution in this chapter.

Problem. For the situation where a nonsinusoidal periodic input function $i(t)$ is applied to a network or system defined by the rational function $N(s) = O(s)/I(s)$, find the steady-state waveform of the output $o(t)$. $O(s)$ and $I(s)$ are the Laplace transforms of $o(t)$ and $i(t)$.

We shall see that the solution of this problem illustrates the application of many of the techniques that we have developed in earlier parts of this book.

10.2 FINDING THE COEFFICIENTS OF THE FOURIER SERIES: THE PROCEDURE FourSeries

The relations for the coefficients of the Fourier series given in (10.4) are readily implemented by a PASCAL procedure that calls the trapezoidal-integration procedure IntegTrpz introduced in Chap. 2. Such a procedure requires the following inputs:

1. The number of harmonic frequencies n_f for which the coefficients are to be computed.
2. The period T of the periodic function $f(t)$.
3. The number of iterations to be used by the trapezoidal-integration procedure.
4. The name for the PASCAL function used to define the periodic function $f(t)$.

As outputs the procedure must supply the values of the coefficients a_i, b_i, c_i, and ϕ_i as defined in (10.4) and (10.6). A procedure named FourSeries (for *Fourier series*) which will implement these operations may be identified by the following heading:

```
PROCEDURE FourSeries (nHarmon:integer; period:real; iter:integer;
                      VAR a0:real; VAR a, b, c, phi:array10;
                      FUNCTION F (t:real):real);
```

where array10 is a user-defined type ARRAY[1..10] of reals. A listing and a flowchart for such a procedure are given in Fig. 10.1. A summary of its characteristics is given in Table 10.1. Example 10.1 illustrates its use.

EXAMPLE 10.1

As an example of the use of the procedure FourSeries, we find the Fourier-series coefficients for the first five harmonic frequencies of the square wave shown in Fig. 10.2. The main program and the output are shown in Fig. 10.3. Direct mathematical analysis of the function using the relations of (10.4) and (10.6) shows that the actual series is

$$
\begin{aligned}
f(t) &= \frac{4}{\pi}\left(\sin \pi t + \frac{1}{3}\sin 3\pi t + \frac{1}{5}\sin 5\pi t + \cdots\right) \\
&= \frac{4}{\pi}\left[\cos\left(\pi t - \frac{\pi}{2}\right) + \frac{1}{3}\cos\left(3\pi t - \frac{\pi}{2}\right) + \frac{1}{5}\left(\cos\left(5\pi t - \frac{\pi}{2}\right) + \cdots\right]
\end{aligned}
$$

which agrees with the numerical values for the coefficients shown in the output from the program.

10.3 PLOTTING THE FOURIER SERIES: THE PROCEDURE FourPlot

In the last section we developed a technique for finding the values of the coefficients of the Fourier series of an arbitrary nonsinusoidal periodic function of time. In this section we treat the inverse problem, namely, given the values of the coefficients of a Fourier series, how we plot the waveform actually generated by the series. A PASCAL procedure that will perform such an operation must be supplied with the following inputs:

1. The number n_t of time steps to be used in plotting the series.
2. The period T of the function (as determined by its fundamental frequency).
3. The value of the coefficient a_0.
4. The number n_c of coefficients of the series which are to be used in constructing the plot.

```
PROCEDURE FourSeries (nHarmon : integer; period : real; iter : integer;
                VAR a0 : real; VAR a, b, c, phi : array10;
                FUNCTION F (t : real) : real);
(* Procedure for finding the coefficients of the Fourier
   series for a periodic function of time f(t)
     nHarmon - Number of harmonic frequencies for which the
          coefficients are to be found
     period - The period T of the periodic function f(t)
     iter - The number of iterations used in the trapezoidal
          integration procedure IntegTrpz
     a0 - The Fourier series coefficient a0
     a - Array of Fourier series cosine coefficients
     b - Array of Fourier series sine coefficients
     c - Array of Fourier series magnitude coefficients
     phi - Array of Fourier series phase coefficients
   Note: A PASCAL function must be provided to define the
   periodic function f(t)
   Note: This procedure calls the function Arg and the
   procedure IntegTrpz                                          *)

   VAR
      (* omega - The fundamental frequency
         h - Number of harmonic currently being evaluated
         k - Index used to change integrand from cosine form
             to sine form                                       *)
      omega, temp : real;
      i, h, k     : integer;
      z           : complex;

   FUNCTION Integrand (t : real) : real;
      BEGIN
         IF k = 0
            THEN Integrand := F(t) * cos(omega * h * t)
            ELSE Integrand := F(t) * sin(omega * h * t)
      END;

   BEGIN

      (* Integrate F(t) to find the coefficient a0 *)

      IntegTrpz(0.0, period, iter, temp, F);
      a0 := temp / period;

      (* Find nHarmon coefficients and store in the arrays
         a, b, c, and phi                                  *)

      omega := 6.2831853 / period;
      FOR i := 1 TO nHarmon DO
         BEGIN
            h := i;

            (* Set k = 0 so integrand contains the
               product of f(t) and the cosine term  *)

            k := 0;
            IntegTrpz(0.0, period, iter, temp, Integrand);
            a[i] := 2.0 * temp / period;

            (* Set k = 1 so integrand contains the
               product of f(t) and the sine term  *)

            k := 1;
            IntegTrpz(0.0, period, iter, temp, Integrand);
            b[i] := 2.0 * temp / period;
            c[i] := sqrt(a[i]*a[i] + b[i]*b[i]);
```

Figure 10.1 Listing and flowchart for procedure FourSeries.

```
            (* Define a[i] and b[i] as elements of a complex
               variable z so that the function Arg can be used
               to define phi[i] in all four quadrants        *)

            z.re := a[i];
            z.im := -b[i];
            phi[i] := Arg(z)
        END
END;
```

(*a*)

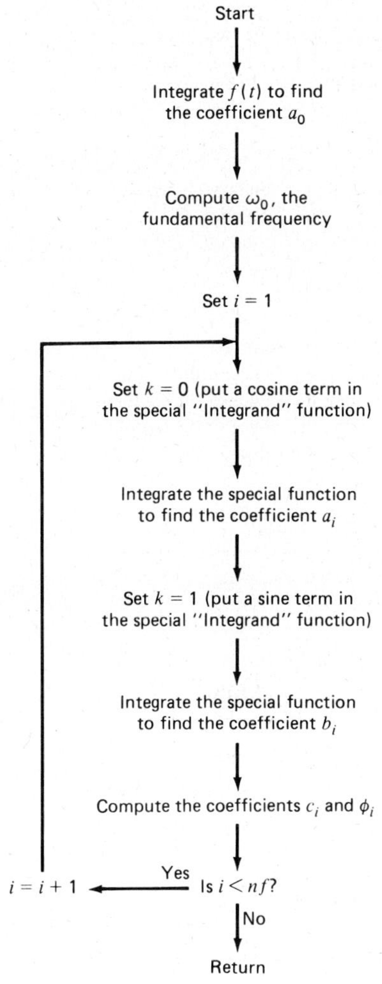

(*b*)

Figure 10.1 (*Continued*)

TABLE 10.1 SUMMARY OF THE CHARACTERISTICS OF THE PROCEDURE FourSeries

Identifying Statement PROCEDURE FourSeries (nHarmon : integer; period : real; iter : integer; VAR a0 : real; VAR a, b, c, phi : array10; FUNCTION F (t : real) : real);

Purpose To determine the coefficients of the Fourier series of a nonsinusoidal periodic function $f(t)$, that is, to find the coefficients a_i, b_i, c_i, and ϕ_i, where

$$F(t) = a_0 + \sum_i (a_i \cos i\omega_0 t + b_i \sin i\omega_0 t)$$
$$= a_0 + \sum_i c_i \cos(i\omega_0 t + \phi_i)$$

and where ω_0 is the fundamental harmonic frequency.

Additional Subprograms Required This procedure calls the procedure IntegTrpz, which in turn calls a function F (t : real) : real; defining the function $f(t)$. It also calls the procedure Arg.

Input Arguments

nHarmon	The number of harmonic components which are to be found
period	The period of the periodic function $f(t)$
iter	The number of iterations to be used by the trapezoidal-integration procedure in determining the coefficients

Output Arguments

a0	The value of the coefficient a_0
a	The array containing the values of the coefficients a_i
b	The array containing the values of the coefficients b_i
c	The array containing the values of the coefficients c_i
phi	The array containing the values of the coefficients ϕ_i

User-Defined Type

array10 = ARRAY[1..10] OF real;

5. The magnitude and phase coefficients of the series (these are easier to use than the sine/cosine ones).
6. The position in the plotting array where the data describing the function is to be stored.
7. Any scaling factor that it is desired to apply to the function prior to plotting.

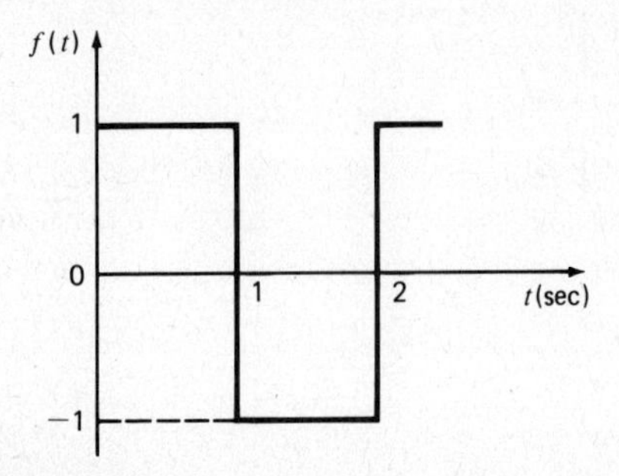

Figure 10.2 Square wave for Example 10.1

```
PROGRAM Main (output);    (* Example 10.1 *)

(* Main program for plotting the Fourier series
   of a square wave
     a0 - The Fourier series coefficient a0
     a - Array of Fourier series cosine coefficients
     b - Array of Fourier series sine coefficients
     c - Array of Fourier series magnitude coefficients
     phi - Array of Fourier series phase coefficients
     plt - Plotting array                                  *)

TYPE
   array10 = ARRAY[1..10] OF real;
   plotArray = ARRAY[1..5,0..100] OF real;
   complex = RECORD
               re, im : real
               END;

VAR
   a, b, c, phi  : array10;
   a0            : real;
   i             : integer;
   plt           : plotArray;

FUNCTION SqWv (t : real) : real;
   BEGIN
      IF t > 1.0
         THEN SqWv := -1.0
         ELSE SqWv := 1.0
   END;

BEGIN
   FourSeries(5, 2.0, 201, a0, a, b, c, phi, SqWv);
   writeln;
   writeln(' a0 = ', a0);
   writeln;
   writeln(' i     a[i]       b[i]       c[i]     ph[i]');
   FOR i := 1 TO 5 DO
      writeln(i:3, a[i]:9:5, b[i]:9:5, c[i]:9:5, phi[i]:9:5)
END.
```

(*a*)

```
A0 =    0.0000000000000E+000

I   A[I]       B[I]       C[I]      PH[I]
1  0.00000   1.27321   1.27321  -1.57080
2  0.00000  -0.00015   0.00016   1.57080
3  0.00000   0.42434   0.42434  -1.57080
4  0.00000  -0.00031   0.00031   1.57080
5  0.00000   0.25452   0.25452  -1.57080
```

(*b*)

Figure 10.3 Program and output for Example 10.1.

The output from the procedure will be stored in a standard plotting array and plotted by the procedure Plot5. From the above discussion, if we call the procedure FourPlot (for *Four*ier series *plot*ting procedure), it can be identified by the following heading:

PROCEDURE FourPlot (nTimes : integer; period, a0 : real; nc : integer;
VAR c, phi : array10; na : integer;
VAR p : plotArray; scale : real);

A listing and a flowchart of a procedure for implementing these operations are shown in Fig. 10.4. A summary of its characteristics is given in Table 10.2. Example 10.2 illustrates the use of this procedure.

EXAMPLE 10.2

As an example of the use of the procedure FourPlot, let us extend Example 10.1 by having the procedure reconstruct the square wave shown in Fig. 10.2 using the coefficients found in that example. A program for doing this through the fifth harmonic is

```
PROCEDURE FourPlot (nTimes : integer; period, a0 : real; nc : integer;

                VAR c, phi : array10; na : integer; VAR p :

                plotArray; scale : real);

(* Procedure for plotting one period of a non-sinusoidal
   periodic function defined by its Fourier series coefficients
     nTimes - Number of time increments to be used in plotting
     period - The period of the non-sinusoidal period function
     a0 - The Fourier series coefficient a0
     c - Array of Fourier series magnitude coefficients
     phi - Array of Fourier series phase coefficient
     na - Index for the plotting array
     p - Plotting array
     scale - Factor by which data is tobe scaled before
          being stored in the plotting array           *)

   VAR
      (* t - Value of time at which computations are made
         dt - Time increment between adjacent values
         wrad - Fundamental frequency (rad/sec)
         f - Value of f(t)                                 *)

      t, dt, wrad, f : real;
      i, j           : integer;

   BEGIN

      (* Initialize and calculate local variables *)

      t := 0.0;
      dt := period / nTimes;
      wrad := 6.2831853 / period;

      (* Make calculations at nTimes+1 different values of time *)

      FOR i := 0 TO nTimes DO
         BEGIN

            (* Determine the effect of the harmonic terms *)

            f := a0;
            FOR j := 1 TO nc DO
               f := f + c[j] * cos(wrad*j*t + phi[j]);
            p[na,i] := f * scale;
            t := t + dt
         END
   END;
```

(a)

Figure 10.4 Listing and flowchart for procedure FourPlot.

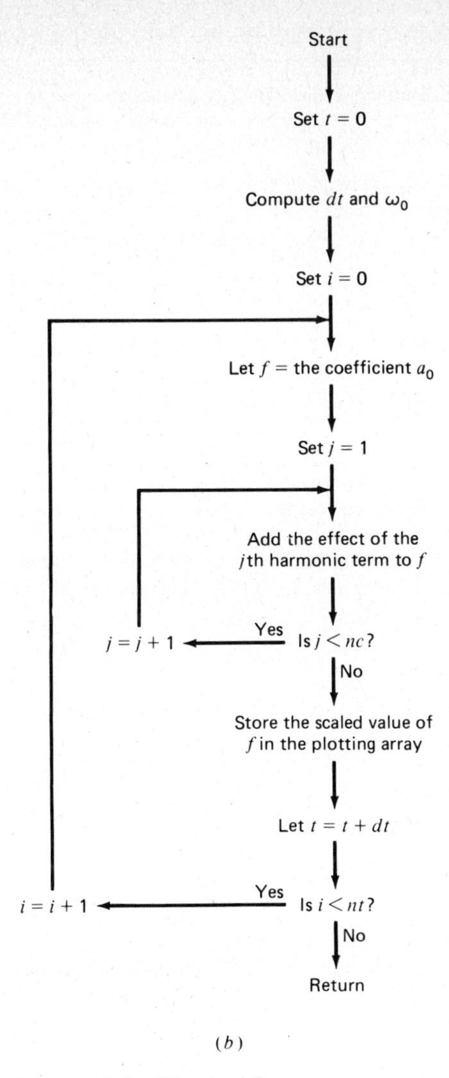

Figure 10.4 *(Continued)*

shown in Fig. 10.5. The output, together with a plot of the square wave, is shown in Fig. 10.6.

10.4 NETWORK AND SYSTEM ANALYSIS USING THE FOURIER SERIES

The Fourier-series techniques introduced in the preceding sections of this chapter are readily applied to the problem of determining the steady-state response to a nonsinusoidal

TABLE 10.2 SUMMARY OF THE CHARACTERISTICS OF THE PROCEDURE FourPlot

Identifying Statement PROCEDURE FourPlot (nTimes : integer; period, a0 : real; nc : integer; VAR c, phi : array10; na : integer; VAR p : plotArray; scale : real);

Purpose To compute and store the data required for plotting the waveform of one period of a nonsinusoidal function $f(t)$ which is specified in the form

$$f(t) = a_0 + \sum_i c_i \cos (i\omega_0 t + \phi_i)$$

where ω_0 is the fundamental frequency.

Additional Subprograms Required None

Input Arguments

nTimes	The number of time steps to be used in plotting the function
period	The period of the periodic function $f(t)$
a0	The value of the coefficient a_0
nc	The number of coefficients, that is, the number of harmonic terms, used for plotting $f(t)$
c	The array containing the values of the coefficients c_i
phi	The array containing the values of the coefficients ϕ_i
na	The position in the plotting array where the data describing the function is to be stored
scale	The scaling factor which is applied to the data stored in the plotting array. If no scaling is desired, this argument must be set to 1.0

Output Argument

p	The plotting array

User-Defined Type

plotArray = ARRAY[1..5,0..100] OF real;

periodic input of a network (or system) which is described by a network function $N(s)$. The procedure is as follows:

1. Use the procedure FourSeries to find the coefficient $a_0^{(\text{input})}$ and the coefficients $c_i^{(\text{input})}$ and $\phi_i^{(\text{input})}$ ($i = 1, 2, 3, \ldots$) for the nonsinusoidal periodic input function.
2. Evaluate the network function $N(s)$ for the dc ($\omega = 0$) case to obtain $N(j0)$ using the procedure SinStdySt of Chap. 9.
3. Modify the coefficient $a_0^{(\text{input})}$ of the input waveform to obtain the corresponding coefficient $a_0^{(\text{output})}$ of the output waveform using the relation

$$a_0^{(\text{output})} = N(j0)\, a_0^{(\text{input})}$$

4. Evaluate $N(s)$ for the ith harmonic frequency (starting at $i = 1$) to obtain $N(ji\omega_0)$ using the procedure SinStdySt.
5. Modify the ith coefficients (starting with $i = 1$) $c_i^{(\text{input})}$ and $\phi_i^{(\text{input})}$ of the input waveform to obtain the coefficients $c_i^{(\text{output})}$ and $\phi_i^{(\text{output})}$ of the output waveform using the relations

$$c_i^{(\text{output})} = |N(ji\omega_0)|\, c_i^{(\text{input})}$$
$$\phi_i^{(\text{output})} = \text{Arg } N(ji\omega_0) + \phi_i^{(\text{input})}$$

```
PROGRAM Main (output);     (* Example 10.2 *)

(* Main program for plotting the Fourier series
   of a square wave
     a0 - The Fourier series coefficient a0
     a - Array of Fourier series cosine coefficients
     b - Array of Fourier series sine coefficients
     c - Array of Fourier series magnitude coefficients
     phi - Array of Fourier series phase coefficients
     plt - Plotting array                                  *)

TYPE
   array10 = ARRAY[1..10] OF real;
   plotArray = ARRAY[1..5,0..100] OF real;
   complex = RECORD
               re, im : real
               END;

VAR
   a, b, c, phi  : array10;
   a0            : real;
   i             : integer;
   plt           : plotArray;

FUNCTION SqWv (t : real) : real;
   BEGIN
      IF t > 1.0
         THEN SqWv := -1.0
         ELSE SqWv := 1.0
   END;

BEGIN
   FourSeries(5, 2.0, 201, a0, a, b, c, phi, SqWv);
   FourPlot(50, 2.0, a0, 5, c, phi, 1, plt, 40.0);
   FOR i := 0 TO 50 DO
      IF i > 25
         THEN plt[2,i] := -40.0
         ELSE plt[2,i] := 40.0;
   Plot5(plt, 2, 50, 50);
END.
```

Figure 10.5 Program for Example 10.2.

6. Repeat steps 4 and 5 above for the other Fourier-series harmonic components ($i = 2, 3, \ldots$).
7. Use the procedure FourPlot to compute and store the values of the output periodic waveform, and the procedure Plot5 to plot it.

A flowchart outlining the method is shown in Fig. 10.7. An example follows.

EXAMPLE 10.3

As an example of the use of the method for analyzing a network presented in this section, let us consider the effect of using the square-wave input waveform shown in Fig. 10.2

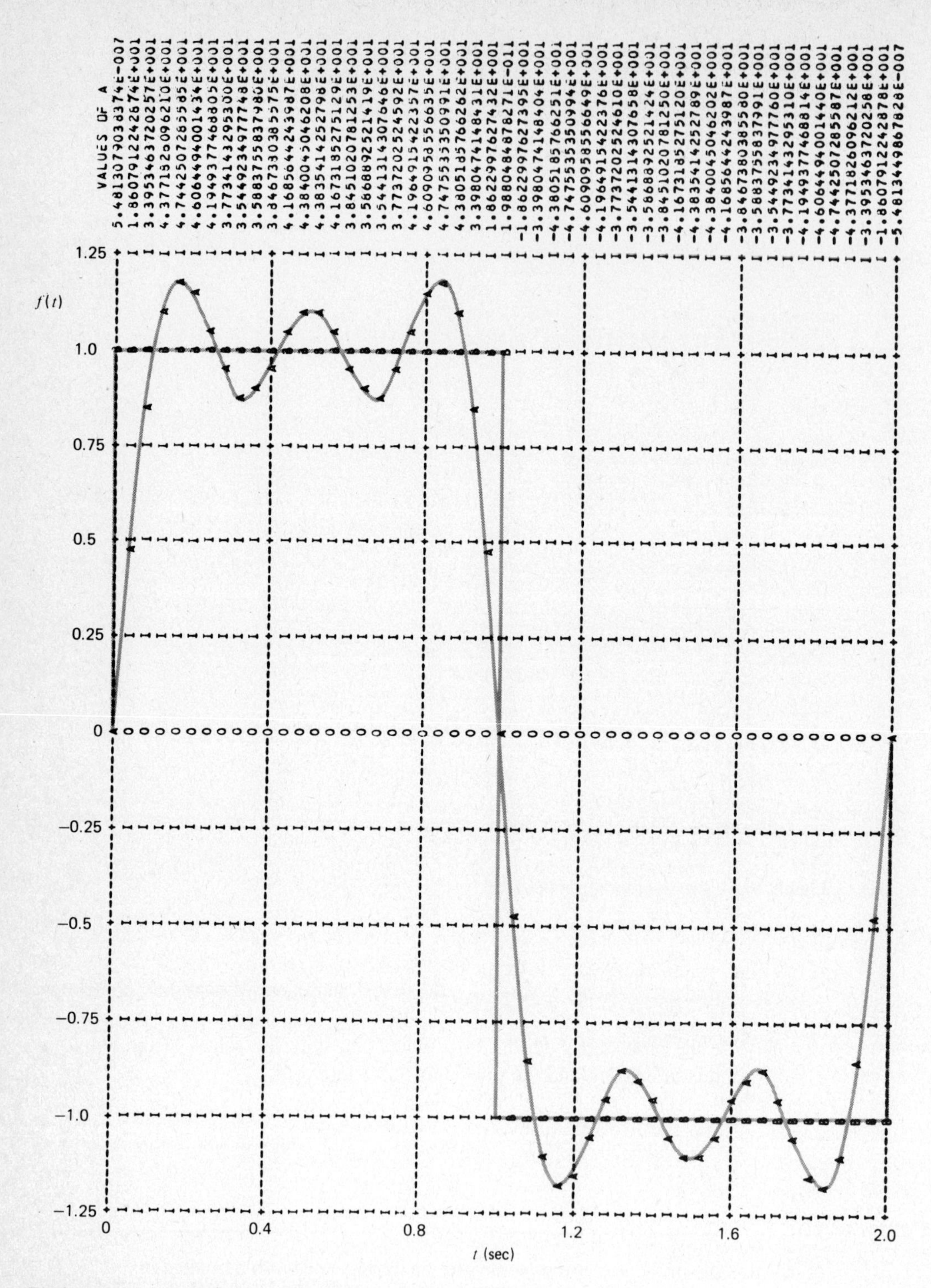

Figure 10.6 Output for Example 10.2

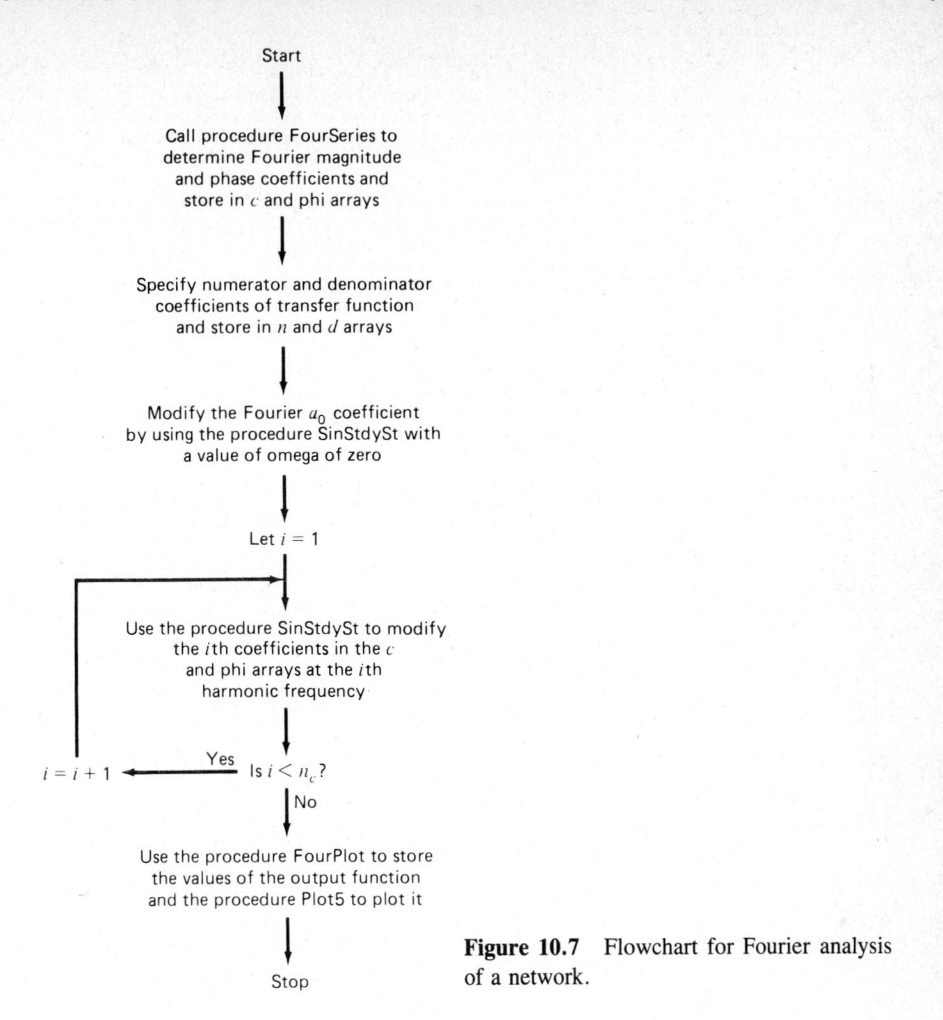

Figure 10.7 Flowchart for Fourier analysis of a network.

as the waveform of the input voltage $v_1(t)$ in the network shown in Fig. 10.8. For this network the voltage transfer function is readily shown to be

$$N(s) = \frac{V_2(s)}{V_1(s)} = \frac{4}{s+4}$$

A program for finding the steady-state output voltage $v_2(t)$ using seven harmonic terms is given in Fig. 10.9. A plot of the output waveform is shown in Fig. 10.10.

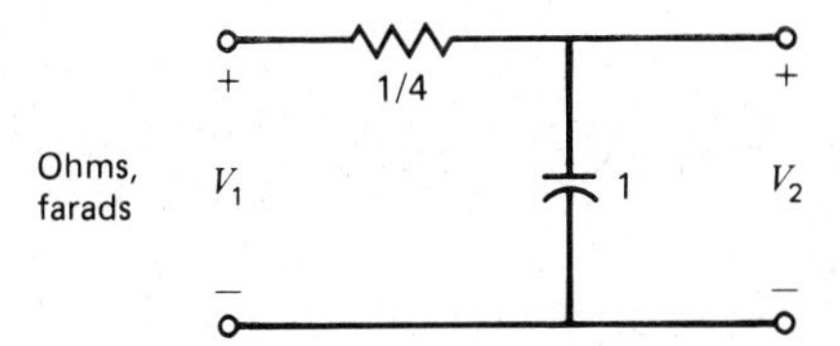

Figure 10.8 Network for Example 10.3.

```
PROGRAM Main (output);    (* Example 10.3 *)

(* Main program for finding the output waveform
   for an RC network with a square wave input
     a0 - The Fourier series coefficient a0
     a - Array of Fourier series cosine coefficients
     b - Array of Fourier series sine coefficients
     c - Array of Fourier series magnitude coefficients
     phi - Array of Fourier series phase coefficients
     plot - Plotting array
     num - Array of numerator coefficients of network function
     den - Array of denominator coefficients of network function
     vmag - Magnitude of network function
     vphase - Phase of network function
     omega - Frequency in rad/sec                                *)

TYPE
   array10 = ARRAY[1..10] OF real;
   array11 = ARRAY[0..10] OF real;
   plotArray = ARRAY[1..5,0..100] OF real;
   complex = RECORD
             re, im : real
             END;

VAR
   a0, omega, vreal, vimag, vmag, vphase : real;
   a, b, c, phi                          : array10;
   num, den                              : array11;
   i                                     : integer;
   plot                                  : plotArray;

FUNCTION SqWv (t : real) : real;
   BEGIN
      IF t > 1.0
         THEN SqWv := -1.0
         ELSE SqWv := 1.0
   END;

BEGIN

   (* Find the Fourier coefficients for the input waveform *)

   FourSeries(7, 2.0, 201, a0, a, b, c, phi, SqWv);

   (* Modify the dc coefficient by evaluating the network
      function at omega = 0                                  *)

   num[0] := 4.0;
   den[0] := 4.0;
   den[1] := 1.0;
   SinStdySt(0, num, 1, den, 0.0, vreal, vimag, vmag, vphase);
   a0 := a0 * vreal;

   (* Modify each of the harmonic coefficients to take
      account of the effect of the network function  *)

   FOR i := 1 TO 7 DO
      BEGIN
         omega := 3.1415927 * i;
         SinStdySt(0, num, 1, den, omega, vreal, vimag, vmag, vphase);
         c[i] := c[i] * vmag;
         phi[i] := phi[i] + vphase
      END;

   (* Compute and plot the output waveform *)

   FourPlot(50, 2.0, a0, 7, c, phi, 1, plot, 40.0);
   Plot5(plot, 1, 50, 50)
END.
```

Figure 10.9 Program for Example 10.3

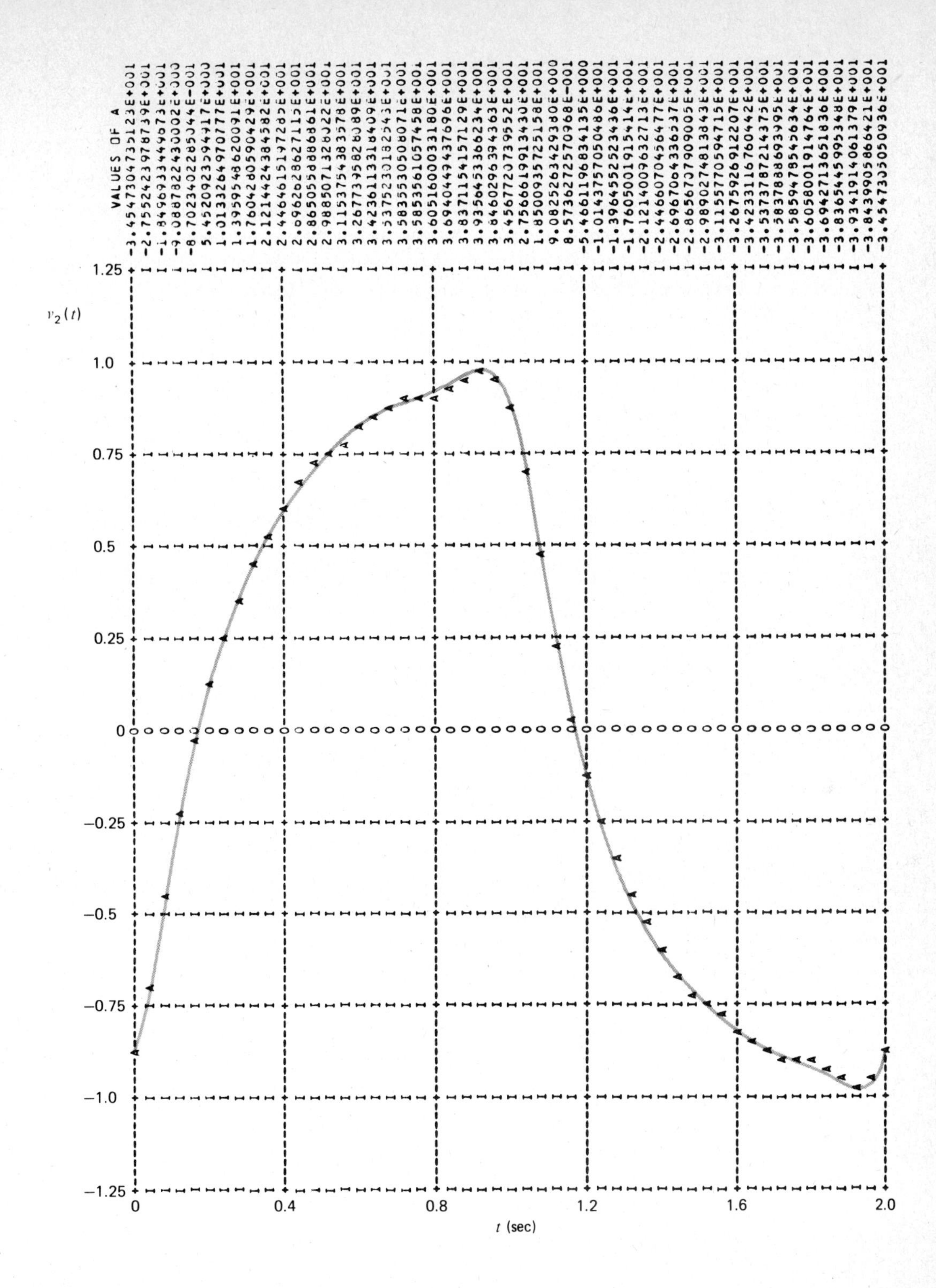

Figure 10.10 Output for Example 10.3

10.5 CONCLUSION

In this chapter we have introduced a powerful and useful technique—the Fourier series. It has been implemented in the PASCAL procedures FourSeries and FourPlot. Using these procedures together with the procedure SinStdySt from Chap. 9, it has been shown that any network or system described by a system function in the complex frequency variable s is readily analyzed to find the steady-state time-domain output waveform under the condition that the input is a nonsinusoidal but periodic function of time.

PROBLEMS

10.1. If a nonsinusoidal periodic waveform has a discontinuity, such as the square wave shown in Fig. 10.2, the use of an odd number of iterations will frequently produce more accurate values for the coefficients than will the use of an even number. To see this, repeat Example 10.1 using 200 iterations (rather than the 201 used in the example).

10.2. One method of avoiding the problem caused when integrating functions with discontinuities (see Problem 10.1) is the use of continuous functions that approximate the discontinuous ones. In Fig. P10.2, an approximation is shown for the square wave of Fig. 10.2. Find the Fourier-series coefficients for this waveform using 200 iterations.

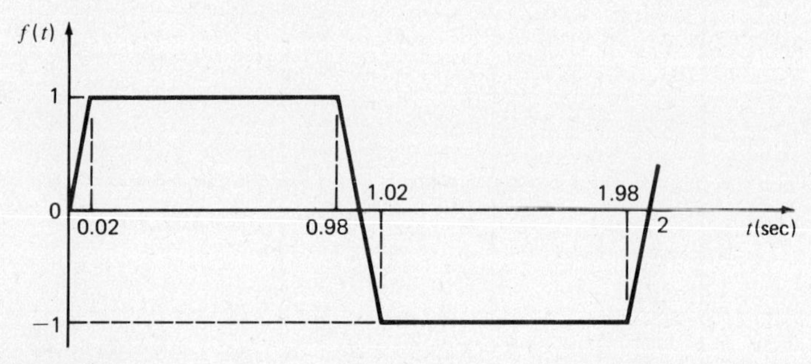

Figure P10.2

10.3. Investigate the effect of using different numbers of iterations by finding the Fourier-series coefficients for the waveform shown in Fig. P10.2 using 50, 100, 200, and 400 iterations.

10.4. Repeat Example 10.2 using seven harmonic terms (rather than the five used in the example). Use 100 increments of time in the plot (rather than the 50 used in the example).

10.5. Find the first seven Fourier-series coefficients for the triangular waveform shown in Fig. P10.5. Plot the original function and the one reconstructed from the series on the same plot.

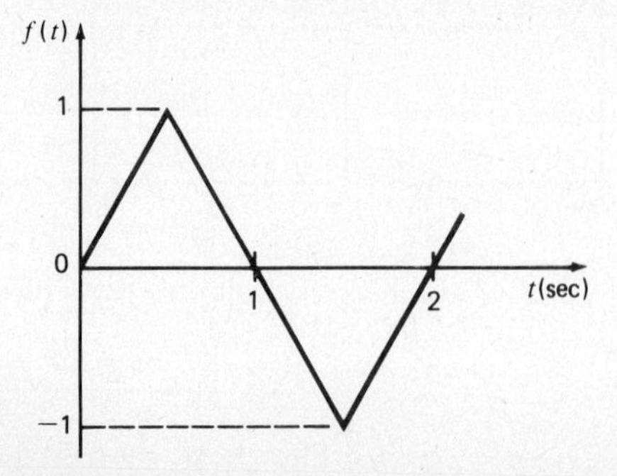

Figure P10.5

10.6. Show how the resonant circuit of Fig. P10.6*a* may be used to reduce the magnitude of the higher-order harmonics of the square-wave input waveform $v_1(t)$ shown in Fig. P10.6*b* by finding the Fourier series of the output waveform $v_2(t)$. Show the plots of the input and output waveforms on the same plot. Use five harmonic terms.

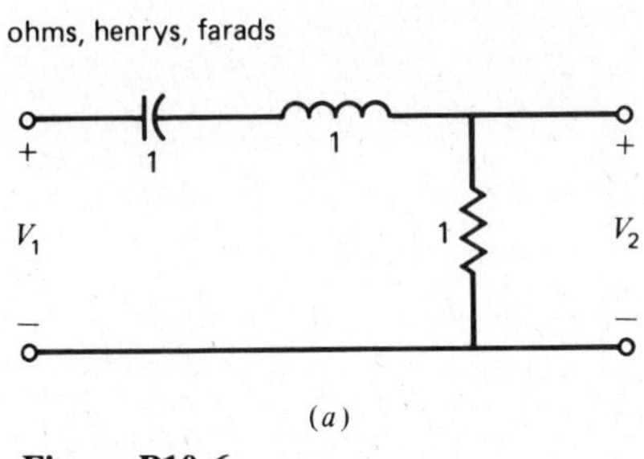

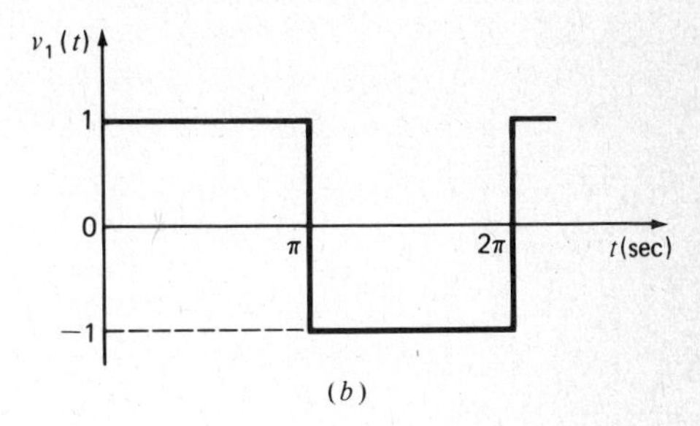

Figure P10.6

10.7. Change the values of the inductor (*L*) and the capacitor (*C*) in the network used in Prob. 10.6 so that the third harmonic component of the input waveform is passed without attenuation. Use the relation $\omega_{\text{3rd harmonic}} = 1/\sqrt{LC}$. Show plots of the input and output waveforms.

10.8. Use the triangular waveform shown in Fig. P10.5 as an input to the network shown in Fig. P10.6*a*, and plot the resulting output waveform.

Chapter 11

Sinusoidal Steady-State Analysis Using the Impedance Matrix

In Chap. 9 we presented methods for sinusoidal steady-state analysis based on the use of the network or system function. In this chapter we introduce a different method for sinusoidal steady-state analysis using the concept of impedance. A comparison of the advantages and disadvantages of the two methods is also presented.

11.1 IMPEDANCE AND THE IMPEDANCE MATRIX

In Chap. 8 we introduced the network function which defines the ratio between some transformed output or response variable and some transformed input or excitation variable. Thus, we may write

$$N(s) = \frac{O(s)}{I(s)} \tag{11.1}$$

where $N(s)$ is a network function and $O(s)$ and $I(s)$ are the Laplace transforms of some output and input variables $o(t)$ and $i(t)$, respectively. Some especially simple network functions result when the single network elements, namely, the resistor, inductor, and capacitor, are excited by sources. This is shown for the use of current-source excitations in Fig. 11.1. In this case the voltage across these elements is the response variable. If we use $Z_i(s)$ (where $i = R, L,$ or C) to designate the various network functions, we may write

$$Z_R(s) = \frac{V_R(s)}{I_R(s)} \qquad Z_L(s) = \frac{V_L(s)}{I_L(s)} \qquad Z_C(s) = \frac{V_C(s)}{I_C(s)} \tag{11.2}$$

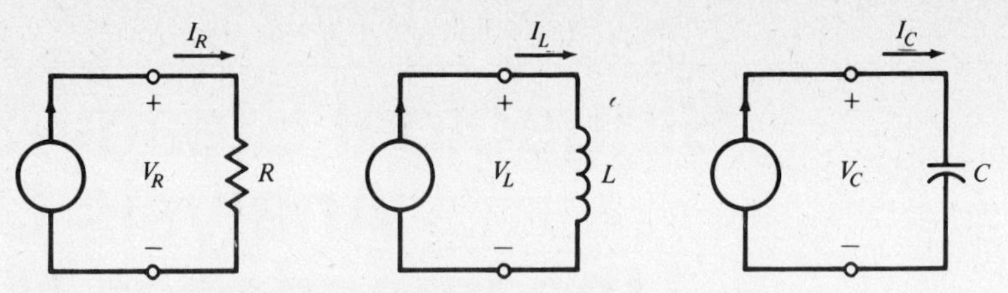

Figure 11.1 *RLC* elements and their variables.

For these network elements, the time-domain equations relating the terminal variables are

$$v_R(t) = Ri_R(t) \qquad v_L(t) = L\frac{di_L}{dt} \qquad v_C(t) = \frac{1}{C}\int i_C(t)\,dt \tag{11.3}$$

Using the methods of Chap. 8 to transform these equations, and comparing the results with the relations of (11.2), we obtain

$$Z_R(s) = R \qquad Z_L(s) = sL \qquad Z_C(s) = \frac{1}{sC} \tag{11.4}$$

Under sinusoidal steady-state conditions these expressions may be written

$$Z_R(j\omega) = R \qquad Z_L(j\omega) = j\omega L \qquad Z_C(j\omega) = \frac{1}{j\omega C} \tag{11.5}$$

The network functions $Z_R(j\omega)$, $Z_L(j\omega)$, and $Z_C(j\omega)$ defined in (11.5) are called the *impedances,* respectively, of a resistor, an inductor, and a capacitor. The unit used for these quantities is the ohm. These impedances have exactly the same role in the analysis of a network under conditions of sinusoidal steady-state excitation that resistors do when a network is analyzed for the dc case. Thus, in performing a mesh analysis of an *RLC* network under sinusoidal steady-state conditions, techniques almost identical with those developed in Chap. 7 may be used. The major difference is that in Chap. 7, the elements of all the matrices were real, while in the sinusoidal steady-state case they will be complex.

From the above considerations we see that applying KVL to a multimesh network consisting of resistors, inductors, and capacitors (this is called an *RLC* network) under sinusoidal steady-state conditions allows us to write a set of simultaneous equations having the form

$$\mathcal{V} = \mathbf{Z}(j\omega)\,\mathcal{I} \tag{11.6}$$

where $\mathcal{V}$ is an $n \times 1$ column matrix of phasor excitation voltages $\mathcal{V}_i$, $\mathbf{Z}(j\omega)$ is an $n \times n$ square matrix called an *impedance matrix* with elements $z_{ij}(j\omega)$, and $\mathcal{I}$ is an $n \times 1$ column matrix of phasor mesh currents $\mathcal{I}_i$. In a manner closely paralleling that done for the resistance network in Chap. 7, the elements of the matrix equation of (11.6) are found to have the following significance:

$\mathcal{V}_i$ The summation of the complex phasor representations of the voltage sources encountered in traversing the ith mesh. The sources are treated as positive if they represent a voltage rise while traversing the mesh in the direction of the mesh current. Otherwise they are treated as negative.

$z_{ij}(j\omega)$ If $i = j$, this element represents the total impedance encountered in traversing the ith mesh. If $i \neq j$, this element is the negative of the impedance that is mutual to the ith and jth meshes (assuming that the reference currents of the ith and jth mesh traverse the mutual impedance in opposite directions).

$\mathscr{I}_i$ This element represents the phasor current in the ith mesh. As such it is an unknown whose value is to be determined.

The first problem that we shall treat in this chapter is that of formulating a set of equations of the form given in (11.6) from a specified *RLC* circuit. Thus, we may define the following problem:

Problem. Given a circuit diagram for an *RLC* network, find the matrix $\mathbf{Z}(j\omega)$ of (11.6) relating the phasor mesh voltages and the phasor mesh currents of the network.

In the following section we discuss a solution to this problem that is easily implemented for the digital computer.

11.2 FORMULATING THE SINUSOIDAL STEADY-STATE EQUATIONS FOR AN *RLC* NETWORK: THE PROCEDURE ImpedMesh

In this section we describe a procedure that performs the function of determining the $\mathbf{Z}(j\omega)$ matrix of (11.6) from input information giving the complex values of the impedances and the identifying numbers of the meshes in which they are located. We make the following assumptions with respect to the topological structure of the network:

1. All the meshes are chosen as shown in Fig. 7.1; that is, the paths followed by the meshes do not enclose any branches not included in the mesh.
2. All mesh currents are given the same reference direction; that is, currents flowing in a clockwise direction around the meshes are considered as positive currents.
3. All *RLC* elements are traversed by not more than two mesh currents.

For the class of networks defined above, each impedance will affect the elements of the matrix $\mathbf{Z}(j\omega)$ of (11.6) in one of two ways:

1. If an impedance of value Z_i Ω occurs only in the jth mesh, then the quantity Z_i must be added to the element $z_{jj}(j\omega)$ of the matrix $\mathbf{Z}(j\omega)$ to take account of the effect of the element.
2. If an impedance of value Z_i Ω occurs in the jth and the kth meshes, then the quantity Z_i must be added to the elements $z_{jj}(j\omega)$ and $z_{kk}(j\omega)$ of the matrix $\mathbf{Z}(j\omega)$ and subtracted from the elements $z_{kj}(j\omega)$ and $z_{jk}(j\omega)$ of that matrix.

The above two possibilities are easily programmed into a procedure. As an input such a procedure requires the following information:

1. The number of impedances n_z that are to be found in the *RLC* network that is being analyzed.
2. The values Z_i of the impedances of the network.
3. The values of the first mesh $m_i^{(1)}$ in which the corresponding impedances Z_i are located.
4. The number of the second mesh (if any) $m_i^{(2)}$ in which the corresponding impedances Z_i are located.

As outputs, the procedure must provide the values of the elements $z_{ij}(j\omega)$ of the matrix $\mathbf{Z}(j\omega)$ for the given network. Consideration of the above inputs and outputs for the procedure leads us to define a parameter list for it that has the form

```
(nMesh, nz: integer; VAR imped: carray20;
 VAR mesh1, mesh2: iarray20; VAR z: cmtrx10)
```

where nMesh is the number of meshes, nz is the variable name for n_z, the number of impedances, imped is a one-dimensional array [1..20] of complex numbers in which are stored the values of the impedances Z_i, mesh1 and mesh2 are one-dimensional arrays [1..20] of integers in which are stored the values $m_i^{(1)}$ and $m_i^{(2)}$ for the impedances Z_i, and z is a two-dimensional [1..10,1..10] array of complex numbers in which are stored the elements of the matrix $\mathbf{Z}(j\omega)$. The various arrays are declared as global user-defined types by the following main program statements:

```
TYPE complex = RECORD re, im : real END;
     carray20 = ARRAY[1..20] OF complex;
     iarray20 = ARRAY[1..20] OF integer;
     cmtrx10 = ARRAY[1..10,1..10] OF complex;
```

Let us use the name ImpedMesh for the procedure. Its heading will be

```
PROCEDURE ImpedMesh (nMesh, nz : integer; VAR imped : carray20;
                     VAR mesh1, mesh2 : iarray20; VAR z : cmtrx10);
```

A listing of a set of PASCAL statements defining the procedure ImpedMesh is given in Fig. 11.2. The operation of the procedure is readily understood in terms of the discussion given above. A summary of the important features of this procedure is given in Table 11.1. An example follows.

```
PROCEDURE ImpedMesh (nMesh, nz : integer; imped : carray20;

                   mesh1, mesh2 : iarray20; VAR z : cmtrx10);

(* Procedure for computing the elements of an impedance
   matrix from the values of the impedances and the
   mesh numbers in which they are located. The algorithm
   is identical to that used for Rmesh,
     nMesh - Number of meshes
     nz - Number of impedances
     imped - Array of complex values of impedances
```

```
      mesh1 - Array of first mesh in which impedances
          are located
      mesh2 - Array of second mesh in which impedances
          are located
      z - Impedance matrix
    Note: This procedure calls the procedures Add and Sub  *)

VAR
   i, j, k : integer;

BEGIN

   (* Set the elements of the impedance matrix to zero *)

   FOR i := 1 TO nMesh DO
      FOR j := 1 TO nMesh DO
         BEGIN
            z[i,j].re := 0.0;
            z[i,j].im := 0.0
         END;
   FOR i := 1 TO nz DO
      BEGIN
         j := mesh1[i];
         k := mesh2[i];
         IF k > 0 THEN
            BEGIN
               Add(z[k,k], imped[i], z[k,k]);
               Sub(z[j,k], imped[i], z[j,k]);
               z[k,j] := z[j,k]
            END;
         Add(z[j,j], imped[i], z[j,j])
      END
END;
```

Figure 11.2 Listing of procedure ImpedMesh.

TABLE 11.1 SUMMARY OF THE CHARACTERISTICS OF THE PROCEDURE ImpedMesh

Identifying Statement PROCEDURE ImpedMesh (nMesh, nz : integer; VAR imped : carray20; VAR mesh1, mesh2 : iarray20; VAR z : cmtrx10);

Purpose To form the complex impedance matrix $\mathbf{Z}(j\omega)$ which represents a network under sinusoidal steady-state conditions.

Additional Subprograms Required This procedure calls the procedures Add and Sub.

Input Arguments

nMesh	The order of the matrix $\mathbf{Z}(j\omega)$
nz	The number of impedances
imped	The complex array whose elements are the values of the impedances Z_i
mesh1	The array of values of the first mesh in which the impedances are located
mesh2	The array of values of the second mesh (if any) in which the impedances are located

Output Argument

z	The two-dimensional complex array whose elements are the impedances $z_{ij}(j\omega)$ of the impedance matrix $\mathbf{Z}(j\omega)$

User-Defined Types

complex = RECORD re, im : real END;
carray20 = ARRAY[1..20] OF complex;
iarray20 = ARRAY[1..20] OF integer;
cmtrx10 = ARRAY[1..10,1..10] OF complex;

EXAMPLE 11.1

As an example of the use of the procedure ImpedMesh to form the impedance matrix for a network under sinusoidal steady-state conditions, consider the three-mesh network shown in Fig. 11.3. The impedance (in ohms) of each of the elements is written alongside it. Let us assume that the circuit is to be analyzed at $\omega = 5$ rad/sec and that the elements have the values and thus the resulting numerical impedances shown below:

Element	Value	Impedance (Ω)
1	1 Ω	$1 + j0$
2	2 Ω	$2 + j0$
3	3 Ω	$3 + j0$
4	4 Ω	$4 + j0$
5	0.3 H	$0 + j1.5$
6	0.1 F	$0 - j2$
7	0.05 F	$0 - j4$

Using the values of impedance given above, a program for finding the impedance matrix for the network, together with the necessary input data, is given in Fig. 11.4. The output from the program is shown in Fig. 11.5.

11.3 SOLVING THE SIMULTANEOUS SINUSOIDAL STEADY-STATE EQUATIONS: THE PROCEDURE CmplxSimEqn

In Secs. 11.1 and 11.2 we formulated the problem of forming a matrix equation of the type

$$\mathcal{V} = \mathbf{Z}(j\omega)\,\mathcal{I} \tag{11.7}$$

where $\mathcal{V}$ is an n-element column matrix with complex elements $\mathcal{V}_i$, $\mathbf{Z}(j\omega)$ is an $n \times n$ square impedance matrix with complex elements $z_{ij}(j\omega)$, and $\mathcal{I}$ is an n-element column matrix with complex elements $\mathcal{I}_i$. We showed how, if this was the matrix equation for

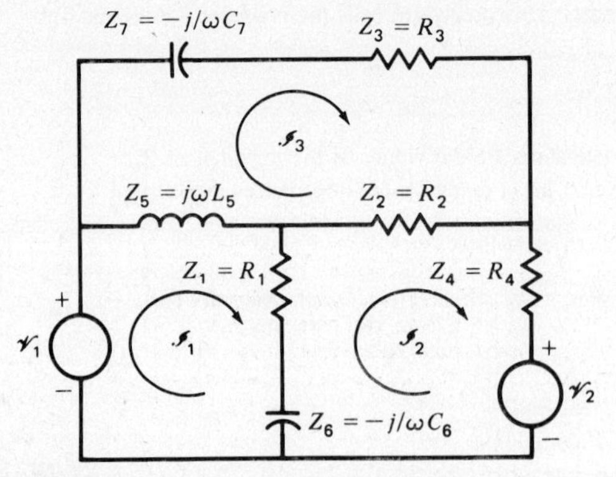

Figure 11.3 Network for Example 11.1.

```
PROGRAM Main (input, output);     (* Example 11.1 *)

(* Main program for finding the impedance matrix of
   an RLC network
     nMesh - Number of meshes
     nz - Number of impedances
     imped - Array of complex values of impedances
     mesh1 - Array of first mesh in which impedances
          are located
     mesh2 - Array of second mesh in which impedances
          are located
     z - Impedance (complex) matrix
   Note: This program calls the procedure ImpedMesh    *)

TYPE
   complex   = RECORD
                re, im : real
                END;
   carray20  = ARRAY[1..20] OF complex;
   cmtrx10   = ARRAY[1..10,1..10] OF complex;
   iarray20  = ARRAY[1..20] OF integer;

VAR
   nMesh, nz, i, j   : integer;
   mesh1, mesh2 : iarray20;
   imped        : carray20;
   z            : cmtrx10;

BEGIN

   (* Input the data defining the network *)

   readln(nMesh, nz);
   writeln('  I Mesh1 Mesh2     Z.re     Z.im');
   FOR i := 1 TO nz DO
      BEGIN
         readln(mesh1[i], mesh2[i], imped[i].re, imped[i].im);
         writeln(i:3, mesh1[i]:4, mesh2[i]:6, imped[i].re:10:3,
                 imped[i].im:8:3)
      END;

   (* Compute and print the impedance matrix *)

   ImpedMesh(nMesh, nz, imped, mesh1, mesh2, z);
   writeln;
   writeln('                        Impedance Matrix');
   FOR i := 1 TO nMesh DO
      BEGIN
         FOR j := 1 TO nMesh DO
            write(z[i,j].re:8:2,' +J ', z[i,j].im:5:2);
         writeln
      END
END.
```

(*a*)

Input Data

```
3 7 <------------- Number of meshes, number of impedances
1 2 1.0 0.0 <----- Z1, mesh1, mesh2, real part, imaginary part
2 3 2.0 0.0 <----- Z2, mesh1, mesh2, real part, imaginary part
3 0 3.0 0.0 <----- Z3, mesh1, mesh2, real part, imaginary part
2 0 4.0 0.0 <----- Z4, mesh1, mesh2, real part, imaginary part
1 3 0.0 1.5 <----- Z5, mesh1, mesh2, real part, imaginary part
1 2 0.0 -2.0 <---- Z6, mesh1, mesh2, real part, imaginary part
3 0 0.0 -4.0 <---- Z7, mesh1, mesh2, real part, imaginary part
```

(*b*)

Figure 11.4 Program and input data for Example 11.1.

```
I MESH1 MESH2     Z.RE      Z.IM
1   1     2       1.000     0.000
2   2     3       2.000     0.000
3   3     0       3.000     0.000
4   2     0       4.000     0.000
5   1     3       0.000     1.500
6   1     2       0.000    -2.000
7   3     0       0.000    -4.000

                     IMPEDANCE MATRIX
  1.00 +J -0.50    -1.00 +J  2.00     0.00 +J -1.50
 -1.00 +J  2.00     7.00 +J -2.00    -2.00 +J  0.00
  0.00 +J -1.50    -2.00 +J  0.00     5.00 +J -2.50
```

Figure 11.5 Output for Example 11.1.

an *RLC* network under sinusoidal steady-state conditions, the square impedance matrix could be found directly from a knowledge of the values of the impedances and the identifying numbers for the meshes in which they were located. In this section we concern ourselves with the solution of such a matrix equation. The problem may be formulated as follows:

Problem. Given the matrices $\mathcal{V}$ and $\mathbf{Z}(j\omega)$ of the matrix equation (11.7), find the matrix $\mathcal{I}$.

Such a problem is frequently spoken of as finding the solution of a set of simultaneous equations.

For a solution to the problem formulated above to exist, the determinant of the square matrix $\mathbf{Z}(j\omega)$ must be nonzero. In the following developments, we assume that this is so. We also assume that the elements on the main diagonal of the matrix are nonzero. Although this is not a necessary condition for a solution, it is a condition that usually exists in physical situations, and it will simplify the resulting computations. The procedure for accomplishing the solution is identical to that given in Sec. 7.3. The only difference is that the elements of the matrices are complex numbers. Thus, all the algebraic operations must be implemented using the complex arithmetic PASCAL procedures introduced in Chap. 8.

A procedure for finding the solution must be provided with the following information:

1. The complex elements $z_{ij}(j\omega)$ of the matrix $\mathbf{Z}(j\omega)$ of (11.7), representing the coefficients of the set of simultaneous equations.
2. The complex elements of the matrix $\mathcal{V}$ of (11.7), representing the known quantities in the set of simultaneous equations.
3. The number n of simultaneous equations, that is the size of the square matrix $\mathbf{Z}(j\omega)$.

As an output, such a procedure should supply the complex values of the $\mathcal{I}_i$, that is, the elements of the matrix $\mathcal{I}$, giving the phasor values of the unknown variables that satisfy the simultaneous equations. The parameter list for the procedure should have the form

```
(VAR v:carray10; VAR z:cmtrx10;
 VAR amps:carray10; n:integer)
```

where v is a one-dimensional array [1..10] of complex numbers in which are stored the (known) values of the phasor mesh voltages, etc. If we use the name CmplxSimEqn (for *complex* Gauss-Jordan *sim*ultaneous *eq*uatio*n* solution) for the procedure, the heading for it will be of the form

PROCEDURE CmplxSimEqn (VAR v: carray10; VAR z: cmtrx10;
VAR amps: carray10; n: integer);

A listing of a set of PASCAL statements for a procedure that will provide the operations described above is given in Fig. 11.6. The operation of the procedure is readily verified by comparing its statements with the flowchart given in Fig. 7.3. A summary of the characteristics of the procedure is given in Table 11.2.

```
PROCEDURE CmplxSimEqn (VAR v : carray10; VAR z : cmtrx10;

                  VAR amps : carray10; n : integer);

(* Procedure for solving a set of simultaneous equations
   with complex coefficients of the form v = z * amps
     v - Array of complex phasor values of independent variables
     z - Matrix of complex coefficients of the set of equations
     amps - Array of (unknown) dependent variables which are
          the solutions to the set of equations
     n - Number of equations
   Note: This procedure calls the procedures Dvd, Mul, and Sub *)

   TYPE
      cmtrx11 = ARRAY[1..10,1..11] OF complex;

   VAR
      i, j, k, np, flag : integer;
      beta, alfa, temp  : complex;
      za                : cmtrx10;

   BEGIN

      (* Enter the z matrix as the first n columns of the
         za matrix and enter the v array as the n+1th
         column of the za matrix                          *)

      FOR i := 1 TO n DO
         BEGIN
            FOR j := 1 TO n DO
               za[i,j] := z[i,j];
            za[i,n+1] := v[i]
         END;

      (* Reduce the first n columns of the augmented matrix
         za to an identity matrix                          *)

      np := n + 1;
      FOR i := 1 TO n DO
         BEGIN

            (* Set the main diagonal elements to unity *)

            alfa := za[i,i];
            IF Mag(alfa) < 0.0001
               THEN flag := 1
               ELSE flag := 0;
            FOR j := 1 TO np DO
               Dvd(za[i,j], alfa, za[i,j]);
```

```
            (* Set the non-diagonal elements of the itn
               column to zero                           *)

            FOR k := 1 TO n DO
               BEGIN
                  IF k <> i THEN
                     BEGIN
                        beta := za[k,i];
                        FOR j := 1 TO np DO
                           BEGIN
                              Mul(beta, za[i,j], temp);
                              Sub(za[k,j], temp, za[k,j])
                           END
                     END    (* End of If/Then *)
               END    (* End of k index loop *)
         END;    (* End of i index loop *)

      (* Store the n+1th column of za as amps *)

   FOR i := 1 TO n DO
      amps[i] := za[i,np];
   IF flag = 1 THEN
      BEGIN
         writeln;
         writeln(' Main diagonal element zero in z matrix');
         writeln(' Results from CmplxSimEqn may not be accurate')
      END;
   END;
```

Figure 11.6 Listing of procedure CmplxSimEqn.

The procedure CmplxSimEqn may be used in conjunction with the procedure ImpedMesh to solve for the phasor mesh currents of an *RLC* network from the known values of the phasor mesh voltages and the impedances. An example follows.

EXAMPLE 11.2

As an example of the use of the procedure CmplxSimEqn, let us analyze the circuit shown in Fig. 11.7 by solving for the phasor mesh currents. A program and input for

TABLE 11.2 SUMMARY OF THE CHARACTERISTICS OF THE PROCEDURE CmplxSimEqn

Identifying Statement PROCEDURE CmplxSimEqn (VAR v : carray10; VAR z : cmtrx10; VAR amps : carray10; n : integer);

Purpose To solve a set of simultaneous equations with complex coefficients having the form $\mathcal{V} = \mathbf{Z}(j\omega)\,\mathcal{I}$.

Additional Subprograms Required This procedure calls the procedures Dvd, Mul, and Sub.

Input Arguments

v The one-dimensional complex array of phasor voltages $\mathcal{V}_i$

z The two-dimensional complex array of impedances $z_{ij}(j\omega)$ which form the impedance matrix $\mathbf{Z}(j\omega)$

n The number of simultaneous equations

Output Argument

amps The one-dimensional complex array of phasor mesh currents $\mathcal{I}_i$

User-Defined Types

complex = RECORD re, im : real END;
carray10 = ARRAY[1..10] OF complex;
cmtrx10 = ARRAY[1..10,1..10] OF complex;

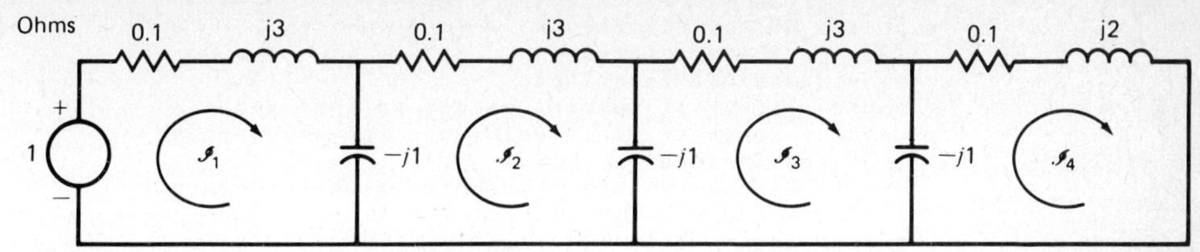

Figure 11.7 Network for Example 11.2.

doing this are shown in Fig. 11.8. The output is shown in Fig. 11.9. It is readily verified that the substitution of the indicated values of phasor mesh currents into the original equation [of the form given in (11.7)] satisfies the equation.

11.4 PLOTTING THE NETWORK FUNCTION DIRECTLY FROM THE NETWORK

In order to apply the techniques of Chap. 9 to make magnitude and phase plots of a network function, it is necessary to first find the network function. For large networks consisting of many elements, this usually proves to be a difficult and error-prone task. The techniques of this chapter provide an alternate approach, since the same magnitude and phase plots may be generated *without having to find the network function* by using the procedures ImpedMesh and CmplxSimEqn. The method for accomplishing this is as follows:

1. Identify the meshes and the impedances of the network, assign reference numbers to each of them, and enter the mesh data in arrays mesh1 and mesh2 as was done in Examples 11.1 and 11.2.
2. Use complex variables imped[i] to represent the impedances and set the real parts of these variables to the correct values.
3. Choose a starting value for ω equal to the first value to be used in the plots.
4. Calculate the values of the imaginary part of the variables imped[i] at the specified value of ω.
5. Call the procedures ImpedMesh and CmplxSimEqn to solve for the phasor mesh currents of the network at the specified value of ω.
6. Calculate the phasor value of the output variable for the network and store its magnitude and/or phase in plotting arrays.
7. Increase the value of ω and repeat steps 4 to 6.

A flowchart of the process is shown in Fig. 11.10. An example follows.

EXAMPLE 11.3

As an example of the use of the procedures ImpedMesh and CmplxSimEqn to plot the magnitude of a network function without actually finding the function, consider the network shown in Fig. 11.11. It is desired to plot the magnitude of the voltage transfer function for this network over the frequency range of 0 to 2.5 rad/sec. The identifying

```
PROGRAM Main (input, output);    (* Example 11.2 *)

(* Main program for finding the complex phasor mesh
   currents of an RLC network
     nMesh - Number of meshes
     nz - Number of impedances
     imped - Array of complex values of impedances
     mesh1 - Array of first mesh in which impedances
          are located
     mesh2 - Array of second mesh in which impedances
          are located
     z - Impedance (complex) matrix
     v - Array of phasor excitation voltages
     amps - Array of phasor mesh currents
   Note: This program calls the procedures ImpedMesh and CmplxSimEqn *)

TYPE
   complex   = RECORD
                 re, im : real
               END;
   carray10 = ARRAY[1..10] OF complex;
   carray20 = ARRAY[1..20] OF complex;
   cmtrx10  = ARRAY[1..10,1..10] OF complex;
   iarray20 = ARRAY[1..20] OF integer;

VAR
   nMesh, nz, i, j  : integer;
   mesh1, mesh2 : iarray20;
   imped        : carray20;
   z            : cmtrx10;
   v, amps      : carray10;

BEGIN

   (* Input the data defining the network *)

   readln(nMesh, nz);
   writeln('1');
   writeln('  I Mesh1 Mesh2     Z.re    Z.im');
   FOR i := 1 TO nz DO
      BEGIN
         readln(mesh1[i], mesh2[i], imped[i].re, imped[i].im);
         writeln(i:3, mesh1[i]:4, mesh2[i]:6, imped[i].re:10:3,
                 imped[i].im:8:3)
      END;

   (* Input the data defining the excitation voltages *)

   writeln;
   writeln('  i          Voltage');
   FOR i := 1 TO nMesh DO
      BEGIN
         readln(v[i].re, v[i].im);
         writeln(i:3, v[i].re:10:3, ' +J', v[i].im:7:3)
      END;

   (* Compute and print the impedance matrix *)

   ImpedMesh(nMesh, nz, imped, mesh1, mesh2, z);
   writeln;
   writeln('                              Impedance Matrix');
   FOR i := 1 TO nMesh DO
      BEGIN
         FOR j := 1 TO nMesh DO
            write(z[i,j].re:8:2,' +J ', z[i,j].im:5:2);
         writeln
      END;

   (* Use CmplxSimEqn to solve for the mesh currents *)
```

Figure 11.8 Program and input data for Example 11.2

```
    CmplxSimEqn(v, z, amps, nMesh);
    writeln;
    writeln('  i           Amps.re                      Amps.im');
    FOR i := 1 TO nMesh DO
       writeln(i:3, amps[i].re, amps[i].im);
END.
```

(*a*)

Input Data

```
4 7 ←———————————————— Number of meshes, number of impedances
1 0 0.1 3.0 ←———————— Z1, mesh1, mesh2, real part, imaginary part
1 2 0.0 -1.0 ←——————— Z2, mesh1, mesh2, real part, imaginary part
2 0 0.1 3.0 ←———————— Z3, mesh1, mesh2, real part, imaginary part
2 3 0.0 -1.0 ←——————— Z4, mesh1, mesh2, real part, imaginary part
3 0 0.1 3.0 ←———————— Z5, mesh1, mesh2, real part, imaginary part
3 4 0.0 -1.0 ←——————— Z6, mesh1, mesh2, real part, imaginary part
4 0 0.1 2.0 ←———————— Z7, mesh1, mesh2, real part, imaginary part
1.0 0.0 ←———————————— Phasor voltage V1, real part, imaginary part
0.0 0.0 ←———————————— Phasor voltage V2, real part, imaginary part
0.0 0.0 ←———————————— Phasor voltage V3, real part, imaginary part
0.0 0.0 ←———————————— Phasor voltage V4, real part, imaginary part
```

(*b*)

Figure 11.8 (*Continued*)

numbers for the various elements are shown in circles in the figure. The corresponding impedances for these elements are

$$Z_1(j\omega) = Z_5(j\omega) = 1 + j0$$
$$Z_2(j\omega) = Z_4(j\omega) = 0 + j\omega$$
$$Z_3(j\omega) = 0 - \frac{j}{2\omega}$$

```
I MESH1 MESH2    Z.RE     Z.IM
1   1     0      0.100    3.000
2   1     2      0.000   -1.000
3   2     0      0.100    3.000
4   2     3      0.000   -1.000
5   3     0      0.100    3.000
6   3     4      0.000   -1.000
7   4     0      0.100    2.000

I          VOLTAGE
1      1.000 +J   0.000
2      0.000 +J   0.000
3      0.000 +J   0.000
4      0.000 +J   0.000

                          IMPEDANCE MATRIX
  0.10 +J  2.00     0.00 +J  1.00     0.00 +J  0.00     0.00 +J  0.00
  0.00 +J  1.00     0.10 +J  1.00     0.00 +J  1.00     0.00 +J  0.00
  0.00 +J  0.00     0.00 +J  1.00     0.10 +J  1.00     0.00 +J  1.00
  0.00 +J  0.00     0.00 +J  0.00     0.00 +J  1.00     0.10 +J  1.00

I           AMPS.RE                 AMPS.IM
1  7.3357892344934E-002 -4.9626497563172E-001
2 -9.7087287126692E-002 -9.4259502066762E-005
3  2.3738820731968E-002  4.8667050642112E-001
4  2.4681415752618E-002 -4.8420236484585E-001
```

Figure 11.9 Output for Example 11.2.

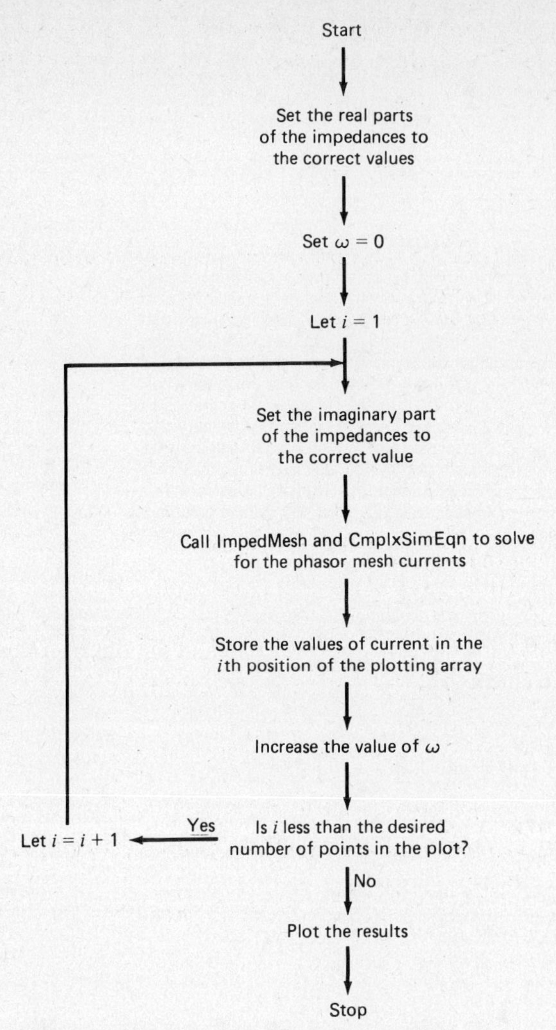

Figure 11.10 Flowchart for network analysis.

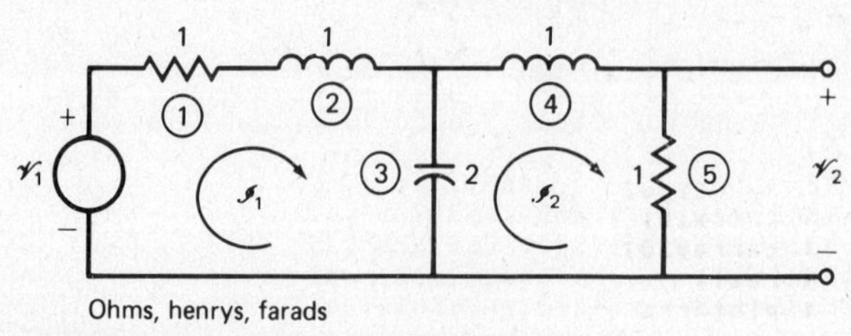

Figure 11.11 Network for Example 11.3.

Since the mesh current $\mathscr{I}_2$ flows through a 1-Ω resistor, it is equal to the output voltage $\mathscr{V}_2$. In addition, since the input voltage is unity, we see that the network function $N(j\omega)$ may be defined as

$$N(j\omega) = \frac{\mathscr{V}_2}{\mathscr{V}_1} = \frac{\mathscr{I}_2}{\mathscr{V}_1} = \frac{\mathscr{I}_2}{1} = \mathscr{I}_2$$

Thus, plotting the magnitude of the mesh current $\mathscr{I}_2$ gives the same result as plotting the magnitude of the network function. A program for making the necessary calculations is shown in Fig. 11.12. The output is shown in Fig. 11.13.

```
PROGRAM Main (output);    (* Example 11.3 *)

(* Example of the use of the CmplxSimEqn procedure to generate
   a plot of the magnitude of a network function from
   an analysis of the network
     n - Number of meshes
     nz - Number of impedances
     imped - Array of complex values of impedances
     mesh1 - Array of first mesh in which impedances
            are located
     mesh2 - Array of second mesh in which impedances
            are located
     z - Impedance (complex) matrix
     v - Array of phasor excitation voltages
     amps - Array of phasor mesh currents
     w - Frequency (rad/sec)
     dw - Change in frequency between plotted points
     p - Plotting array
   Note: This program calls the procedures ImpedMesh, CmplxSimEqn,
            Mag, and Plot5                                          *)

CONST
   n = 2;
   nz = 5;
   scale = 100.0;

TYPE
   complex    = RECORD
                re, im : real
                END;
   carray10   = ARRAY[1..10] OF complex;
   carray20   = ARRAY[1..20] OF complex;
   cmtrx10    = ARRAY[1..10,1..10] OF complex;
   iarray20   = ARRAY[1..20] OF integer;
   plotArray  = ARRAY[1..5,0..100] OF real;

VAR
   i              : integer;
   mesh1, mesh2   : iarray20;
   imped          : carray20;
   z              : cmtrx10;
   v, amps        : carray10;
   w, dw          : real;
   p              : plotArray;
```

Figure 11.12 Program for Example 11.3

```
BEGIN
   (* Initialize the values of the constant parts of the
      impedances and the excitation voltages *)

   imped[1].re := 1.0;     imped[1].im := 0.0;
   imped[2].re := 0.0;     imped[2].im := 0.0;
   imped[3].re := 0.0;     imped[3].im := 0.0;
   imped[4].re := 0.0;     imped[4].im := 0.0;
   imped[5].re := 1.0;     imped[5].im := 0.0;
   mesh1[1] := 1;     mesh2[1] := 0;
   mesh1[2] := 1;     mesh2[2] := 0;
   mesh1[3] := 1;     mesh2[3] := 2;
   mesh1[4] := 2;     mesh2[4] := 0;
   mesh1[5] := 2;     mesh2[5] := 0;
   v[1].re := 1.0;     v[1].im := 0.0;
   v[2].re := 0.0;     v[2].im := 0.0;
   w := 0.00001;
   dw := 0.05;

   (* For each value of frequency, compute the values of
      the impedances, form the impedance matrix, solve for
      the mesh currents and store the magnitude of the
      currents in a plotting array                      *)

   FOR i := 0 TO 50 DO
      BEGIN
         imped[2].im := w;
         imped[3].im := -1.0 / (2.0 * w);
         imped[4].im := w;
         ImpedMesh(n, nz, imped, mesh1, mesh2, z);
         CmplxSimEqn(v, z, amps, n);
         p[1,i] := Mag(amps[2]) * scale;
         w := w + dw
      END;
   Plot5(p, 1, 50, 100)
END.
```

Figure 11.12 (*Continued*)

11.5 CONCLUSION

In this chapter we have presented some additional techniques for the analysis of networks and systems under sinusoidal steady-state conditions based on mesh analysis and the use of the impedance matrix. These techniques are readily extended to node analysis and also to the case where controlled sources are present. Some examples of these cases may be found in the problems.

PROBLEMS

11.1. The techniques developed in this chapter for finding the elements of the impedance matrix $\mathbf{Z}(j\omega)$ and solving the equations $\mathcal{V} = \mathbf{Z}(j\omega)\mathcal{I}$ to find the phasor mesh currents may be directly applied to finding the *admittance* matrix $\mathbf{Y}(j\omega)$ and solving the equations $\mathcal{I} = \mathbf{Y}(j\omega)\,\mathcal{V}$ to find the phasor node voltages. Illustrate the method by using the procedures

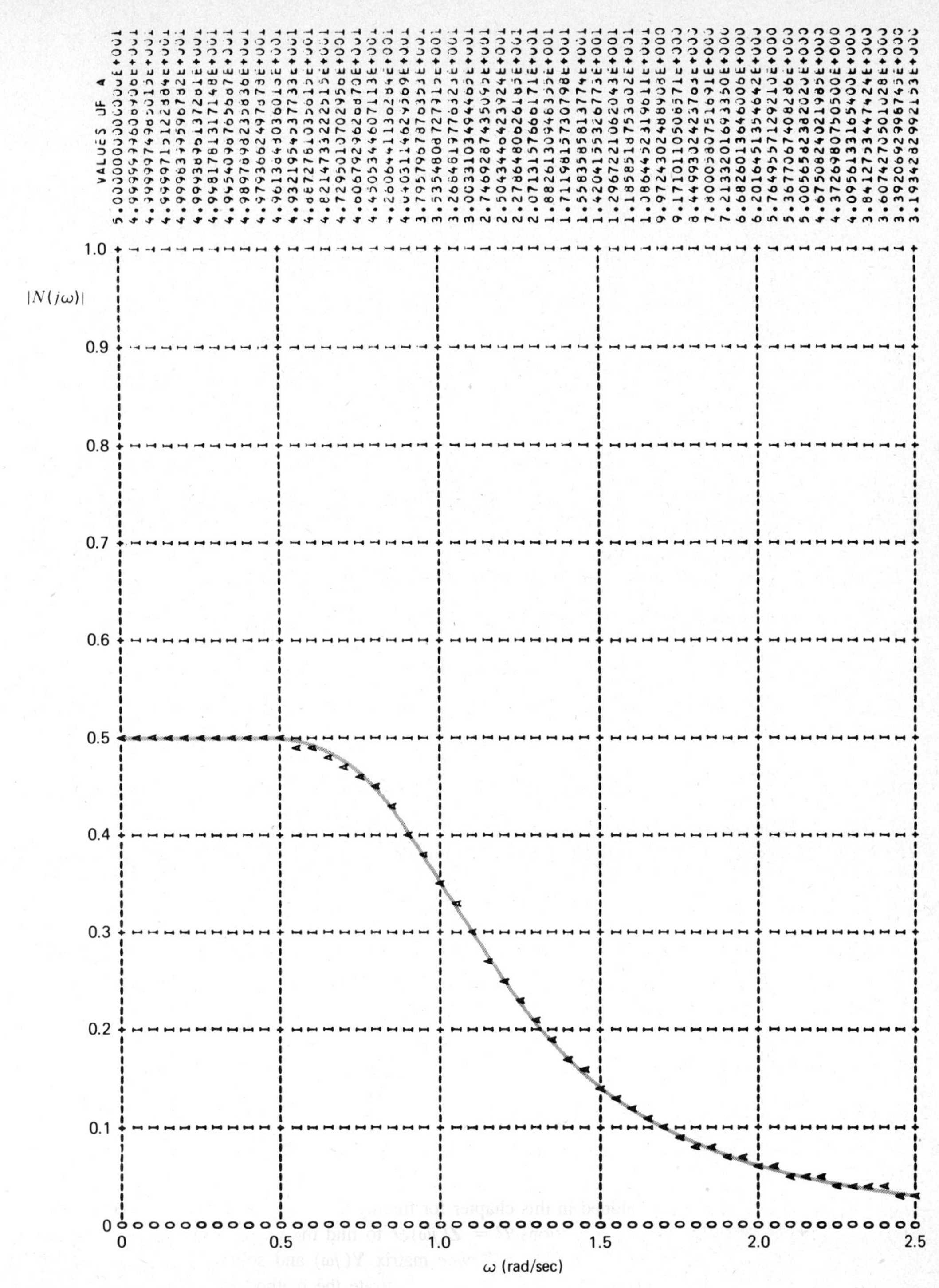

Figure 11.13 Output for Example 11.3.

ImpedMesh and CmplxSimEqn to find the values of the real and imaginary parts of the phasor node voltages for the network shown in Fig. P11.1.

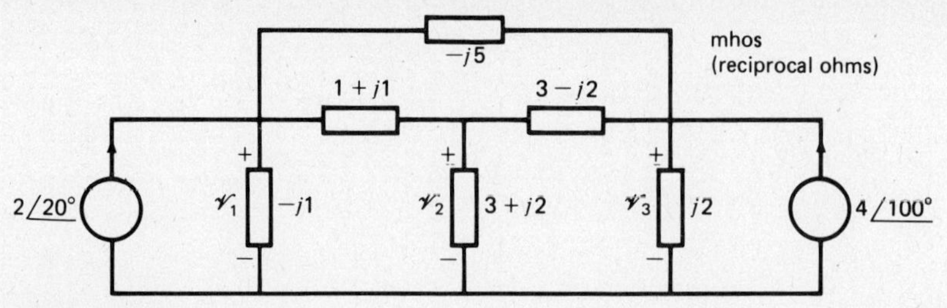

Figure P11.1

11.2. Two independent sinusoidal voltage sources (of different frequencies) are used to excite the network shown in Fig. P11.2. Use the procedures ImpedMesh and CmplxSimEqn to find the *time-domain* expressions for the sinusoidal steady-state behavior for the mesh currents.

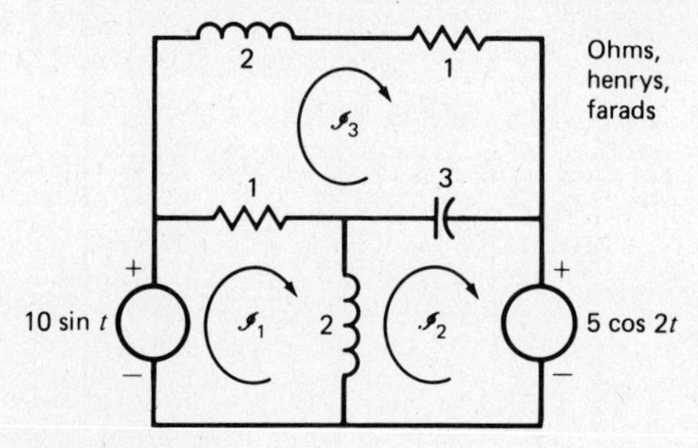

Figure P11.2

11.3. The effect of controlled sources in sinusoidal steady-state analysis may be treated using the same method as was discussed in Sec. 7.4 for the resistance network. Use the procedures ImpedMesh and CmplxSimEqn to solve for the real and imaginary parts of the phasor mesh currents for the network shown in Fig. P11.3.

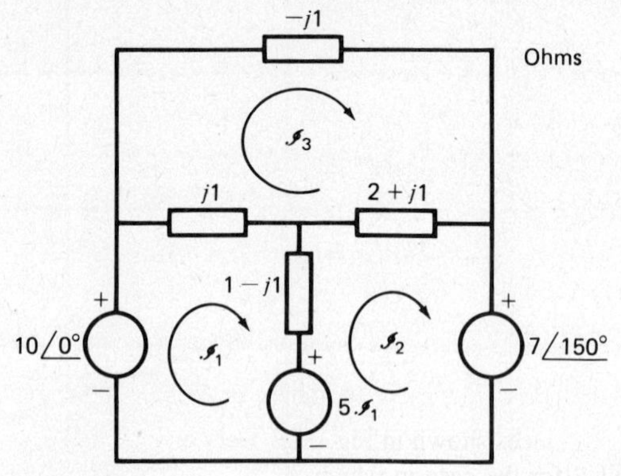

Figure P11.3

11.4. Using the techniques introduced in Sec. 11.4, make a plot of the magnitude of the network function $N(j\omega) = \mathcal{V}_2/\mathcal{V}_1$ for the passive filter circuit shown in Fig. P11.4 over a frequency range of 0 to 1.5 rad/sec. Use 61 points and scale the data by 200.

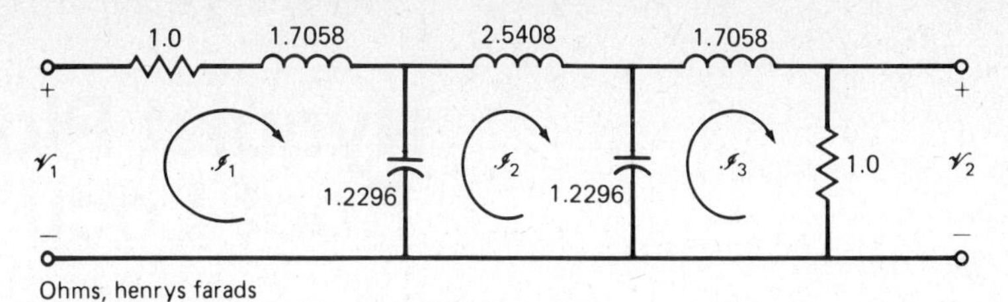

Figure P11.4

11.5. Using an admittance-matrix format (see Prob. 11.1), make a plot of the magnitude of the network function $N(j\omega) = \mathcal{V}_2/\mathcal{V}_1$ for the active-*RC* filter circuit shown in Fig. P11.5 over a frequency range of 0 to 2.5 rad/sec. Use 51 points and scale the data by 50. The triangle shown in the figure represents a VCVS (voltage-controlled voltage source) for which the output voltage $\mathcal{V}_2$ is 1.586 times the input voltage $\mathcal{V}_4$. The effects of this source on the admittance matrix are: (1) since $\mathcal{V}_2 = 1.586\,\mathcal{V}_4$, the elements of the fourth column of the admittance matrix may be divided by 1.586 and added to the elements of the second column; and (2) the second row of the matrix may be eliminated. Thus, the original admittance equations

$$\begin{bmatrix} \mathcal{I}_1 \\ \mathcal{I}_2 \\ \mathcal{I}_3 \\ \mathcal{I}_4 \end{bmatrix} = \begin{bmatrix} y_{11} & y_{12} & y_{13} & y_{14} \\ y_{21} & y_{22} & y_{23} & y_{24} \\ y_{31} & y_{32} & y_{33} & y_{34} \\ y_{41} & y_{42} & y_{43} & y_{44} \end{bmatrix} \begin{bmatrix} \mathcal{V}_1 \\ \mathcal{V}_2 \\ \mathcal{V}_3 \\ \mathcal{V}_4 \end{bmatrix}$$

become

$$\begin{bmatrix} \mathcal{I}_1 \\ \mathcal{I}_3 \\ \mathcal{I}_4 \end{bmatrix} = \begin{bmatrix} y_{11} & y_{12} + y_{14}/1.586 & y_{13} \\ y_{31} & y_{32} + y_{34}/1.586 & y_{33} \\ y_{41} & y_{42} + y_{44}/1.586 & y_{34} \end{bmatrix} \begin{bmatrix} \mathcal{V}_1 \\ \mathcal{V}_2 \\ \mathcal{V}_3 \end{bmatrix}$$

Assuming an arbitrary input current $\mathcal{I}_1 = 1$ and letting $\mathcal{I}_3 = \mathcal{I}_4 = 0$, we may solve for $\mathcal{V}_2$ and $\mathcal{V}_1$ at each frequency and thus find their ratio, namely, $N(j\omega)$.

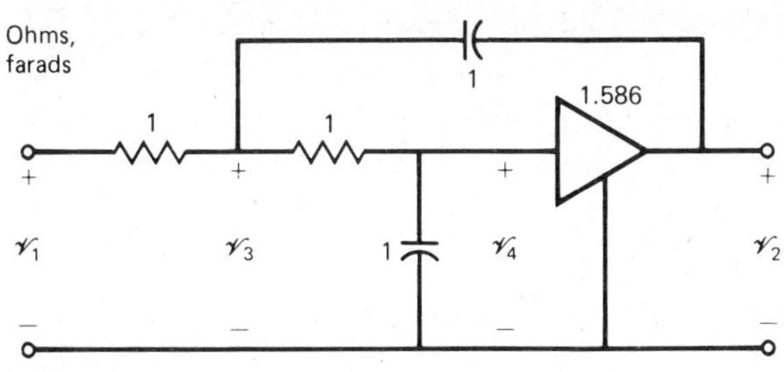

Figure P11.5

11.6. Repeat Prob. 11.4, but make a plot of the phase of $N(j\omega)$.

11.7. Repeat Prob. 11.5, but make a plot of the phase of $N(j\omega)$.

11.8. Assume that the impedances shown in Fig. 11.7 were evaluated at a frequency of 1 rad/sec. Repeat Example 11.2 for the case in which the frequency is 2 rad/sec.

Chapter 12

Nyquist Plots and Root-Locus Plots

In Chap. 9 we discussed some methods for describing the characteristics of network and system functions under conditions of sinusoidal steady-state excitation. Specifically, we developed methods for constructing plots of the magnitude and phase of the network functions. In this chapter we present some other means of describing the characteristics of network functions. To do this we introduce Nyquist plots and root-locus plots. In addition to their use in displaying the characteristics of networks, such plots have considerable application in the field of control theory.

12.1 THE xy PLOT: THE PROCEDURE Xyplt

In Chap. 3, we introduced the concept of using a procedure (the procedure Plot5) to plot several functions of an independent variable. It was assumed, in making the plot, that the values of the functions were calculated for a series of values of the independent variable. The values of the variable were chosen to satisfy the following characteristics: (1) successive values were assumed to be in increasing order, and (2) the difference between successive values was chosen to be a constant. We found that such a procedure was well adapted to plotting the results of many network studies both in the time domain and in the frequency domain. Frequently, in engineering studies, the need arises for a different type of plot, which is referred to as an *xy* plot. The data plotted in such a case has the following characteristics:

1. Each point on the plot gives the values of two functions for some value (which is not plotted) of a third (independent) variable. Thus, we see that such a plot displays the manner in which two functions vary with respect to each other, for different values of their independent variable.

2. The values of the independent variable are, in general, not evenly spaced; that is, in some portions of the plot it may be desirable to use closely spaced values of the independent variable, in others to use widely spaced values, to produce the values of the functions that are plotted. Frequently, such plots will cover a range of the independent variable from 0 to ∞.

In this section we introduce a procedure designed to produce a plot of the type described above. We label the functions that are being plotted x and y. We assume that both of these are functions of some independent variable ω; thus, they may be written $x(\omega)$ and $y(\omega)$. The points that are to be plotted will be (x_i, y_i), where $x_i = x(\omega_i)$ and $y_i = y(\omega_i)$. The procedure has the following features:

1. *Number of points*. From 1 to 200 values of x_i and the same number of values of y_i may be entered as input data.
2. *Identification of data points*. All the data points are identified on the plot by printing the letter X.
3. *Orientation of the plot*. In order to simplify the plotting routine, the scale for the function $x(\omega)$ is plotted in an increasing direction downward by the printer; thus, the scale that we should normally refer to as the abscissa increases in a downward direction on the plot, while the scale for the function $y(\omega)$ increases from left to right across the printed line. This is the same convention that was used for the procedure Plot5 and is shown in Fig. 3.4. The more conventional orientation of the plot is easily achieved by rotating it 90° counterclockwise.
4. *Scaling the ordinate*. The scale for the ordinate on which the values of the y_i are plotted can be adjusted by the user to include any desired range of 100 units. Thus, it may be set to plot values of the y_i from -100 to 0, or from -50 to 50, or from 0 to 100, or for any other convenient range of 100 units.
5. *Scaling the abscissa*. The scale of the abscissa on which the values of the x_i are plotted may be specified by the user with respect to the following: (*a*) the total range of values that it is desired to include on the abscissa scale, and (*b*) the maximum value of the abscissa scale that is desired. Both of these numbers are most conveniently chosen to be some multiple of 10. The minimum value of the x_i that will be plotted is the maximum value minus the range. If it is desired to keep the plot on a single page of printer output, the maximum range that should be used is 90.
6. *Automatic scaling*. If automatic scaling of the data of the plot is desired, the parameter specifying the range of the abscissa should be set to 999. When this option is selected, the values of the quantities x_i and y_i will be independently scaled to fit into an abscissa range of 90 points and an ordinate range of 100.
7. *Coordinate lines*. Coordinate lines are provided every 10 units in both the ordinate direction and the abscissa direction. To avoid distortion in the plot, such a coordinate grid should consist of squares. Since most printers provide 10 characters per inch across a printed line but 6 lines per inch down the printed page, the scale for the y_i (printed across the line) is divided into 10 parts per decade, while the scale for the x_i (printed down the page) is divided into 6 parts per decade. Thus, a range of 90 units of x_i data for the function $x(\omega)$ requires

9 in of vertical page dimension and also represents 55 lines (an extra line is required so that both the beginning coordinate and the ending coordinate are included) of printing.

8. *Data that exceed scale*. Any of the values of the y_i that exceed the maximum or minimum values of the ordinate scale are indicated by printing the symbol $ on the upper or lower edge of the plot at the corresponding value of the x_i. Any values of the x_i which exceed the maximum or minimum values of the abscissa scale are printed along the edge of the plot at the corresponding value of the y_i.
9. *Ordering of data*. Although the plot and the data points must actually be printed in order of increasing values of the abscissa scale, it is not necessary for the points to be arranged in any special order before the procedure is called. Specifically, they need *not* be arranged in order of increasing values of the x_i. The procedure provides for automatic internal rearranging of the data to provide for the correct printing of the data points.

The procedure described above, which presents an *xy* plot, is named Xyplt. We shall not take time at this point to describe the operation of the procedure, since such a description would take us too far afield from our primary goal of applying digital-computational techniques to basic theory. A brief description of the procedure, however, together with a flowchart and a listing, is given in Appendix B. This information should be sufficient to permit interested readers to produce their own working copies of the procedure. The procedure Xyplt has the identifying statement

```
PROCEDURE Xyplt (nPoints: integer; VAR x, y: array200;
                 abRng, abMax, ordMax: integer);
```

where array200 is a user-defined type ARRAY [1..200] OF real, and where the various arguments have the following significance:

nPoints The number of data points that are being used in constructing the *xy* plot (nPoints must have a value from 1 to 200).

x The one-dimensional array in which are stored the values x_i of the function $x(\omega)$.

y The one-dimensional array in which are stored the values y_i of the function $y(\omega)$.

abRng The range of values of the x_i to be included along the abscissa scale (abRng must have a value from 10 to 990). The minimum value for the scale of the abscissa will be abMax − abRng (see 5 above). This minimum value must not contain more than three digits.

abMax The maximum value used for the scale of the abscissa, for plotting the values of the x_i (abMax must not contain more than three digits).

ordMax The maximum value used for the scale of the ordinate, for plotting the values of the y_i. The value of ordMax must be chosen so that it and the minimum value of the ordinate (which is ordMax − 100) do not contain more than three digits. Practically this represents a range of −890 to 990 for ordMax.

TABLE 12.1 SUMMARY OF THE CHARACTERISTICS OF THE PROCEDURE Xyplt

Identifying Statement PROCEDURE Xyplt (nPoints : integer; VAR x, y : array200; abRng, abMax, ordMax : integer);

Purpose To plot pairs of numbers x_i and y_i, that is, to make an xy plot of the data points (x_i, y_i).

Additional Subprograms Required None

Input Arguments

nPoints	The number of data points to be plotted, that is, the number of values of x_i and y_i
x	The one-dimensional array of variables x[i] in which are stored the values of the quantities x_i
y	The one-dimensional array of variables y[i] in which are stored the values of the quantities y_i
abRng	The total abscissa range used in the plot for the x_i quantities
abMax	The maximum abscissa value used in the plot for the x_i quantities
ordMax	The maximum ordinate value used in the plot for the y_i quantities

Output The program provides a plot of the data points (x_i, y_i) for a range of values of y_i from ordMax − 100 to ordMax along the ordinate, and for a range of values of x_i from abMax − abRng to abMax along the abscissa. Coordinate lines are printed every 10 units in both the ordinate and the abscissa direction. The direction of increasing values on the abscissa is vertically downward on the printed page. An automatic scaling option is called by setting abRng to 999. In this case the values of the other ordinate and abscissa parameters are ignored.

User-Defined Type array200 = ARRAY[1..200] OF real

A summary of the characteristics of the procedure Xyplt is given in Table 12.1. An example of the use of the procedure will be given in connection with the development of the Nyquist plot in the next section.

12.2 THE NYQUIST PLOT

One of the most frequently encountered examples of an xy plot in engineering studies, especially those involving circuits or control systems, is the *Nyquist* plot. Such a plot describes the behavior of a given system under conditions of sinusoidal steady-state excitation by plotting the real and imaginary parts of the network function (evaluated for $s = j\omega$) describing the system for a given set of values of ω. Typically, a range of ω from 0 to ∞ is used, except, of course, for those cases where the network function has a pole at the origin or at infinity (or on the $j\omega$ axis), in which case the range of ω is suitably restricted. The procedure SinStdySt introduced in Chap. 9 can be used to provide the real and imaginary values of the network function, since these quantities are already listed in the output arguments specified for the procedure. Thus, to make a Nyquist plot, we need merely select a value of ω (in radians per second), use SinStdySt to find the real and imaginary parts of the network function, store the real value as x[i] and the imaginary value as y[i], and continue this process for as many values of ω as are desired. When the real and imaginary values have been stored as the single-subscripted variables x[i] and y[i], we need simply call the procedure Xyplt to produce the desired Nyquist plot. An example follows.

EXAMPLE 12.1

It is desired to produce a Nyquist plot of the low-pass network function

$$N(p) = \frac{60}{s^2 + s + 1}$$

over a range of frequency from 0 to ∞. This may be done by using the procedure SinStdySt to calculate the desired values of the real and imaginary parts of the function (under conditions of sinusoidal steady-state excitation) and the procedure Xyplt to plot the resulting data. A listing of a main program for accomplishing this is shown in Fig. 12.1. The output plot produced by the procedure Xyplt is shown in Fig. 12.2.

```
PROGRAM Main (output);     (* Example 12.1 *)

(* Main program for constructing a Nyquist plot
   for a second-order low-pass network function
     num - Array of numerator coefficients
     den - Array of denominator coefficients
     x - Array of values of the real part of
          the network function
     y - Array of values of the imaginary part
          of the network function
     omega - Frequency (rad/sec)                   *)

TYPE
   array200 = ARRAY[1..200] OF real;
   array11  = ARRAY[0..10] OF real;
   complex = RECORD
                re, im : real
                END;

VAR
   omega, vmag, vphase : real;
   x, y                : array200;
   num, den                : array11;
   i                   : integer;

BEGIN
   (* Initialize the variables and the arrays for the
      numerator and the denominator coefficients of the
      network function                                   *)

   omega := 0.0;
   num[0] := 60.0;
   den[0] := 1.0;
   den[1] := 1.0;
   den[2] := 1.0;

   (* Calculate and store the values of the real and imaginary
      parts of the network function                            *)

   FOR i := 1 TO 100 DO
      BEGIN
         SinStdySt(0, num, 2, den, omega, x[i], y[i], vmag, vphase);
         omega := (omega + 0.03) * 1.008
      END;
   Xyplt(100, x, y, 90, 60, 20)
END.
```

Figure 12.1 Program for Example 12.1.

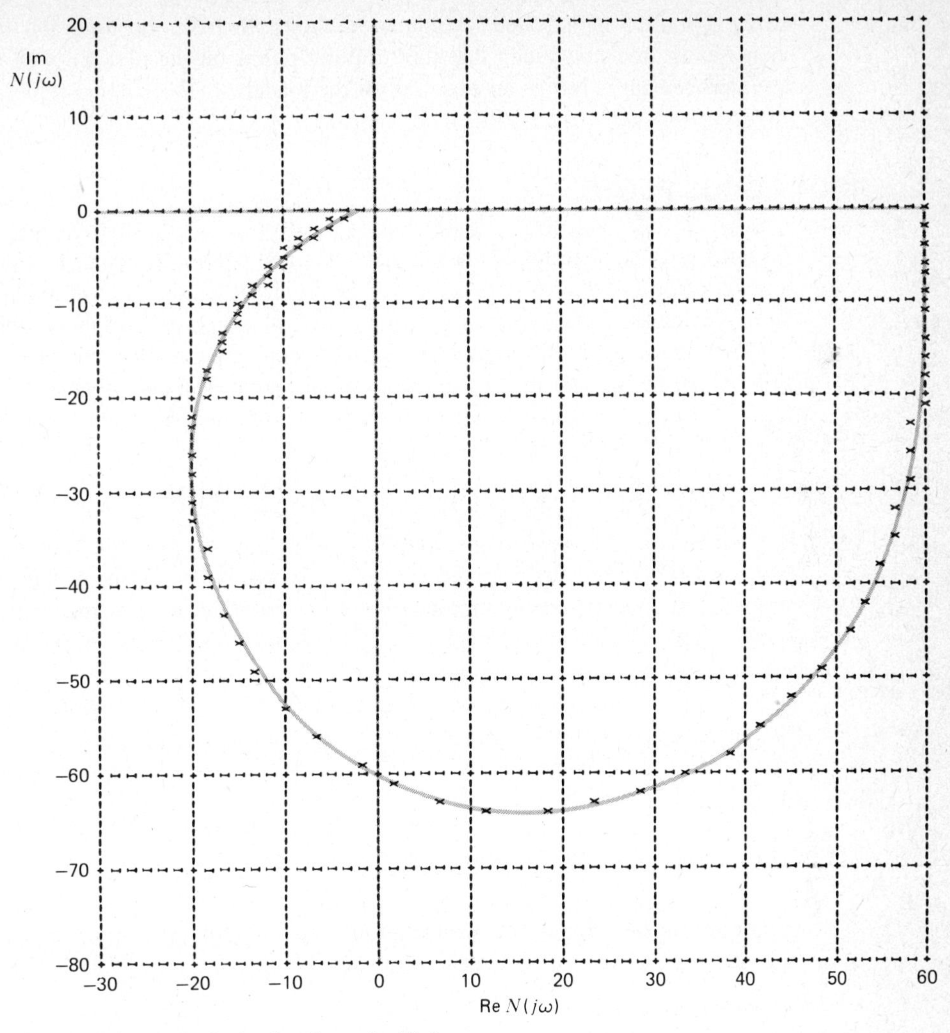

Figure 12.2 Output plot for Example 12.1.

It should be noted that one of the problems in producing a plot of the type shown in Fig. 12.2 is choosing the values of ω to provide acceptable spacing of the points on the plot. In the above example, we have used the relation

$$\text{omega} := (\text{omega} + 0.03) * 1.008 \tag{12.1}$$

to provide both an arithmetic increase and a geometric increase of ω at each step. The arithmetic increases provide good spacing of the frequency values for low values of frequency, and the geometric increases provide good spacing for higher values of frequency. For very large values of frequency, even the geometric increases are not sufficient to provide for evenly spaced points on the plot, as may be readily observed from Fig.

12.2. A more sophisticated technique would be to have the program itself calculate increases in frequency such that the resulting points on the plot are well spaced. Such an improvement is left as an exercise for the reader.

12.3 ROOT-LOCUS PLOTS

Frequently in engineering studies of networks and control systems, it is desirable to be able to determine the manner in which the poles of a given network function vary as some parameter in the system varies. One method of approach to this problem would be to select successively different values of the specified parameter, determine the resulting polynomial coefficients that are functions of this parameter, and use a root-solving routine such as the procedure Root to determine the roots of the polynomial. The root locations as determined by the above procedure could then be plotted (by using the procedure Xyplt) to determine the locus that the roots follow as the parameter is varied. Such a plot is appropriately referred to as a root-locus plot.

In practice, the procedure outlined above (using a root-solving routine) is inefficient because of the need for repeated use of the root-solving routine, which is a complex program. A more effective approach is to use the knowledge of the root locations, for some initial value of the parameter which is being varied, to aid in the determination of the new root locations for a given (small) change of the parameter. Such a procedure is readily programmed for the computer and requires relatively little computing time. As one example of such a procedure, consider a polynomial $P(s)$ which has a root at $s = p_i$. Let $p_i^{(0)}$ be some approximation to the root. We may improve the approximation to p_i by the relation

$$p_i^{(1)} = p_i^{(0)} - \frac{P(p_i^{(0)})}{P'(p_i^{(0)})} \tag{12.2}$$

where $p_i^{(1)}$ is the improved approximation, $P(p_i^{(0)})$ is the given polynomial evaluated at the original approximation for the root, and $P'(p_i^{(0)})$ is the derivative of $P(s)$ evaluated at the original approximation. Obviously, the process may be applied repeatedly to further reduce the error between the actual (unknown) root and its approximation. This iterative technique for improving the approximation to a root of a given polynomial is called the *Newton-Raphson method*. For approximate roots which are close to an actual root, convergence is usually quite rapid. This is exactly the situation which occurs in our determination of the points for a root-locus plot; therefore, we implement the Newton-Raphson method on the digital computer to serve as our primary tool for constructing a root-locus plot.

The operations of evaluating a polynomial and differentiating a polynomial that are required by (12.2) may be accomplished by the procedures ValPoly and DifPoly, which were introduced in Chap. 8. A main program is easily constructed to implement the other operations necessary for determining a set of numbers to construct an xy plot showing the movements of the roots of a given polynomial as some coefficient of the polynomial is varied. As an illustration of how this may be done, consider the polynomial

$$P(s) = a_0 + a_1 s + a_2 s^2 + a_3 s^3 + \cdots + a_n s^n \tag{12.3}$$

which has n roots located at p_i ($i = 1, 2, \ldots, n$). A flowchart of the process by means of which values of x_j and y_j giving the locations of these roots are determined for the variation of some coefficient a_k is shown in Fig. 12.3. It will be noted in the flowchart that use is made of a variable iter to count the number of times the relation given in (12.2) is applied in the determination of each new root location after the coefficient a_k has been changed in value. If this variable reaches a value of 10, it is assumed that the method is not converging, and the program is stopped. Thus, the variable iter provides

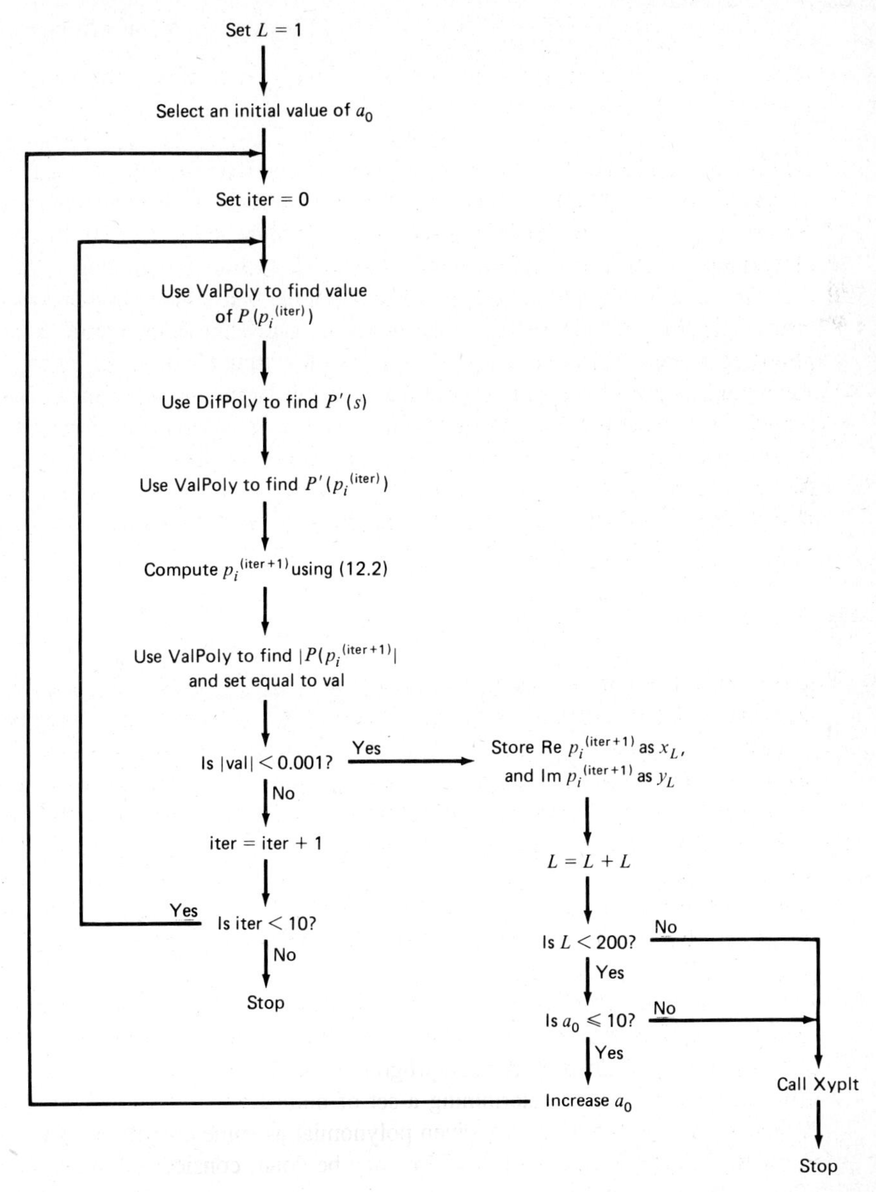

Figure 12.3 Flowchart for making a root-locus plot.

a safeguard against the possibility that the program will run indefinitely (until stopped by the computer) and use up prohibitive amounts of computer time uselessly. Such safeguards should always be made a part of any program that uses iterative methods. Example 12.2 illustrates the use of the flowchart given in Fig. 12.3 to set up a main program to make a root-locus plot.

EXAMPLE 12.2

We may use the flowchart shown in Fig. 12.3 as a guide in writing a program for constructing a root-locus plot for the polynomial

$$P(s) = a_0 + 2s + 2s^2 + s^3$$

for a range of values of a_0 from 0 to 10. For $a_0 = 0$, the initial root locations are at zero and at $-1 \pm j1$. Such a program is shown in Fig. 12.4. It uses the procedure Dvd to compute the quotient of the two complex numbers resulting from the polynomial evaluations given in (12.2). It also includes a provision for multiplication of the root-location data by a scale factor before storing in the one-dimensional arrays x and y. The statement a[0] := (a[0] + 0.2) * 1.1 used for increasing the value of the coefficient a_0 was determined by trial and error. A plot of the output is shown in Fig. 12.5. Only the upper half-plane is plotted. The lower half-plane is similar. A more sophisticated program could easily be constructed to determine automatically increases in the coefficient a_0 such

```
PROGRAM Main (output);     (* Example 12.2 *)

(* Main program for plotting the locus of the complex
   root of a third-order polynomial as the zero-degree
   coefficient is varied
     a - Array of polynomial coefficients
     L - Index giving the number of the point being
           plotted
     iter - Counter for the number of iterations
           required for the convergence of the Newton
           Raphson algorithm
     p - Complex current value of the root location
     x - Array for storing the real value of the points
     y - Array for storing the imaginary value of the points *)

CONST
   scale = 40.0;

TYPE
   complex = RECORD
               re, im : Real
               END;
   array200 = ARRAY[1..200] OF real;
   array11  = ARRAY[0..10] OF real;

VAR
   L, iter, flg, m : integer;
   p, val, dval    : complex;
   magpol          : real;
   a, b            : array11;
   x, y            : array200;
```

Figure 12.4 Program for Example 12.1.

```
BEGIN

   (* Initialize the values of the polynomial coefficients,
      the root location, and the counters                     *)

   a[0] := 0.0;
   a[1] := 2.0;
   a[2] := 2.0;
   a[3] := 1.0;
   L := 1;
   iter := 0;
   p.re := -1.0;
   p.im := 1.0;
   WHILE (L <= 200) AND (a[0] <= 10.0) AND (iter <= 10) DO
      BEGIN

         (* Apply the Newton-Raphson algorithm *)

         ValPoly(3, a, p, val);
         DifPoly(3, a, m, b);
         ValPoly(m, b, p, dval);
         Dvd(val, dval, val);
         Sub(p, val, p);

         (* Check to see if the new root is accurate *)

         ValPoly(3, a, p, val);
         IF Mag(val) > 0.001
            THEN    (* Repeat the Newton-Raphson algorithm *)
               BEGIN
                  iter := iter + 1;
                  flg := 0
               END
            ELSE    (* Store root and increase a[0] *)
               BEGIN
                  x[L] := p.re * scale;
                  y[L] := p.im * scale;
                  iter := 0;
                  a[0] := (a[0] + 0.2) * 1.1;
                  L := L + 1;
                  flg := 1
               END;
      END; (* End of While *)

   (* Check the number of points and plot the locus *)

   IF flg = 1
      THEN L := L - 1;
   Xyplt(L, x, y, 90, 20, 100)
END.
```

Figure 12.4 *(Continued)*

that the points on the root-locus plot are evenly spaced. Such a project is left to the reader as an exercise.

It should be noted that the relation for the Newton-Raphson method, given in (12.2), cannot take account of a situation such as that occurring when a root locus "breaks away" from a second-order (or higher-order) zero on the real axis, since it cannot produce imaginary values when the polynomials are evaluated at purely real arguments. In such a case, the method will not converge. One means of resolving such a situation might be to have the main program perturb the approximate root by an imaginary quantity and

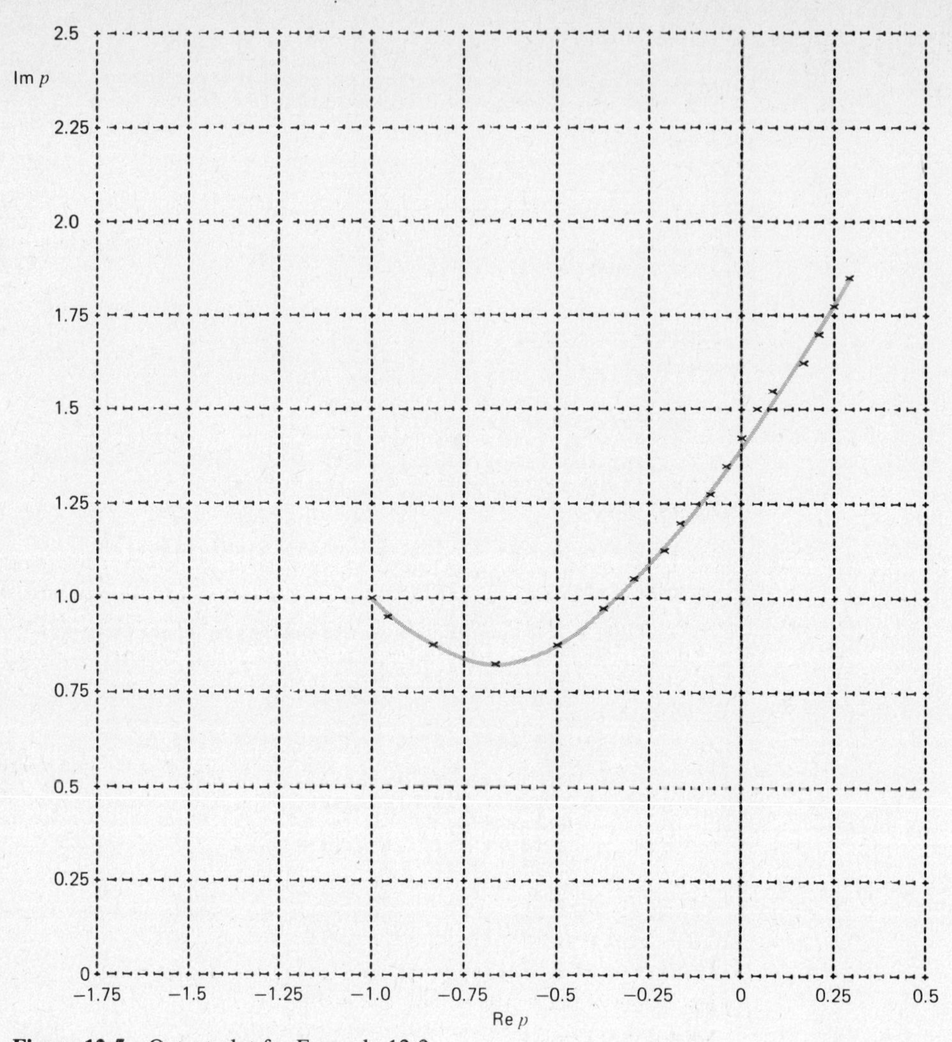

Figure 12.5 Output plot for Example 12.2.

then perform additional iterations of the basic relation of (12.2) to determine the complex root more accurately. Such an improvement is left to the reader as an exercise.

12.4 CONCLUSION

In this chapter we have introduced the use of another type of computational tool, a procedure for making an *xy* plot. In addition to the Nyquist plots and root-locus plots discussed here, the *xy* plot may also be readily adapted for use in plotting phase-plane characteristics for the analysis of nonlinear systems. Many other engineering applications for this procedure will suggest themselves to the reader.

PROBLEMS

12.1. The network shown in Fig. P12.1 has the transfer impedance

$$\frac{V(s)}{I(s)} = \frac{1}{s^3 + 2s^2 + 2s + 1}$$

Use the procedures SinStdySt and Xyplt to make a Nyquist plot of the network function (that is, the transfer impedance) for a range of frequency from 0 to 6 rad/sec. Scale the data by a factor of 50 before plotting.

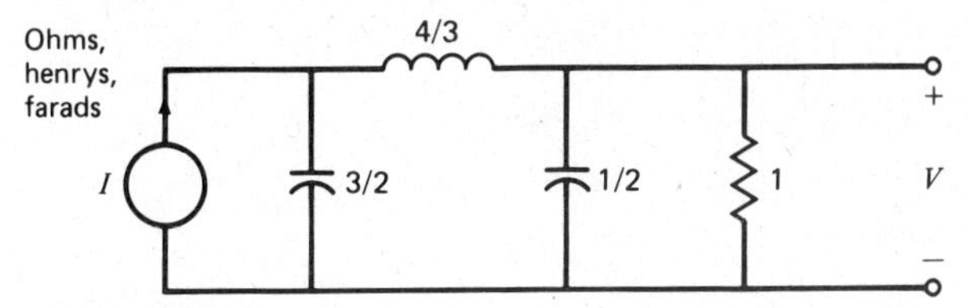

Figure P12.1

12.2. System functions that have zeros on the $j\omega$ axis will have a Nyquist contour that passes through the origin of the complex plane. As an example, consider the system function given below, which has a zero at $\sqrt{1/2}$ rad/sec. Use the procedures SinStdySt and Xyplt to make a Nyquist plot of the function for a range of frequency from 0 to 6 rad/sec.

$$N(s) = \frac{40(s^2 + 1/2)}{s^3 + 1.2s^2 + 1.2s + 1}$$

12.3. System functions that have a pole at the origin will have a root locus starting at an angle of $k(\pi/2)$ rad, where k is the order of the pole. As an illustration of this, use the procedures SinStdySt and Xyplt to make a Nyquist plot of the function

$$N(s) = \frac{(s + 1)(s + 2)}{s^2}$$

for a frequency range from 0.27 to 10 rad/sec.

12.4. An all-pass function (see Prob. 8.6) has a constant magnitude that is independent of frequency. The Nyquist plot for such a function will be a circle centered at the origin of the complex plane. As an example of this, use procedures SinStdySt and Xyplt to construct a Nyquist plot for the all-pass function

$$N(s) = \frac{40(s^3 - 2s^2 + 2s - 1)}{s^3 + 2s^2 + 2s + 1}$$

over a frequency range from 0 to 5 rad/sec.

12.5. Use the procedures SinStdySt and Xyplt to construct Nyquist plots for the following functions on a single graph. The functions have maximally flat magnitude characteristics of second, third, and fourth order.

(a) $N_a(s) = \dfrac{50}{s^2 + 1.4141s + 1}$

(b) $N_b(s) = \dfrac{50}{s + 2s^2 + 2s + 1}$

(c) $N_c(s) = \dfrac{50}{s^4 + 2.6131s^3 + 3.4142s^2 + 2.6131s + 1}$

12.6. Use the procedures SinStdySt and Xyplt to construct Nyquist plots for the following functions on a single graph. The functions are second-order and have low-pass, bandpass, and high-pass magnitude characteristics.

(a) $N_a(s) = \dfrac{40}{s^2 + s + 1}$ **(b)** $N_b(s) = \dfrac{40s}{s^2 + s + 1}$ **(c)** $N_c(s) = \dfrac{40s^2}{s^2 + s + 1}$

12.7. In the use of Nyquist plots for stability studies of control systems, a plot of the quantity $1 - N(j\omega)$, where $N(j\omega)$ is the system function, is made for $-\infty < \omega < \infty$. The plot is then evaluated to determine whether or not it encircles the xy point $(-1, 0)$. Use this criterion on the following system functions:

$$N_a(s) = \frac{1}{(s+1)(s^2+s+1)} = \frac{1}{s^3 + 2s^2 + 2s + 1}$$

$$N_b(s) = \frac{1}{(s-1)(s^2+s+1)} = \frac{1}{s^3 + 1}$$

Scale the quantity $1 - N(j\omega)$ by a factor of 20 to make the plots more readable [the xy point to test thus becomes $(-20, 0)$].

12.8. **(a)** An active RC circuit consisting of two resistors, two capacitors, and a voltage-controlled voltage source (a device with infinite input impedance, zero output impedance, and a voltage gain of K), as shown in Fig. P12.8 has the transfer voltage ratio

$$\frac{V_2(s)}{V_1(s)} = \frac{K}{s^2 + (3 - K)s + 1}$$

Use the techniques described in Sec. 12.3 together with the procedure Xyplt to make a plot of the root locus for the poles of the network function for a range of the gain K from 1.01 to 4.99. Scale the data by 40.

(b) What can you conclude about the stability of this circuit for values of gain greater than 3?

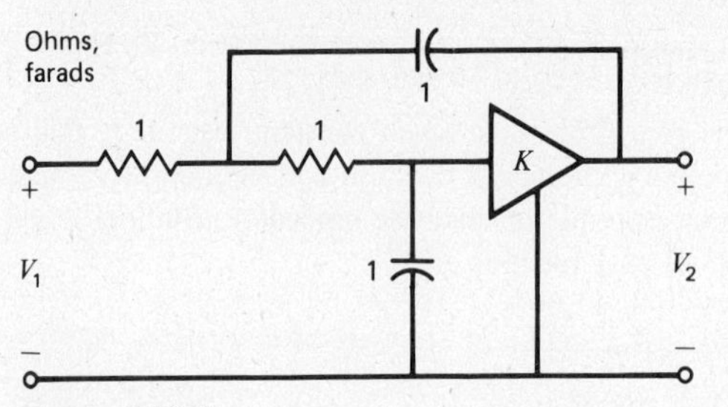

Figure P12.8

12.9. We may use the root-locus techniques introduced in Sec. 12.3 to illustrate the variation of the poles of a given polynomial with respect to the variation of any of its coefficients. As an example of this, plot the zeros of the polynomial

$$P(s) = s^3 + s^2 + ds + 1$$

for a range of the constant d from -2.00738 to 7. Plot only the complex zero with a positive imaginary part. Scale the data by a factor of 40. For the given initial value of d, the polynomial has zeros at -2.15 and $0.575 \pm j.366$.

Chapter 13

Some Large-Scale Programs

In the preceding chapters of this text we have (1) illustrated many specific digital-computational techniques, (2) developed procedures for implementing these techniques, and (3) given examples of how they could be used to solve basic circuit and system-analysis problems. The reader who has followed through the developments to this point, and who has worked a representative number of the problems, should have a fairly respectable working knowledge both of PASCAL programming and of the ways in which numerical techniques may be applied to an engineering discipline such as circuit theory. The method of approach has been to develop procedures and use one or more of these procedures as part of a main program to accomplish some desired computational goal. The procedures and programs that have been used up to this point have been relatively short ones. The method of approach, however, is readily applicable to much larger programs. In this chapter, therefore, we discuss the development of some larger-scale programs. Although we do not actually write the programs, we indicate the method of attack that could be used in developing them and illustrate the manner in which an individual or a group of individuals might undertake to break the overall development problem down into separate subproblems, each being capable of implementation by a subprogram. The results would then be combined to form a useful large-scale program.

13.1 A PROGRAM FOR THE SOLUTION OF NONLINEAR RESISTANCE NETWORKS

A problem that is frequently encountered in the design of electronic circuits is the determination of network variables such as loop currents and/or node voltages under conditions of dc excitation, when the circuit contains two- and three-terminal elements with nonlinear characteristics. An example of such a two-terminal element is a diode. It can be modeled (for static dc analysis) as a nonlinear two-terminal resistor. Similarly, a transistor can be modeled by a three-terminal resistive network including a controlled source with a nonlinear characteristic. The solution for the currents and/or voltages of a network containing such nonlinear elements in effect determines the operating points for the nonlinear devices. Such operating points can then be used to create small-signal equivalent models for use in representing the nonlinear elements under sinusoidal steady-state conditions. The problem in finding the operating points is complicated by the fact that a network that includes nonlinear two- and three-terminal elements may have more than one solution. In general, analytic techniques cannot be used except in the most trivial cases (such as those which include only a single nonlinear element). Therefore, a program for providing the solution or solutions for a network of this type can be a valuable analysis tool.

The method of approach that might be used to form a digital-computational solution to such a problem is similar to that used in solving Prob. 7.1, that is, to represent each of the nonlinear elements by a piecewise-linear characteristic consisting of as many segments as are desirable for each of the given nonlinearities. The computer can then be used to construct linear representations for every element for each section of this piecewise-linear characteristic. If the analysis is being made on a loop basis, such a representation will consist of a resistor and a voltage source. A set of such representations can then be used to define a linear-resistance-network problem such as that discussed in Chap. 7. For a given nonlinear element, if the values of the variables found in the solution are within the range of the values of the variables determining the portion of the nonlinear characteristic being used, then the solution is valid for the nonlinear element as well as for the linear equivalent circuit. For the case where there are several nonlinear elements, the computer simply needs to be programmed to perform the above operation for all combinations of possible linear equivalent circuits. The problem might be broken down into the following parts:

1. A subprogram to provide for the input of data such as the following:
   (a) The number of linear and nonlinear resistors
   (b) The number of loops (or nodes)
   (c) The data that characterizes each of the nonlinear elements
   (d) The loops in which the various linear and nonlinear resistors are located
   (e) The nature of the elements, that is, whether they are two-terminal elements or whether they are controlled sources
   (f) The magnitude, value, and location of the dc voltage sources

   This subprogram could also be used to provide a listing of the input data with appropriate output statements.

2. A subprogram to compute the various linear equivalent circuits for the different sections of each of the piecewise-linear equivalent characteristics. This subprogram could also provide an output listing of the values of the voltages and resistances associated with each of these equivalent representations.

3. A subprogram that selects the various combinations of linear equivalent representations and provides a means of generating the data for the corresponding linear equivalent resistance networks.

4. A subprogram that formulates the basic matrix equation from the data determined by the subprogram specified in 3 and that solves for the unknown variables for such a network. This subprogram would be similar to the procedure ResisMesh discussed in Chap. 7, except for the inclusion of the effects of controlled sources. The procedure GjSimEqn discussed in Chap. 7 could then be used to provide a solution for the unknowns of the matrix equations.
5. A subprogram that checks the solutions obtained in 4 with the limiting values of the variables that define each of the equivalent linear representations of the piecewise-linear segments. If all the values of the variables are within the correct limits, a solution exists, and the subprogram can record and print the pertinent data.

The main program would simply call the subprograms in the given order. After reaching the subprogram described in 5, it would return to the subprogram described in 3, repeating this cycle until all the possible combinations of linear and equivalent representations for the nonlinear elements had been evaluated. A flowchart of the program is shown in Fig. 13.1.

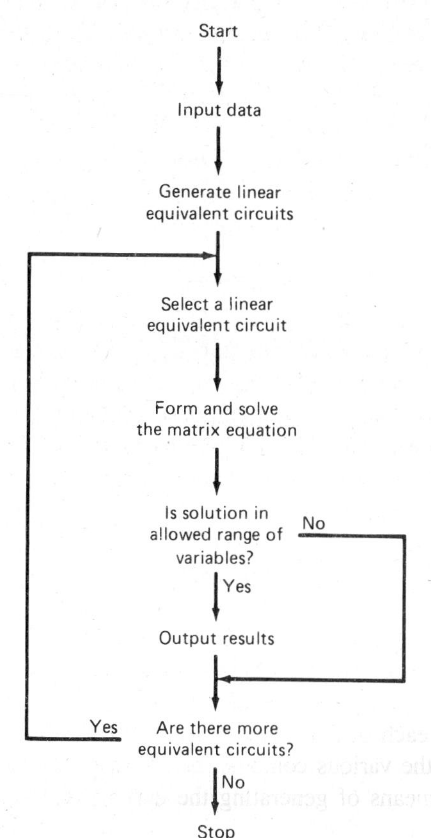

Figure 13.1 Flowchart for solution of nonlinear resistance networks.

13.2 A PROGRAM FOR NETWORK FUNCTION ANALYSIS

Many problems in network (and system) theory are concerned with the time- and frequency-domain analysis of the function $N(s)$ representing the input-output characteristics of a given network. Desirable time-domain information includes step response and impulse response as well as the response to an arbitrary input. Important frequency-domain information includes pole-zero locations, and Bode (log-log) or linear plots of sinusoidal steady-state magnitude, phase, and delay. A very useful capability which such a program can provide is a multicase one that allows the user to create families of loci displaying the way in which the properties of the network function change as some of its parameters are varied. Such a capability provides a convenient method for displaying sensitivity information. A program that implements these various types of analysis can be a most useful design tool.

A basic program for providing the time- and frequency-domain analysis capabilities described above can be created by implementing the techniques outlined in Chaps. 8 and 9. Such a program could be broken down into the following subprograms:

1. A subprogram to provide the input of data such as the following:
   (a) The order of the numerator and denominator polynomials
   (b) The values of the coefficients of the numerator and denominator polynomials
   (c) The type of output desired (time-domain, frequency-domain, etc.) and any parameters for controlling the output (time range, frequency range, scaling factors for plots, etc.)
   (d) Any information needed to specify a multicase situation

2. A subprogram to apply the techniques developed in Chap. 8 to perform a time-domain analysis and store the results for plotting. This subprogram could include a provision for accommodating various types of input, for example, a step function, by appropriately modifying the coefficients of the network function.

3. A subprogram to apply the techniques of Chap. 9 to a frequency-domain analysis and store the results for plotting. This subprogram could include a provision for computing the delay response $D(\omega)$ using the relation

$$D(\omega) = -\frac{d}{d\omega} \operatorname{Arg} N(j\omega) \tag{13.1}$$

The function of the main program would be simply to call the subprograms given in 1, 2, and 3 above as required. In addition, it would provide recalls of the subprograms given in 2 and 3 as necessary to provide a multicase capability. Finally, the main program would call a plotting subprogram, such as the procedure Plot5 to display the results. An option could be provided to also display a Nyquist plot. A flowchart for the program is shown in Fig. 13.2.

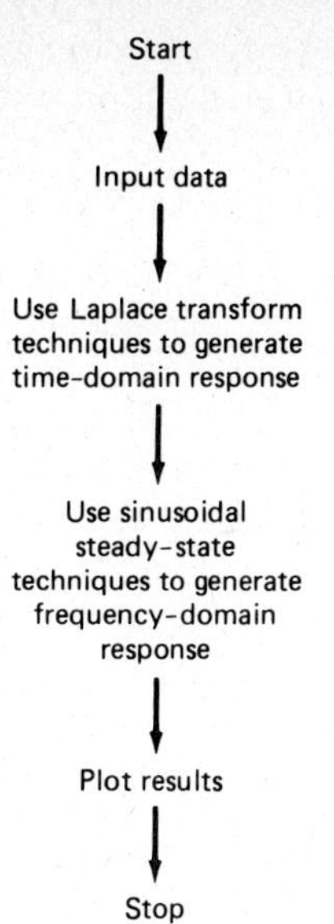

Figure 13.2 Flowchart for network function analysis.

13.3 A PROGRAM FOR NETWORK SINUSOIDAL STEADY-STATE ANALYSIS

One of the most frequently recurring problems of circuit theory is the analysis of networks under conditions of sinusoidal steady-state excitation. In this section we discuss the development of a program for the determination of the magnitude and phase of the voltage transfer function of a linear lumped *RLC* network that includes controlled sources as active elements. We assume node 1 is the input node and node 2 is the output one. The first step in obtaining a solution to this problem could consist of a determination of the (complex) admittance matrix $\mathbf{Y}(j\omega)$ representing the network at a given frequency. If the network has $n + 1$ nodes, $\mathbf{Y}(j\omega)$ will be an $n \times n$ matrix. The admittance matrix equations $\mathcal{I} = Y(j\omega)\ \mathcal{V}$ will have the form

$$\begin{bmatrix} \mathcal{I}_1 \\ \cdot \\ \cdot \\ \mathcal{I}_i \\ \mathcal{I}_j \\ \cdot \\ \cdot \\ \mathcal{I}_n \end{bmatrix} = \begin{bmatrix} y_{11} & \cdots & y_{1j} & \cdots & y_{1n} \\ \cdot & & \cdot & & \cdot \\ \cdot & & \cdot & & \cdot \\ y_{i1} & \cdots & y_{ij} & \cdots & y_{in} \\ y_{j1} & \cdots & y_{jj} & \cdots & y_{jn} \\ \cdot & & \cdot & & \cdot \\ \cdot & & \cdot & & \cdot \\ y_{n1} & \cdots & y_{nj} & \cdots & y_{nn} \end{bmatrix} \begin{bmatrix} \mathcal{V}_1 \\ \cdot \\ \cdot \\ \mathcal{V}_i \\ \mathcal{V}_j \\ \cdot \\ \cdot \\ \mathcal{V}_n \end{bmatrix} \tag{13.2}$$

where we have indicated two arbitrary rows and columns by the subscripts i and j. For convenience these are shown adjacent to each other. Now let us consider how a VCVS (voltage-controlled voltage source) constrains such a matrix. We assume that the VCVS has gain K, input at node j, and output at node i. Thus the constraint equation is $\mathcal{V}_i =$

$K \mathcal{V}_j$ and the variable $\mathcal{V}_j$ can be eliminated from the equations by substituting $\mathcal{V}_i/K$. This effectively reduces the equations to an $n \times (n-1)$ set with the form

$$\begin{bmatrix} \mathcal{I}_1 \\ \cdot \\ \cdot \\ \mathcal{I}_i \\ \mathcal{I}_j \\ \cdot \\ \cdot \\ \mathcal{I}_n \end{bmatrix} = \begin{bmatrix} y_{11} & \cdots & y_{1i} + y_{1j}/K & \cdots & y_{1n} \\ \cdot & & \cdot & & \cdot \\ \cdot & & \cdot & & \cdot \\ y_{i1} & \cdots & y_{ii} + y_{ij}/K & \cdots & y_{in} \\ y_{j1} & \cdots & y_{ji} + y_{jj}/K & \cdots & y_{jn} \\ \cdot & & \cdot & & \cdot \\ \cdot & & \cdot & & \cdot \\ y_{n1} & \cdots & y_{ni} + y_{nj}/K & \cdots & y_{nn} \end{bmatrix} \begin{bmatrix} \mathcal{V}_1 \\ \cdot \\ \cdot \\ \mathcal{V}_i \\ \cdot \\ \cdot \\ \mathcal{V}_n \end{bmatrix} \tag{13.3}$$

Since the output impedance of the VCVS is zero, node i is effectively short-circuited to ground (the reference node). As a result, the equation for $\mathcal{I}_i$ is not an independent one, and it may be eliminated. The result is a set of equations with an $(n-1) \times (n-1)$ admittance matrix having the form

$$\begin{bmatrix} \mathcal{I}_1 \\ \cdot \\ \cdot \\ \mathcal{I}_j \\ \cdot \\ \cdot \\ \mathcal{I}_n \end{bmatrix} = \begin{bmatrix} y_{11} & \cdots & y_{1i} + y_{1j}/K & \cdots & y_{1n} \\ \cdot & & \cdot & & \cdot \\ \cdot & & \cdot & & \cdot \\ y_{j1} & \cdots & y_{ji} + y_{jj}/K & \cdots & y_{jn} \\ \cdot & & \cdot & & \cdot \\ \cdot & & \cdot & & \cdot \\ y_{n1} & \cdots & y_{ni} + y_{nj}/K & \cdots & y_{nn} \end{bmatrix} \begin{bmatrix} \mathcal{V}_1 \\ \cdot \\ \cdot \\ \mathcal{V}_i \\ \cdot \\ \cdot \\ \mathcal{V}_n \end{bmatrix} \tag{13.4}$$

The procedure outlined above can, of course, be repeated for each VCVS in the network.

The final step in obtaining a solution to the problem is to solve the set of simultaneous equations that remain at the end of the constraint procedure described above. For convenience we redefine n as the number of remaining equations and assume that the constrained admittance matrix $\mathbf{Y}(j\omega)$ is an $n \times n$ one. In addition we partition it as follows:

$$\mathbf{Y} = \begin{bmatrix} \mathbf{Y}_{11} & \mathbf{Y}_{12} \\ \mathbf{Y}_{21} & y_{nn} \end{bmatrix} \tag{13.5}$$

where $\mathbf{Y}_{11}$ is an $(n-1) \times (n-1)$ square submatrix, $\mathbf{Y}_{12}$ is a column matrix with $n-1$ elements, $\mathbf{Y}_{21}$ is a row matrix with $n-1$ elements, and y_{nn} is the indicated element of the matrix. We may now reduce the $n \times n$ matrix $\mathbf{Y}$ to an $(n-1) \times (n-1)$ matrix $\mathbf{Y}^{(1)}$ by the relation

$$\mathbf{Y}^{(1)} = \mathbf{Y}_{11} - \mathbf{Y}_{12}\mathbf{Y}_{21}\frac{1}{y_{nn}} \tag{13.6}$$

If the matrix $\mathbf{Y}^{(1)}$ is now also partitioned in a manner similar to that shown in (13.5), a relation similar to that shown in (13.6) may be applied to produce an $(n-2)$

$\times$ (n − 2) matrix $\mathbf{Y}^{(2)}$. Continuing in this fashion, we may derive a 2 × 2 matrix $\mathbf{Y}^{(n-2)}$ that has the form

$$\mathbf{Y}^{(n-2)} = \begin{bmatrix} y'_{11} & y'_{12} \\ y'_{21} & y'_{22} \end{bmatrix} \tag{13.7}$$

where the elements y'_{ij} are, of course, complex. The voltage transfer function between node 1 and node 2 is given by the relation

$$\frac{\mathcal{V}_2}{\mathcal{V}_1} = \frac{-y'_{21}}{y'_{22}} \tag{13.8}$$

This process may be repeated at other frequencies to determine the data for plotting a magnitude or phase curve for any specified frequency range.

The basic problem and method outlined above might well be broken into the following subprograms:

1. A subprogram to provide for the input of data such as the following:
  (a) The number of elements and the number of loops (or nodes) in the network
  (b) The location and value of the elements in the network
  (c) The specifications for any controlled sources
  (d) The frequency range over which the magnitude and phase are desired and the number of points in this frequency range (and an indication of whether logarithmically or linearly spaced frequency values are desired)

  This subprogram should also provide a listing of the input data with appropriate output statements identifying the data.

2. A subprogram to form the **G**, **Γ**, and **C** matrices, where the **G** matrix takes account of the values of conductance (reciprocal resistance) of the resistors connected to the various nodes, the **Γ** matrix takes account of the reciprocal values of the inductors connected to the various nodes, and the **C** matrix takes account of the values of the capacitors connected to the various nodes. Thus, these three matrices effectively constitute a description of the network elements and the nodes to which they are connected.

3. A subprogram to provide a sequence of frequency values for which it is desired to determine the magnitude and phase of the network. The function LogIncrem discussed in Chap. 9 could be used as part of this program to provide logarithmically spaced frequency values.

4. A subprogram to form the real part and the imaginary part of the $n \times n$ admittance matrix **Y** from the **G**, **Γ**, and **C** matrices found in 2 and the frequency found in 3 using the relation

$$\mathbf{Y} = \mathbf{G} + \frac{1}{j\omega}\mathbf{\Gamma} + j\omega\mathbf{C} \tag{13.9}$$

5. A subprogram to iteratively apply the relation of (13.6) to reduce the $n \times n$ matrix **Y** to a 2 × 2 matrix and to compute the resulting magnitude and phase of the transfer function by using (13.8). This subprogram could use the procedures Dvd, Mag, and Arg presented in Sec. 8.2.

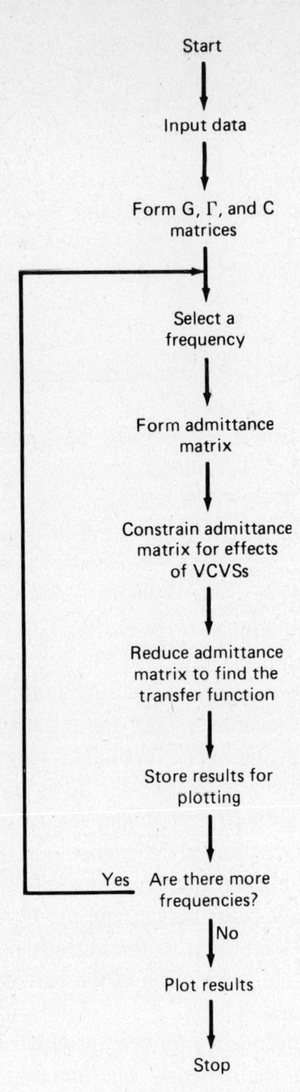

Figure 13.3 Flowchart for network sinusoidal steady-state analysis.

The function of the main program would be simply to call the various subprograms in the order given and to repeat the sequence of subprograms given in 3, 4, and 5 as often as needed to cover the specified frequency range. Its final role would be to call a plotting subprogram such as the procedure Plot5. A flowchart for the program is shown in Fig. 13.3.

13.4 CONCLUSION

In the learning of a new discipline, it is an old maxim that doing engenders learning. This is perhaps even more true of digital-computational processes. The more use that is made of such processes, the more facile users become with the techniques, and the more likely they are to find additional applications for them. Many problems to which digital-computational processes may be applied have necessarily been omitted from these chapters. For example, the use of optimization processes in the synthesis of networks and systems has not been discussed. It is hoped, however, that this text will serve as an introduction to illustrate how much can be accomplished by even the simplest of techniques, and that readers will be motivated to continue their explorations into the wonderful world of digital-computational techniques. It is a journey they will find richly rewarding.

Appendix A

The PASCAL Language

A.1 INTRODUCTION

This appendix presents an introduction to PASCAL. It is written to provide both an easily followed introduction and a readily usable reference to the language. To accommodate its role as an introduction, the appendix starts off with a very simple program consisting of a single write statement. This is used to illustrate the basic block structure and the forms of the initial and final statements used in PASCAL. Each succeeding section covers additional features of the language, emphasizing their relation and application to the material that has gone before. Thus, the reader is led through declarations, data types, structured statements, arrays, functions, procedures, block structures, user-defined types, records, and sets. Many simple example programs are included to illustrate the correct usage of the various features of the language. To accommodate its role as a reference, this material has been organized to facilitate the easy and rapid location of a given topic. For example, each section is titled with a key word or phrase such as ''Structured Statements'' or ''Records.'' In the sections, subheadings and a standard format are used to identify each new type of statement or declaration and to define its component parts.

To keep the appendix as brief as possible, only minimal references to programming style have been included. A treatment of this topic could well fill an entire book, and indeed, many such texts (such as Kernighan and Plauger, *The Elements of Programming Style*) are available. Instead, the emphasis is to allow the reader to develop a knowledge of the syntax of the PASCAL language as rapidly and efficiently as possible.

A.2 A SIMPLE PASCAL PROGRAM

To introduce PASCAL, consider the simple program shown as Program A2.1.

Program A2.1

```
PROGRAM PrintOneLine (output);
{This is a simple PASCAL program}
BEGIN
    write ('We have now written a PASCAL program.')
END.
```

The first line is called the *program heading*. The second line contains a *comment*. The third line specifies the action to be performed by the program. The last line terminates the program. Now let us consider each of the lines in Program A2.1 in more detail.

Program Heading

The first line of every PASCAL program is the *program heading*. Its component parts are named as follows:

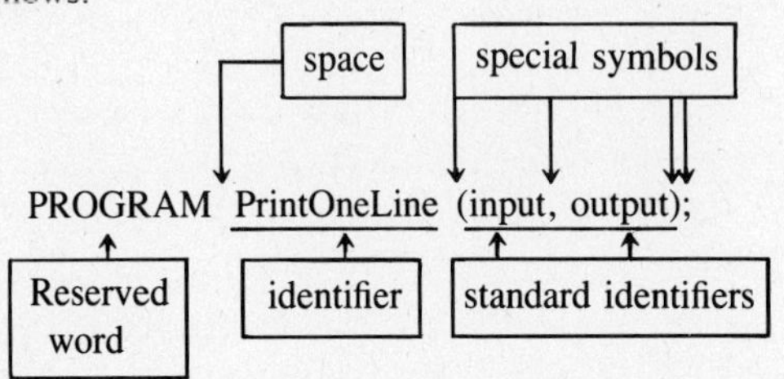

PROGRAM The word PROGRAM is one of a set of words that have a special meaning to the PASCAL compiler. They are called *reserved words*. A list of them is given in Table A2.1. In order to emphasize their character, we show these words in capitals. For typing convenience when actually entering a program into the computer, they are frequently entered as lowercase (this avoids having to hold down the shift key). In general, the compiler will accept them in either lower- or uppercase.

Space The space is one of a set of *special symbols* that are recognized by the PASCAL Compiler. These symbols may consist of a single character or a pair of characters. They are shown in Table A2.2. The special symbol *space* is required between reserved words and identifiers (see below). Even when they are not required, the use of spaces before and after many of the other special symbols greatly improves the readability of the program.

Identifier An identifier is a name consisting of alphanumeric characters (letters and numbers). The first character *must* be a letter. Special symbols (including the space) are *not* permitted.[1] Identifiers are used for naming, that is, identifying, programs, special procedures, functions, files, constants, variables, etc. In selecting an identifier, the programmer can enhance the quality and readability of the program by making it as descriptive as possible. To this end, PASCAL permits identifiers of almost unlimited length. Only the first eight characters, however, are actually rec-

TABLE A2.1 PASCAL RESERVED WORDS[a]

AND	DOWNTO	IF	OR	THEN
ARRAY	ELSE	IN	PACKED	TO
BEGIN	END	LABEL	PROCEDURE	TYPE
CASE	FILE	MOD	PROGRAM	UNTIL
CONST	FOR	NIL	RECORD	VAR
DIV	FUNCTION	NOT	REPEAT	WHILE
DO	GOTO	OF	SET	WITH

[a]Other reserved words that may not be available at every installation are FORWARD, EXTERN, FORTRAN.

[1]The underscore "_" is permitted in some implementations.

TABLE A2.2 PASCAL SPECIAL SYMBOLS

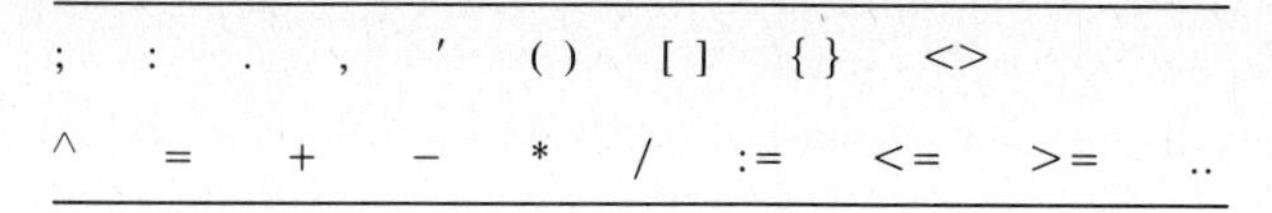

;	:	.	,	'	( )	[ ]	{ }	<>	
^	=	+	−	*	/	:=	<=	>=	..

ognized by the compiler. Thus, the name of Program A2.1 is treated as PrintOne. In identifiers, upper- and lowercase characters are treated the same. In PASCAL, in addition to the names used by the programmer, there are standard identifiers that are used for various purposes. A list of these is given in Table A2.3. Identifiers will be shown here in lowercase; however, identifiers used for program and subprogram names will begin with a capital letter. In addition, in any identifier that consists of multiple words, to provide easy identification, each word after the first will have its first letter capitalized. Some examples of correct and incorrect identifiers follow.

Correct identifiers	Incorrect identifiers
A123456	2abc (starts with a number)
CalcSimProgram	Program (reserved word)
stopRun	try again (contains space)
index	a*b (contains special symbol)

Files Accessed by the Program In the program heading, the identifier giving the name of the program is followed by a list of the files accessed by the program. These are enclosed in parentheses and separated by commas. The word *file* is a standard computer term used to indicate how data transfers into and out of the program are made. Thus, if a read operation is included in the program, it must access an *input* file. Similarly, a write operation must go to some type of *output* file. The exact terminology, meaning, and function of these files depend on the particular system on which the program is being run. In a batch system, for example, the input might be from a card reader and the output to a line printer. In a personal microcomputer, the input might be from a keyboard and the output to a cathode-ray-tube (CRT) display screen. The general subject of file handling is considerably beyond the scope of our introductory treatment of PASCAL. Thus, we simply use the "standard" file names shown in Table A2.3.

Semicolon The special symbol semicolon ";" is used as a separator between statements such as the program heading and the "write" statement in Program A2.1. It is also useful in that it permits

TABLE A2.3 PASCAL STANDARD IDENTIFIERS

Files	Constants	Types	Procedures		Functions		
input	false	integer	get	readln	abs	exp	sin
output	true	real	new	reset	arctan	ln	sqr
	maxint	boolean	pack	rewrite	chr	odd	sqrt
		char	page	unpack	cos	ord	succ
			put	write	eof	pred	trunc
			read	writeln	eoln	round	

the programmer to write several statements on a single line, thus keeping the text of the program in a more compact form.[2]

Newline At the end of the program heading, if the program is being entered at a terminal, a CR (carriage return) would be sent. This is considered as a special symbol and is usually referred to as a *newline*. In general, in this text we shall *not* refer to the CR specifically, assuming that it is always used at the end of each line of program text.

Comments

The second line of Program A2.1 contains a comment. Comments are enclosed in braces ({ }) and may contain any collections of characters. In some computers, the characters "(*" are used to initiate a comment and the characters "*)" to terminate it. Comments may occur anywhere in the program and may be continued on as many lines as is desired. The liberal use of comments is a necessity in the writing of programs to ensure that they are easily understood and readily modified.

Program Statements

The collection of lines of program text starting with the reserved word BEGIN and ending with END forms the statement portion of the PASCAL program. Good programming practice which enhances readability is to put BEGIN and END at the left margin on separate lines and to indent all the statements (in this case there is only one) in between. In Program A2.2, Program A2.1 is rewritten to illustrate this, and also to show some additional uses of comments. Note that no semicolon is used after BEGIN as before END.

Program A2.2

```
PROGRAM PrintOneLine (output);
{This is a simple PASCAL program.
    Its output is a single line, and
    there is no input.}
BEGIN {The statements portion of the program starts
        here.}
  write('We have now written a PASCAL program.')
END.
```

Write Statement

The write statement shown in Programs A2.1 and A2.2 is called a *procedure statement*. It instructs the program to send information to the output file. The desired output is placed inside a pair of parentheses. As shown in the program, text strings must be enclosed in single quotes. A single quote may be included in the string by writing it twice, as in 'Don''t stop yet'. In the "Procedures" section of the standard identifiers list of Table A2.3, two write operations are shown, namely, write and writeln. The writeln operation shifts the output to a new line *after* its output operation. The write one stays on the same line.

[2] A semicolon is *not required* at the end of the last statement preceding the reserved word END. If one is used, however, the compiler will allow it.

Period

The single period after END signifies the end of the program to the compiler.

A.3 DECLARATIONS

In the preceding section we showed that a PASCAL program started with a program heading and finished with a period. Everything in between is referred to collectively as the *block* of the program. In the simple program (Program A2.1) used for an example, the block consisted of a single procedure statement (the write statement) preceded by the reserved word BEGIN and followed by END. In this section we introduce another component of the block. It is called a *declaration*. In all but trivially simple programs, one or more declarations are required. They must precede the statement portion of the block; thus they appear directly after the program heading. The major use of declarations is to define any variables used in the block. PASCAL requires that all variables be so defined. As an example of the use of a declaration, consider Program A3.1. The lines in the program that illustrate new ideas are described below.

Program A3.1

```
PROGRAM AddTwoNumbers (input, output);
{This program inputs two integer numbers
and outputs their sum}
VAR {Declarations portion of the block}
   number1, number2, sum : integer;
BEGIN {Statements portion of the block}
   read(number1);  {These two statements can have}
   read(number2);  {several different forms}
   sum := number1 + number2;
   writeln('The sum is', sum)
END.
```

Variable Declaration

The variable declaration portion of the block is initiated by the reserved word VAR. It lists the variables (and their type) that are be used in the program. The variables are separated by commas and the type is separated from the variables by a colon. Only one use of the VAR declaration is permitted in a block, but as many entries and/or lines as needed may follow it. As an example, other equivalent forms of the variable declaration of Program A3.1 are

```
VAR number1:integer; number2:integer; sum:integer;
VAR
   number1, number2 : integer;
   sum : integer;
```

Note that a semicolon is used to separate statements and also to mark the end of the declaration. Other types of variables are listed in the "Types" section of the standard identifiers given in Table A2.3. These will be discussed in more detail shortly.

Read Statement

The read statement shown in Program A3.1 enables the program to accept input data. Note that an *input file* is specified in the program heading. The input data establishes the value of the program variable enclosed in parentheses. There are two ways of performing the read operation. These are shown in the ''Procedures'' section of the standard identifiers given in Table A2.3. The readln operation shifts to a new input line *after* reading the data. The read operation stays on the same input line. Since the read operation is a free-format one, if there is not enough data to satisfy it on a given line, it will automatically proceed to the next line. This is also true for the readln operation. A single read statement can also be used to read several variables by listing them, separated by commas, inside the parentheses. Thus, the two read statements shown in Program A3.1 can be combined into a single one having the form read(number1, number2); or readln(number1, number2);. The different possibilities and the format of the data required for each are summarized below.

Statements	Data

read(number1);
read(number2);
OR
read(number1, number2);
OR
readln(number1, number2);
	17 24 (both on same line)
	OR
	17 (one number per line)
	24

readln(number1); readln(number2);	17 (one number per line) 24

Note that one or more spaces must be used to separate the numbers if they are on the same line.

Assignment Statement

The statement in which the addition of number1 and number2 is performed in Program A3.1 is called an *assignment statement*. The expression on the right of the special symbol '':='' is called an *arithmetic expression*. Its value is computed and assigned to the variable on the left of the symbol.

Write Statement

In the write statement used in the program there are two outputs, namely, the string of text (enclosed by single quotes) and the numerical value of sum. Alternately, the same output could be produced (less efficiently) by the two statements

```
write('The sum is');
writeln(sum);
```

Write statements may not only contain text strings and variables, but may also contain arithmetic expressions. Thus, Program A3.1 can also be written in the form shown in Program A3.2 (for brevity the comments have been deleted).

Program A3.2

```
PROGRAM AddTwoNumbers (input, output);
VAR number1, number2 : integer;
BEGIN read(number1); read(number2);
   writeln('The sum is', number1 + number2)
END.
```

In general, if a program such as Program A3.1 is to be run on an interactive computer, it is desirable to add prompting statements to tell users when the program is ready for them to enter data. This could be done by the additional statements shown in Program A3.3. In general, to avoid making our example programs too lengthy or hard to follow, we will not include such statements. They are, however, a most important part of any program written for interactive usage.

Program A3.3

```
PROGRAM AddTwoNumbers (input, output);
VAR number1, number2, sum : integer;
BEGIN writeln('Enter first number');
   readln(number1); writeln('Enter second number');
   readln(number2); sum := number1 + number2;
   writeln('The sum is', sum)
END.
```

Constant Declaration

Another type of declaration is frequently useful in the block of a PASCAL program. This is the *constant declaration,* which is used to define *symbolic constants,* that is, to assign a fixed numerical value or an alphanumeric string to an identifier. Such a usage helps to make the program more readable, since a name has far more meaning than a number. The use of declared constants also makes the program easier to modify, since the single action of changing the value of a constant in the declarations obviously changes it in every usage throughout an entire program. The constant declaration, if one is used, precedes the variable declaration. An example of a constant declaration is

```
CONST first = 1; last = 10;
```

where CONST is a reserved word. Note that, in PASCAL usage, the term *constant* refers to an identifier to which a value (such as a number or alphanumeric character) has been assigned. *Numerical quantities* such as 1, 27, and −3 will be referred to as *numbers* rather than as constants. Note that the value of a constant cannot be changed by the program any more than the value of a number can.

A.4 DATA TYPES

In the last section we introduced variables and showed how arithmetic expressions involving them could be written. As shown in the "Types" section of Table A2.3, there are four types of variables—integer, real, boolean, and char. We now consider those types in detail.

Integer

An *integer variable* is one whose value is restricted to being that of an integer number, that is, one with no decimal point. Such a variable is defined by a variable declaration having the form identifier : integer (see Programs A3.1 and A3.2) where the identifier can be a single variable or a set of them separated by commas. An *integer constant* is defined by a constant declaration having the form identifier = integer number. Here, the use of an integer number (rather than one with a decimal point) after the "=" symbol is sufficient to characterize the identifier as being integer in type. The range of values that integer quantities may have depends mainly on the word length of the computer on which the program is being run. In general, for a word length of n (bits), the range is $= \pm 2^{n-1} - 1$. This (positive) value is available as the standard identifier maxint. If integer variables, constants, and numbers are combined in an arithmetic expression, the result is integer for the arithmetic operations of addition (+), subtraction (−), and multiplication (*). If two integer quantities are divided using the operator (/), however, the result will be real; that is, it will contain a decimal portion (which may, of course, be zero). Thus, an arithmetic expression which contains this operation can only be assigned to a real variable (see below). If, however, two integer quantities are divided using the operator DIV, a truncated integer quotient is produced. The remainder can be determined by using the operator MOD. Thus,

$$7/4 = 1.75 \qquad 7\,\text{DIV}\,4 = 1 \qquad 7\,\text{MOD}\,4 = 3 \tag{1}$$

Note that DIV and MOD are reserved words (see Table A2.1). In an arithmetic expression that contains more than a single operator, just as in other high-level languages, the order of operations is first from left to right for multiplication and division, then from left to right for addition and subtraction. Parentheses may be used to clarify the order of operation. If parentheses are nested, the expressions contained in the innermost pair are evaluated first. For example, the following steps illustrate the evaluation of an expression:

$$\begin{aligned} &3 + 8\,\text{DIV}\,2 * (4 + 6\,\text{DIV}\,(4\,\text{DIV}\,2))\\ &3 + 8\,\text{DIV}\,2 * (4 + 6\,\text{DIV}\,2)\\ &3 + 8\,\text{DIV}\,2 * (4 + 3)\\ &3 + 8\,\text{DIV}\,2 * 7\\ &3 + 4 * 7\\ &3 + 28\\ &31 \end{aligned} \tag{2}$$

There are two standard mathematical functions (see Table A4.1) that operate on integer quantities. These are abs and sqr. The assignment statement a := abs (b), where a and b are type integer, will assign to a the absolute value of b. Similarly, the statement a := sqr(b) will assign to a the square of the value of b. Note that, as is true for any defined function, the argument b can be an integer number, constant, variable, or arithmetic expression. When a write or writeln statement is used to output an integer quantity, its value is normally right-justified in a fixed-width (implementation-dependent) field. Frequently this is 10 columns wide. It may be controlled by following the integer quantity with : w, where w is the desired number of columns. Thus, if b is integer and has a value of 37

```
                          eight spaces
write ('B =',b); B =         37

write ('B =',b:3); B = 37                (3)
                      ↑
                   one space
```

Real

A *real variable* is one whose value is treated by the computer as a *floating-point (decimal) number*. Such a number may be written using *fixed-point notation* (32.7, 0.0037, 94.0) or *scientific notation* (43.1E1, −29.0E-1, 0.423E-8) in which the number following "E" gives the power to which the multiplier 10 is raised. A real variable is defined by a variable declaration having the form identifier : real; where either a single identifier or a set of them separated by commas may be used. A *real constant* is defined by a constant declaration having the form identifier = floating-point number. The range of value that floating-point numbers can take on varies with different computers. Typically, such numbers can be + or − and may have an absolute value as small as 10^{-38} and as large as 10^{38}. It should be noted that if a decimal point is used, PASCAL requires that the number begin and end with a digit—0, 1, 2, etc. Thus, 0.745 and 91.0 are acceptable floating-point numbers, while .745 and 91. are not. If real variables, constants, and numbers are combined in an arithmetic expression, the result is real for the arithmetic operations of addition (+), subtraction (−), multiplication (*), and division (/). If there is a mixture of real and integer quantities in an arithmetic expression, the result will be real. An arithmetic expression containing only integer quantities can be converted to real if the variable to which it is assigned (the one on the left of the ":=" symbol in the assignment statement) is real. The converse, however, is not permitted; that is, *an expression containing only reals cannot be assigned to an integer variable*. Instead, one of the standard identifiers trunc and round (see the "Functions" section of Table A2.3) must be used. As an example, if a has been declared an integer, then a := trunc(3.2 + 1.7) yields 4 for a, while a := round(3.2 + 1.7) yields 5. The order of operations and the use of parentheses are the same for real quantities as for integer ones.

Standard functions for real quantities (see Table A2.3) are given in Table A4.1. The arguments of these functions can be any real number, constant, variable, or arithmetic expression.

When a write or writeln statement is used to output a real *variable,* its value normally appears in scientific notation and fills some implementation-dependent field width such as 20 columns. To change this width, the identifier should be followed with :w, where w is the desired width in columns. To output the real variable in fixed-point notation, follow its identifier by :w:d, where d is the number of digits to the right of the decimal point. For example, if the value of b is 123.7245 and the default width is 16 columns,

```
write ('B =', b);      → B = 1.237245000E+02
write ('B =', b:12);   → B = 1.23724E+02          (4)
write ('B =', b:7:2);  → B = 123.72
```

Note that in the above formats, one column is reserved for the possible use of a "−" sign.

TABLE A4.1 STANDARD FUNCTIONS FOR REAL QUANTITIES

Function	Purpose
abs(r)	absolute value of r
arctan(r)	arctangent (in radians) of r
cos(r)	cosine of r where r is in radians
exp(r)	natural logarithm base raised to the power r (e^r)
ln(r)	natural logarithm of r
sin(r)	sine of r where r is in radians
sqr(r)	square of r
sqrt(r)	square root (positive) of r

TABLE A4.2 LOGICAL OPERATIONS

Inputs		Outputs		
x	y	NOT x	(x)AND(y)	(x)OR(y)
false	false	true	false	false
false	true	true	false	true
true	false	false	false	true
true	true	false	true	true

Boolean

A *boolean variable* is one whose value is restricted to being *true* or *false*. Such variables are defined by a variable declaration having the form identifier : boolean; where either a single identifier or a set of them separated by commas may be used. A *boolean constant* is defined by a constant declaration having the form identifier = value; where value is either true or false. There are two types of logical operations that can be performed on boolean quantities, the *unary* (single operand) *operation* NOT and the *binary* (two operands) *operations* AND and OR. These are defined by the truth table shown in Table A4.2. The most common use of boolean quantities is in expressions containing *relational operators*. A list of these is given in Table A4.3. When these operators are used with integer, real, or char (see below) variables, the resulting expressions are called *boolean expressions* since they have a boolean value. Usually this value is directly acted upon by a program statement rather than being assigned to a boolean variable. As an example of such a usage, consider the statement

```
IF j < 1   THEN a := 17.3                                        (5)
```

where j has been declared as a type integer variable and a as a type real one. Note that the boolean expression j < 1 will have a value of true if j is less than 1 and a value of false if j is greater than or equal to 1. In general, the variables, constants, and numbers that appear in an expression using a relational operator must be of the same type. Simple boolean expressions can be combined using the logical operations shown in Table A4.2 to generate more complex ones. In this case parentheses are used to enclose the arguments of the logical operations. As an example, consider

```
IF (j < 1) AND (b < 25.2)   THEN a := 17.3                       (6)
```

where j is integer and a and b are real. Note that the various components of such a complex boolean expression can be of different types. There are two standard functions that have a boolean output. These are eoln, which has a value of true when the end of a line is reached (otherwise it is false), and eof, which is true only when the end of a file, such as an input file, is reached. As such, the latter function provides a useful test to determine when the end of input data is reached.

TABLE A4.3 PASCAL RELATIONAL OPERATORS

Symbol	Meaning
=	Equal to
<	Less than
>	Greater than
<=	Less than or equal to
>=	Greater than or equal to
<>	Not equal to

The statements used in the two preceding examples illustrate a type of statement called a *structured statement*. These will be covered in more detail in the following chapter. The procedure statements write and writeln can be used directly on boolean type variables. For example, if aa is defined as boolean, the following program statements will print the word "true" in the output:

```
aa := true                                                  (7)
write(aa)
```

The procedure statements read and readln, on the other hand, cannot be used to directly input data establishing the value of a boolean variable. For example, for aa defined as boolean, we cannot use read(aa). Instead, we might declare k as a variable of type char (this type is defined below) which can be used to input a single letter, and use the following:

```
IF k = 'T'  THEN aa := true;
IF k = 'F'  THEN aa := false;                               (8)
```

Other similar methods are readily established.

Char

A *char* (character) *variable* is one which can represent any of the set of characters used in communicating with the computer. Almost every computer manufacturer[3] uses a standard 7-bit encoding of these characters specified by the ASCII (American Standard Code for Information Interchange) code. A partial listing of the code is given in Table A4.4. A char variable is defined by a variable declaration having the form identifier : char; where either a single identifier or a set of them separated by commas may be used. A char constant is defined by a constant declaration having the form identifier = 'character';. The same format, namely, single quotes (apostrophes),

TABLE A4.4 ASCII CODE (7-BIT)

<table>
<tr><th></th><th colspan="6">$b_6b_5b_4$</th></tr>
<tr><th>$b_3b_2b_1b_0$</th><th>010</th><th>011</th><th>100</th><th>101</th><th>110</th><th>111</th></tr>
<tr><td>0000</td><td>SP</td><td>0</td><td>@</td><td>P</td><td></td><td>p</td></tr>
<tr><td>0001</td><td>!</td><td>1</td><td>A</td><td>Q</td><td>a</td><td>q</td></tr>
<tr><td>0010</td><td>"</td><td>2</td><td>B</td><td>R</td><td>b</td><td>r</td></tr>
<tr><td>0011</td><td>#</td><td>3</td><td>C</td><td>S</td><td>c</td><td>s</td></tr>
<tr><td>0100</td><td>$</td><td>4</td><td>D</td><td>T</td><td>d</td><td>t</td></tr>
<tr><td>0101</td><td>%</td><td>5</td><td>E</td><td>U</td><td>e</td><td>u</td></tr>
<tr><td>0110</td><td>&</td><td>6</td><td>F</td><td>V</td><td>f</td><td>v</td></tr>
<tr><td>0111</td><td>'</td><td>7</td><td>G</td><td>W</td><td>g</td><td>w</td></tr>
<tr><td>1000</td><td>(</td><td>8</td><td>H</td><td>X</td><td>h</td><td>x</td></tr>
<tr><td>1001</td><td>)</td><td>9</td><td>I</td><td>Y</td><td>i</td><td>y</td></tr>
<tr><td>1010</td><td>*</td><td>:</td><td>J</td><td>Z</td><td>j</td><td>z</td></tr>
<tr><td>1011</td><td>+</td><td>;</td><td>K</td><td>[</td><td>k</td><td>{</td></tr>
<tr><td>1100</td><td>,</td><td><</td><td>L</td><td>\</td><td>l</td><td>|</td></tr>
<tr><td>1101</td><td>-</td><td>=</td><td>M</td><td>]</td><td>m</td><td>}</td></tr>
<tr><td>1110</td><td>.</td><td>></td><td>N</td><td>^</td><td>n</td><td>~</td></tr>
<tr><td>1111</td><td>/</td><td>?</td><td>O</td><td>_</td><td>o</td><td>DEL</td></tr>
</table>

[3]Exceptions are IBM and Burroughs, which use an EBCDIC coding.

is used to define char type ''numbers'' which can be used in arithmetic or relational expressions. For example, if keybrd has been defined as a char variable, then, in a statement such as

IF keybrd < 'P' THEN i := −3 (9)

the boolean expression keybrd < 'P' will be true for any character assigned to keybrd that has a lower ASCII value than (capital) P, and false for P or for any character with a higher ASCII value. Thus, $, 3, N, etc., when stored in keybrd will produce a boolean value of true in the boolean expression, while P, Z, a, w, etc., will produce a value of false. When a write or writeln statement is used to output a char quantity, the appropriate single character is printed. The standard function ord (ordinal) applied to char data gives the numerical value representing the data. For example, in ASCII code, ord ('1') = 31 (hexadecimal). Another useful function is chr, which provides the inverse function, that is, chr (31) = '1'.

In addition to the types of integer, real, boolean and char defined above, PASCAL also permits users to define their own types. For example, they might define a type called COLOR, whose elements could be black, brown, red, etc. We defer our treatment of user-defined types to Sec. A.9.

A.5 STRUCTURED STATEMENTS

The ''heart'' of a PASCAL block of programming occurs between the reserved words BEGIN and END. This part of the block is where the statements defining the actions taken by the program are located. We have already seen two examples of such statements—the *assignment statement,* which assigns a value to a variable, and the *procedural statements* starting with read, readln, write, and writeln. Both assignment and procedural statements are examples of a class of statements called *simple statements*. These statements may be defined as ones which *do not contain other statements* as part of them. Here we introduce a second type of statement called a *structured statement* which will include some other statement as one of its components. Sets of one or more statements, either simple and/or structured, can be grouped together in such a way as to be treated as a single statement. In this case they are delineated by the reserved words BEGIN and END and referred to as a *compound statement.* The differences among simple, structured, and compound statements are illustrated in Fig. A5.1. In general, throughout this text, when we use the word *statement* we imply any *simple, structured, or compound statement.*

IF/THEN

One of the most powerful structured statements in the PASCAL language is the IF statement. In general it has the form

IF condition THEN statement (10)

where IF and THEN are reserved words (see Table A2.1). The components of this statement are described as follows:

Condition A boolean expression containing a single relational operator or a set of such expressions enclosed in parentheses and connected by boolean operators (see Table A4.2).

Statement Any simple, structured, or compound statement.

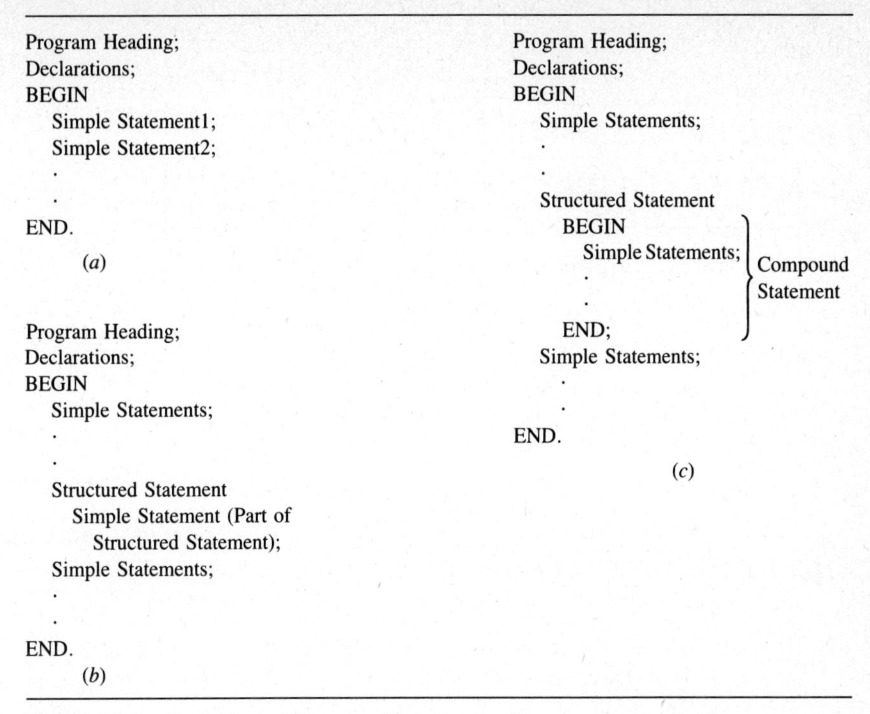

Figure A5.1 Examples of the use of (*a*) simple, (*b*) structured, and (*c*) compound statements.

If the condition portion of the IF statement is true, the statement portion is executed. If it is not true, the statement portion is skipped and the program continues with the statement following the IF statement. A flowchart of the operation of the IF statement is given in Fig. A5.2. Examples are given in (5) and (6).

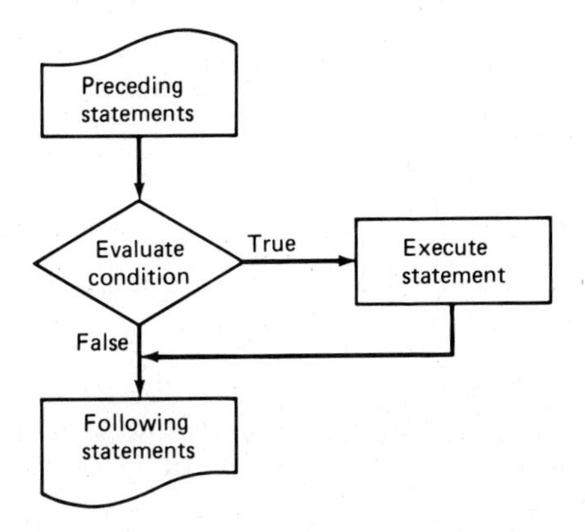

Figure A5.2 The IF/THEN structured statement.

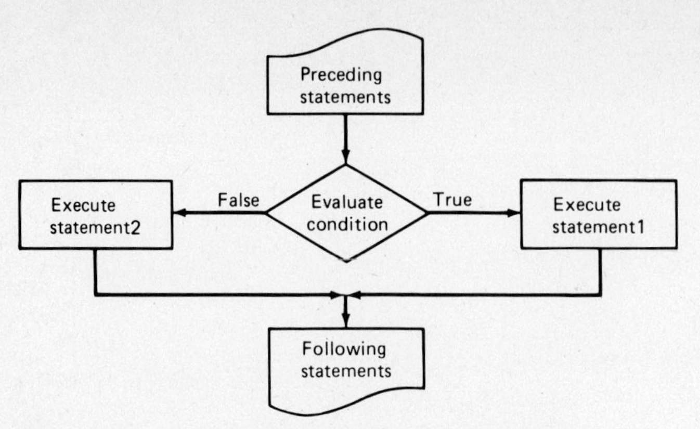

Figure A5.3 The IF/THEN/ELSE structured statement.

IF/THEN/ELSE

A more versatile form of the IF statement allows the programmer to choose between two alternatives. This statement has the form[4]

IF condition THEN statement1 ELSE statement2 (11)

If condition is true, statement1 is executed and the program proceeds with the statement following the IF statement. If condition is not true, statement2 is executed and the program then proceeds with the statement following the IF statement. A flowchart is shown in Fig. A5.3. As an example of the use of the IF/THEN/ELSE statement, consider the determination of the square root of a real variable that may have a positive or negative value. A program for doing this is shown in Program A5.1.

Program A5.1
```
PROGRAM SignSqRt (input, output);
VAR number, root : real;
BEGIN
   read (number);
   IF number >= 0.0 THEN writeln('square root =', sqrt(number))
   ELSE writeln('square root = J', sqrt(-number))
END.
```

Note that in this program we have used the sqrt function listed in Table A4.1.

FOR

In PASCAL there are three statements which may be used to repeat a statement (or a set of statements) a number of times. The first of these is the FOR statement. It has the form:

FOR control := first TO last DO statement (12)

[4]The same effect may be obtained by using two IF/THEN statements, the second one with the negation of the condition specified by the first. For example, in Program A5.1 the ELSE line could be replaced with

```
IF number < 0.0 THEN writeln('square root = J',
  sqrt(-number));
```

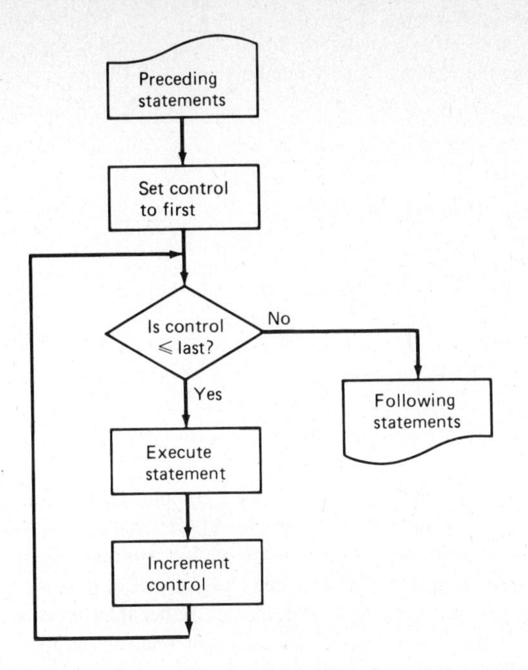

Figure A5.4 The FOR structured statement.

where FOR, TO, and DO are reserved words. The components of this that have not already been defined are now defined.

Control A declared integer variable. Its value is defined only while the FOR statement is being executed.

First An integer variable, constant, number, or expression which is evaluated at the beginning of the execution of the FOR statement and which is then assigned as the initial value given to *control*.

Last An integer variable, constant, number, or expression which is evaluated at the beginning of the execution of the FOR statement and which is used as the last value of *control*.

Note that control, first, and last must all be of the same type.[5] The operation of the FOR statement is as follows:

1. The value of first is computed and assigned to control.
2. The value of control is compared with the value of last. If (control <= last) is true, statement is executed once and step 3 (below) is followed. If it is false, the FOR statement is terminated and the program proceeds to the following statement.
3. The value of control is advanced to its next higher value (an integer variable is incremented by 1) and step 2 is repeated.

A flowchart of the operation is shown in Fig. A5.4. Note that it forms a "loop" around statement. Also note that the first test of the value of control is made before the first execution of statement.

[5]In a later section we see that these quantities may be of types other than integer, namely, any scalar-defined type.

Thus, it is possible for statement to be executed zero times. As an example of the use of a FOR statement, a program for printing the values of binary numbers with 1 to 16 bits is given in Program A5.2.

Program A5.2

```
PROGRAM BinNumb (output);
VAR bits, value : integer;
BEGIN
   writeln('     bits      number');
   value := 2;
   FOR bits := 1 TO 16 DO
      BEGIN writeln(bits, value);
      value := 2 * value
      END
END.
```

Note that the statement portion of the FOR statement uses a compound statement consisting of two simple statements separated by a semicolon, starting with BEGIN and ending with END. Note also how the use of indentations emphasizes the portions of the program which comprise the statement portion of the FOR statement. A variation of the FOR statement uses the reserved word DOWNTO in place of TO. In this case control is decremented rather than incremented, and the iterations continue as long as (control >= last) is true. As two final observations on the FOR statement, note that (1) the number of iterations it will make of statement is fixed once the execution of the FOR is begun, and (2) control is undefined once the loop is terminated.

WHILE

The second type of PASCAL statement which may be used to repeat a statement (or set of statements) a number of times is the WHILE statement. It has the form

WHILE condition DO statement (13)

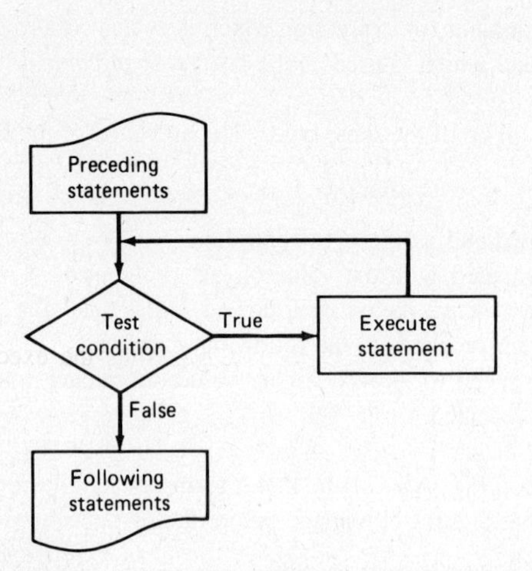

Figure A5.5 The WHILE structured statement.

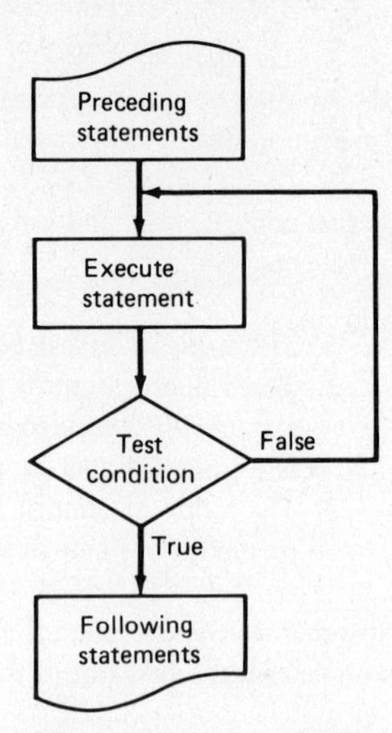

Figure A5.6 The REPEAT structured statement.

where WHILE and DO are reserved words. The operation of the WHILE is as follows:

1. The condition portion of the statement is evaluated.
2. If condition is true, statement is executed and step 1 is repeated. If condition is false, the program proceeds to the statement following the WHILE.

Obviously the statement portion of the WHILE must modify the variable(s) of condition in some way, or an ''endless loop'' will result. A flowchart of the operation is shown in Fig. A5.5. As an example of the use of the WHILE statement, Program A5.3 will print a table similar to the one printed by Program A5.2; however, it will continue printing entries until the maximum integer range of the computer is reached.

Program A5.3

```
PROGRAM MaxBin (output);
VAR bits, value : integer;
BEGIN
   writeln('     bits      number');
   value := 2; bits := 1;
   WHILE value < maxint DIV 2 DO
      BEGIN writeln(bits, value);
      value := 2 * value; bits := bits + 1
      END;
    writeln(bits, value) {Print last entry}
END.
```

The program uses the standard identifier maxint (see constants in Table A2.3) to determine the condition in the WHILE statement. In general, the FOR statement may be considered as a special case of the WHILE statement.

REPEAT

The third type of PASCAL structured statement that may be used to repeat a statement (or a set of statements) is the REPEAT statement. It has the general form

```
REPEAT statement UNTIL condition                                    (14)
```

where REPEAT and UNTIL are reserved words. It is similar to the WHILE statement except that condition is tested after statement is executed, and that statement is repeated as long as condition is false. Since statement is bracketed by the reserved words REPEAT and UNTIL, a compound statement may be used here without using the delimiting BEGIN and END. A flowchart of the REPEAT statement is given in Fig. A5.6.

CASE

One other useful structured statement is the CASE statement. It permits the execution of one particular statement from a set of statements. It has the form

```
CASE index OF
   list1 : statement1;
   list2 : statement2;
     .    .
     .    .
END;                                                                (15)
```

where CASE, OF, and END are reserved words. The components of this structured statement are as follows.

Index A variable or expression of any scalar type.[6] *Caution:* ALL possible values that index may take on in the particular program in which the CASE statement is used must be given in the lists that follow. If index should take on a value not in the lists, the operation of the program is undefined.

List1, List2, etc. Each of these lists contains one or more (separated by commas) values that index may take on. Each value may appear only once. The values must be of the same type as index.

Statement1, Statement2, etc. The statement (simple, structured, or compound) that is executed when index has the value given in the corresponding list.

As an example of the use of the CASE statement, consider Program A5.4. It is designed to appropriately change a set of x and y coordinate values by 1 each time the user enters U, D, R, or L (for up, down, right, and left). The starting point is the origin, $x = 0$, $y = 0$. The program is restarted at the origin by entering S (start) or B (begin).

Program A5.4

```
PROGRAM XY Coordinates (input, output);
VAR inp:char; x, y:integer;
BEGIN x:= 0; y:= 0;
   readln (inp);
   WHILE (inp = 'U') or (inp = 'D') OR (inp = 'R') OR
              (inp = 'L') OR (inp = 'S') OR (inp = 'B') DO
     BEGIN {The statement part of the WHILE begins here}
         CASE inp OF
               'U':y:= y + 1;
               'D':y:= y - 1;
               'R':x:= x + 1;
               'L':x:= x - 1;
           'S', 'B':BEGIN x:= 0; y:= 0 END;
         END; {End of CASE statement}
     writeln('x = ', x, ' y = ', y);
     read(inp);
     END; {END of WHILE statement}
   writeln('Unrecognizable character') {This statement is
              reached when the program exits from the WHILE}
END.
```

Note that in the program a test is made of the input to make certain that it conforms to the available entries in the lists. The program is stopped by entering any other character.

GOTO

The final statement to be considered in this section is the GOTO statement. It is not a structured statement since it does not formally include another statement. It may, however, be used to transfer control from one part of the program to another. The general form of the GOTO statement is

GOTO statement-label (16)

[6]Scalar types are integer, char, and boolean but not real. They may also be user-defined (see Sec. A.9).

where GOTO is a reserved word and statement-label is an unsigned integer (usually not exceeding four digits) that is used to identify some program statement. To perform such an identification, statement-label is placed before the statement and followed by a colon. In addition, any statement label used in the program must have been previously declared using a label declaration which precedes all other declarations. It has the form

LABEL label-list (17)

where LABEL is a reserved word and label-list is a list of one or more (separated by commas) statement-labels. As an example of the use of the GOTO statement, Program A5.2 has been rewritten as shown in Program A5.5.

Program A5.5

```
      PROGRAM BinNumb (output);
      LABEL 23;
      VAR bits, value : integer;
      BEGIN
         writeln('     bits      number');
         bits := 1; value := 2;
23:      writeln(bits, value);
         bits := bits + 1; value := * 2;
         IF bits <= 16 THEN GOTO 23
      END.
```

In general, the use of the GOTO statement is discouraged, since its usage obscures the block-structure properties of PASCAL (more on this in Sec. A.8) which make programs written in it so easy to understand and to maintain.

A.6 ARRAYS

A task that is part of many programming projects is the storage of large quantities of data. Methods for providing such storage are treated under the general heading of *data structures*. In this section we consider one of the simplest and most easily used data structures, the array.

Variable Declarations

An *array* is an ordered list of a set of data items of the same type. In a *one-dimensional array* (also referred to as a vector), the individual data items are specified by the array name and a single index variable. Although many high-level languages require that the range of the index variable start at 1, PASCAL is considerably more flexible, allowing the programmer to pick both the lower and upper limits of the index. Thus an array of 10 items could have index ranges of 0 to 9, 1 to 10, 101 to 110, −5 to 4, etc. We restrict our attention here to cases where the index takes on only integer values.[7] A one-dimensional array is defined using an *array declaration* in the variable declaration section of the PASCAL block. This declaration has the form

VAR identifier : ARRAY[integer1..integer2] OF type; (18)

where ARRAY and OF are reserved words. The components of this declaration are now defined.

[7] In later sections we see that the index is permitted to be relatively unrestricted including even user-defined types of variables.

Identifier The name that identifies the array. It is used in accessing any individual element of the array.

Integer1, Integer2 The possible values of the index for the array are the set of integers starting with integer1 and going through integer2 (where integer1 is less than integer2). This is called the *range* of the index.

Type The type (real, integer, etc.) of the data stored in the individual elements of the array.

Any individual elements of the array may be accessed, either for storing data or for retrieving it, by using the array name followed by a pair of brackets in which is contained the numerical value of the index or the identifier or an expression for the numerical value desired for the index. Thus it has the form

array-name[index-value] (19)

As an example of the use of an array, in Program A6.1 the array fctrl is used to store the factorial values of the integers from 1 to 10. These values are then printed in the form of a table.

Program A6.1

```
PROGRAM Factorial (output);
VAR number, factor : integer;
   fctrl : ARRAY[1..10] OF integer;
BEGIN factor := 1;
FOR number := 1 TO 10 DO
   BEGIN factor := factor * number;
     fctrl[number] := factor   END;
writeln('      number              factorial');
FOR number := 1 TO 10 DO
writeln(number, fctrl[number])
END.
```

Multidimensional arrays can also be defined in PASCAL. For example, a *two-dimensional array* (also referred to as a matrix) would be defined by a variable declaration having the form

VAR identifier : ARRAY[range of row index, range of column
index] OF type; (20)

where each of the ranges has the form integer1..integer2. A component of the array is accessed by

array-name[value of row index, value of column index] (21)

As an example of a two-dimensional array, consider the problem of storing the set of coefficients defining the numerator and denominator polynomials of a rational function $N(s)$ having the form

$$N(s) = \frac{a_{10} + a_{11}s + a_{12}s^2 + a_{13}s^3}{a_{20} + a_{21}s + a_{22}s^2 + a_{23}s^3} \tag{22}$$

Using the first subscript of each coefficient for the row index and the second subscript for the column index, the array will have the form

$$\begin{matrix} a_{10} & a_{11} & a_{12} & a_{13} \\ a_{20} & a_{21} & a_{22} & a_{23} \end{matrix} \tag{23}$$

If we give this array the name a, it will be defined by the variable declaration

```
VAR a : ARRAY[1..2, 0..3] OF real;                                        (24)
```

Any individual coefficient stored in the array is identified by an identifier having the form a[i,j].

The general form of the format specified in (20) and (21) is readily extended to arrays with more than two dimensions. For example, a $2 \times 5 \times 6$ array called b might be defined by the variable declaration

```
VAR b : ARRAY[1..2, 0..4, 11..16] OF real;                                (25)
```

An element of this array would be identified by b[i,j,k].

Type Declarations

Another method for defining arrays is through the use of a *type declaration*. This is another form of declaration similar to the constant declaration and variable declaration described in Sec. A.3. As used to define an array, the type declaration has the general form

```
TYPE type-name = ARRAY[range] OF type;                                    (26)
```

where TYPE, ARRAY, and OF are reserved words. The components of this declaration are now described.

Type-name This is a name defining a type of variable that consists of an array. Note that it does not define a variable, but a variable type (like integer, real, etc.). As such the new type is usually referred to as a *user-defined type*. To use this to define a variable, in the variable declarations a statement of the form

```
identifier : type-name;                                                   (27)
```

must be used. Such a statement establishes identifier as an array variable having the dimensions and type indicated in the type declaration.

Range This parameter gives the range or ranges used to define a one- or higher dimensional array. It has the same format at was described for arrays earlier in this section.

Type This parameter specifies the type of the elements of the array, that is, real, integer, etc.

The TYPE declaration must be placed between the constant declaration and the variable declaration. Its format is the same as the other declarations; namely, it consists of one use of TYPE followed by as many statements as are needed to specify the desired types. As an illustration of the use of a type statement, consider the following:

```
TYPE matrix23 = ARRAY[1..2, 1..3] OF real;
VAR a, b, c : matrix23;                                                   (28)
```

This pair of statements defines three two-dimensional arrays a, b, and c, each of size 2×3. This example illustrates one of the advantages of the use of a type declaration to define an array rather than defining it in a variable declaration; namely, a single declaration can define a large number of arrays. Clearly, this makes it far easier to change the size of the arrays if in the course of subsequent program modification such action should become necessary.

A.7 FUNCTIONS AND PROCEDURES

One of the most important techniques in programming is the structuring of the overall program task into separate components, each of which is realized by a *subprogram*. This technique is especially useful when identical sequences of instructions are to be repeated at different points in the program. In this case, it is only necessary to write the instructions one time and declare them as a subprogram. They can then be "called" by the main program as needed. The subprogram technique also makes it possible to develop sections of code that can be used in more than one program. In this chapter we present a discussion of the two types of subprograms that are available in PASCAL, the function and the procedure.

Functions

In PASCAL, a *function* is a subprogram, that is, an independent collection of statements, which is defined in the declarations section of the main program. A function has the property that when it is called, it returns a value to the program or subprogram in which the call occurred. The declaration for each function starts with a *function heading* having the form

FUNCTION identifier (parameter list) : type (29)

where FUNCTION is a reserved word. The individual components of this heading are defined as follows.

Identifier The name of the function. In addition to identifying the function, this name is also treated as a variable. Its value is determined by the statements of the function. Thus, somewhere in the function there must be an assignment statement having the form

identifier := some constant, variable, or arithmetic expression (30)

The value assigned to the name of the function is considered as one of the outputs from the function.

Parameter List The variables that are used as inputs (we shall shortly see that they may also be used as outputs) to the function. These are listed in the form

identifier1 : type; identifier2 : type; . . . (31)

where either a single identifier or a set of them separated by commas may be used for each type that is declared.

Type The type assigned to the output variable defined by the name of the function.

The body of the function, following the function heading, is a block (see Sec. A.3) similar to that of the main PASCAL program. It consists of declarations followed by a statement or set of statements beginning with BEGIN and ending with END; (note that a semicolon is used after END). The function is called; that is, the statements in its block are processed, by using its name and parameter list in an arithmetic or boolean expression in the main program or in some other function or procedure. An example of a typical overall program structure is given in Fig. A7.1. Note the following points concerning this structure:

1. The statements comprising the function declaration follow the main program variable declaration and precede the BEGIN defining the statements in the main program block.

```
PROGRAM MainProgram (input, output);
 {A constant declaration, if one is used, goes here}
VAR i:integer; r:real; b:boolean; c:char;
 {Other variable declarations, as needed}
FUNCTION Fun (j:integer; s:real; c:boolean; d:char) : integer;
  {If variables other than j, s, c, and d are used in the function, they must be defined in a VAR statement
   here}
   BEGIN  {The first statements of the function go here} ..;
     Fun := . . .; {Some defining expression}
     {Other function statements if needed}
   END;  {End of the function subprogram}
BEGIN  {The main program starts here}
  {Some main program statements defining values of i, r, b, and c would go here}
   ka := Fun(i, r, b, c) * 17.0      {An example of a statement which calls the function}
  {Some other main program statements changing the values of i, r, b, or c might go here}
   IF Fun(i, r, b, c) < 3 THEN. . .;      {Another example of a statement which calls the function}
  {Some other main program statements might go here}
END.  {End of the main program}
```

Figure A7.1 Typical program structure including a function.

2. The variables used in the function are defined in two ways—by the parameter list and also (if required) by a variable declaration in the function block.
3. The order, type, and number of variables in the parameter list of the function when it is called in the main program must be exactly the same as in the function heading.
4. The variables defined in the parameter list of the function heading and used in the statements comprising the body of the function are called *formal parameters*. They are defined only inside the function and bear no relation to other variables (even ones having the same name) in the main program.
5. When the function is called by using its name in the main program, the variables shown in the parameter list are called *actual parameters*. The *current actual values* of these variables are used by the function to determine its output value, which is then used in the main program. Even though the formal parameter values may be changed by the statements of the function, the actual parameters of the main program are not changed when the function is called.
6. In the main program, two examples of the use of the function are shown, one in an arithmetic expression and another in a boolean expression. If the values of the main program variables i, r, b, and c have changed between the two uses of the function, then a different value will be output by the function in the second call.

An illustration of the way in which a function is used is given in Program A7.1.

Program A7.1

```
PROGRAM TestFun (output);
VAR a, b, c, d : integer;
FUNCTION Sum (f, g, h : integer) : integer;
   BEGIN g := g * g;
      sum := f + g + h  END;
BEGIN a := 1; b := 2; c := 3;
   d := Sum(a, b, c);
   writeln('Inputs', a, b, c, 'Output', d)
END.
```

The output for the program will be (the spacing is only approximate)

```
Inputs   1   2   3   Output   8                    (32)
```

As another example of the difference between formal parameters and actual parameters, assume that the function shown in Program A7.1 is replaced with the following one:

```
FUNCTION Sum (a, b, c : integer) : integer;
  BEGIN b := b * b;
    sum := a + b + c   END;                        (33)
```

In this case, the *a, b,* and *c* variables in the function are formal parameters *and are completely different* from the a, b, and c variables in the main program. The output from the program will be unchanged.

Procedures

In PASCAL, a *procedure,* like a function, is a subprogram defined in the declarations section of the main program. When a procedure is called, it performs a set of operations as specified by its statements. Unlike a function, however, there is no output value associated with its name. The declaration for a procedure starts with a *procedure heading* having the form

```
PROCEDURE identifier (parameter list)              (34)
```

where PROCEDURE is a reserved word. The components of this heading include the following.

Identifier The name of the procedure. Since no value is assigned to this name, it will not usually appear in any of the statements of the procedure block.[8]

Parameter List The variables used as inputs for the procedure. These have the same form as the ones in a function, and we shall shortly see that they may also be used as outputs. Many procedures may not require a parameter list. In such a case, the parentheses are also omitted.

The block of a procedure is identical with that of a function.[9] Thus, it consists of a declarations section and a statements section starting with BEGIN and ending with END;. If both procedures and functions are present in a program, they may be listed in any order unless one calls the other. A discussion of this situation is given in Sec. A.8. The procedure is called by using its name and parameter list *as a statement* (not *in* a statement) in the main program. As an example of the use of a procedure, in Program A7.2, a procedure called labeler is used to print a 10-character name stored in the (formal parameter) array tenchar. A line of asterisks is printed above and below the name. Note that, following the discussion given in Sec. A.6, a TYPE statement is used in the declaration section of the main program to define a type named alfab consisting of a one-dimensional array of 10 char-type variables. This user-defined type is used in the procedure heading to assign the correct properties to tenchar. The main program reads 10 characters into the (actual parameter) array name (also of type alfab), and calls the procedure.

[8]An exception is when the procedure calls itself. This is referred to as *recursive* usage.

[9]One exception is that no assignment of a value to the procedure name is made.

Program A7.2

```
PROGRAM TestLabel (input, output);
TYPE alfab = ARRAY[1..10] OF char;
VAR name : alfab; i : integer;
PROCEDURE labeler (tenchar : alfab);
   VAR i : integer;
   BEGIN
      writeln('**********');
      FOR i := 1 TO 10 DO write(tenchar[i]);
      writeln;
      writeln('**********')
   END;  {End of labeler procedure}
BEGIN  {Start of main program}
   FOR i := 1 TO 10 DO read(name[i]);
   Labeler(name)  {This statement calls the procedure}
END.
```

Parameters

The method of passing input information to functions and procedures described in the preceding portion of this section uses a type of parameters called *value parameters* and is named pass-by-value. It has the following properties:

1. At the time the subprogram is called, the values of the main program variables (the actual parameters) are used as the values of the subprogram variables (the formal parameters).
2. Even though the statements of the subprogram may change the values of the formal parameters, the actual parameters in the main program remain unchanged.

A second method of communication between the main program and the subprogram uses a type of parameter called *variable parameters* and is named pass-by-reference. It can be used to input *or output* information to a subprogram. It has the following properties:

1. At the time the subprogram is called, the values of the main program variables (the actual parameters) are used as the values of the subprogram variables (the formal parameters) in the same way that they are for pass-by-value.
2. If the statements of the subprogram change the value of the formal parameters, the corresponding actual parameters in the main program are also changed. Thus, pass-by-reference variables can serve as either input or output quantities.

To identify variable parameters, the reserved word VAR is used before the identifier name in the parameter list of the subprogram heading. All variables that are listed between VAR and type will be treated as variable parameters. As an example of the use of variable parameters, in Program A7.3 we have rewritten Program A7.1 using variable parameters rather than value parameters.

Program A7.3

```
PROGRAM TestFun (output);
VAR a, b, c, d : integer;
FUNCTION Sum (VAR f, g, h : integer) : integer;
   BEGIN g := g * g;
      sum := f + g + h   END;
BEGIN a := 1; b := 2; c := 3;
   d := Sum(a, b, c);
   writeln('Inputs', a, b, c, 'Output', d)
END.
```

The only change between the two programs is in the function heading. The output of the programs, however, will be different. For Program A7.3, it will have the form

Inputs 1 4 3 Output 8 (35)

Note that the value of the main program actual parameter b has been changed by the statements of the function. Other comments on variable parameters and value parameters follow:

1. When one or more output variables must be obtained from a procedure, or more than a single output variable is required from a function, variable parameters must be used.
2. For a scalar (nonarray) input to a function or procedure, value parameters are usually preferred. The reason for this is that not only a variable but also a constant, number, or an arithmetic expression can be used as the actual parameter in the calling statement.
3. If an array, especially a large one, is to be input to a function or procedure, it is usually advantageous to use variable parameters. The reason for this is that, for a variable parameter (whether a scalar or an array), the PASCAL compiler treats both the formal parameter and the actual parameter as the same quantity, avoiding the need for an inefficient transfer of data from memory locations assigned to the main program to memory locations assigned to the subprogram, as is required with value parameters.

In the preceding paragraphs of this section we have introduced user-defined functions and procedures. In addition, we have discussed the two types of parameters used in defining and calling such subprograms—value parameters and variable parameters. In this section we introduce a third type of parameter. It allows a subprogram to call any of several other functions (or procedures), the choice being made at the time the original subprogram is called. The name of this new type of parameter is function parameter.

A *function parameter* is a parameter that specifies the name of a function. By using such a parameter, when the original function or procedure is called, any other function having the proper type and parameter list can be used as an actual parameter. Thus, the calling statement can be used to select any of various functions which may be required in a given situation. The implementation of a function parameter is accomplished by the following steps:

1. Declare the function parameter in the parameter list of the heading of the original function or procedure. This is done by using the reserved word FUNCTION followed by a name, type, and parameter list for the function parameter. Note that, like any formal parameter, the function parameter name is defined only local to the function or procedure in whose heading it appears; thus it has significance only inside the block of that subprogram. Some examples of parameter lists which include function parameters are:

```
(n : integer; FUNCTION Y (t : real) : real)
(FUNCTION Y (x : real) : real; FUNCTION Z (x : real) : real)
(VAR p, q : char; FUNCTION Z : boolean)
```

As a more specific example of the declaration of a function parameter, let us define a function called Arithmetic whose purpose is to perform an arithmetic operation on two real input numbers and give the result as a real output. The specific arithmetic operation performed is to be chosen by having the Arithmetic function call some other function. If we let the name of the formal function parameter be Y, the heading of the Arithmetic function could be written as follows:

```
FUNCTION Arithmetic (a, b : real; FUNCTION Y
                  (p, q : real) : real) : real;          (36)
```

2. In one (or more) of the statements in the block of the function or procedure, the formal function parameter name together with its parameter list must appear. This parameter list is treated as an actual (or calling) parameter list; thus, the types of the parameters are *not* given. For our example Arithmetic function, such a calling statement might have the form

```
Arithmetic := Y(a, b);          (37)
```

3. The functions that are to be called using the function parameter must be declared separately in the main program. The parameters of such functions *can only be value parameters*. More specifically, they *cannot* be variable parameters. For the example we are developing here, let us assume that we require the original function Arithmetic to call either of two functions named Add and Multiply, and that their declarations have the form

```
FUNCTION Add (x, y : real) : real;
    BEGIN Add := x + y END;
FUNCTION Multiply (x, y : real) : real;
    BEGIN Multiply := x * y END;          (38)
```

Note that the parameter list and type of any such functions must agree with the form used in the calling statement of the original function or procedure, in this case as given for Y in (36).

A program that uses the Arithmetic function defined (in part) by (36) and (37), in which the functions given in (38) are used as actual (function) parameters, is given in Program A7.4.

Program A7.4

```
PROGRAM FunParam (output);
VAR c, d : real;
FUNCTION Add (x, y : real) : real;
    BEGIN Add := x + y END;
FUNCTION Multiply (x, y : real) : real;
    BEGIN Multiply := x * y END;
FUNCTION Arithmetic (a, b : real; FUNCTION Y (p, q : real) : real) : real;
    BEGIN
        Arithmetic := Y(a, b)
    END;
BEGIN
    c := 3.0; d := 9.0;
    writeln(' Sum = ', Arithmetic(c, d, Add));
    writeln(' Product = ', Arithmetic (c, d, Multiply))
END.
```

The output from the two write statements of this program will be

Sum = 12.0
Product = 27.0

Obviously, there are far simpler ways of implementing the operations shown in the program. It is included here as a simple medium for illustrating the mechanism of the function parameter.

In this section we have introduced the PASCAL subprograms—the function and the procedure. In addition, we have shown two methods of communicating between the main program and the subprograms—by the use of value parameters and variable parameters. There is one other way in which such communication can take place. This is the use of global variables. It will be covered in the following section.

A.8 BLOCK STRUCTURES

One of the most useful properties of the PASCAL language is that embodied in the concept of a block. The main block of a PASCAL program consists of everything between the program heading and the END statement, that is, the entire program. As such, the block is made up of two parts—declarations and statements. The declarations, however, may include subprograms, namely, functions and procedures. Since these subprograms also consist of declarations and statements, they form their own blocks. Their declarations, in turn, can define other blocks, and so on. Thus, the general structure of a PASCAL program appears as shown in Fig. A8.1. The nesting of blocks shown in the figure is summarized by describing PASCAL as a *block-structured language*. In terms of these blocks we can now indicate a means for establishing communication between the various components of a program, that is, of transferring variable values between the main program and a subprogram or between various subprograms. This method supplements the pass-by-value and pass-by-reference methods described in Sec. A.7.

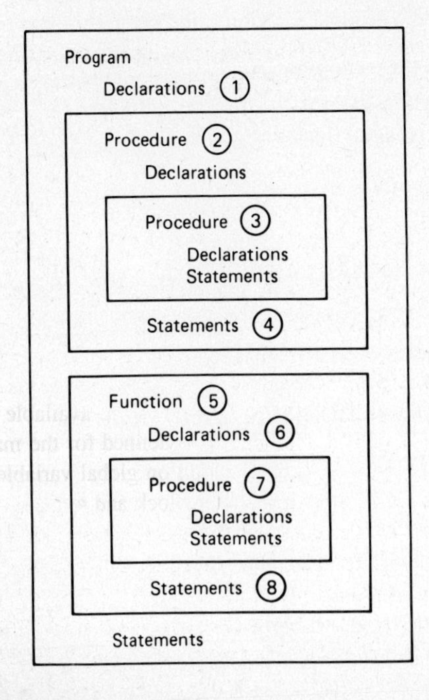

Figure A8.1 The general structure of a PASCAL program.

Global Variables

To begin, let us assume that some variable has been defined in the declarations section of the main program identified by the ① in Fig. A8.1. This variable is considered as *a global variable* in the sense that it is available to every subprogram inside the main program block. As an illustration of the characteristics of such a global variable, consider the function identified by the ⑤ in Fig. A8.1. The variables used in this function can be declared in three ways:

1. The parameter list given in the function heading. These variables may be either pass-by-value or pass-by-reference.
2. The variables specified by the declaration portion of the function and identified by the ⑥ in Fig. A8.1.
3. Any variable specified in the declaration section (identified by ① in Fig. A8.1) of the main program. The latter variables are called global variables. They may be used for output as well as input.

As an illustration of the use of global variables, consider Program A7.1. It has been rewritten as Program A8.1, using global variables to communicate between the main program and the function.

Program A8.1

```
PROGRAM TestFun (output);
VAR a, b, c, d : integer;
FUNCTION Sum : integer;
   BEGIN b := b* b;
      Sum := a + b + c  END;
BEGIN a := 1; b := 2; c := 3;
   d := Sum
   writeln('Inputs', a, b, c, ' Output', d)
END.
```

Note that no parameter list is used either in the function heading or in the statement in the main program. As input *and output,* the function uses the globally defined variables a, b, and c. Specifically, b is redefined by the statements of the function; thus a new value of b is output by the function. The output from the program will be

Inputs 1 4 3 Output 8 (39)

Global variables exist from the main program not only to the subprograms contained in it but also to the subprograms contained in them, etc. They will also exist between a subprogram and its subprograms. A variable declared in a subprogram, however, will not be available to the main program. The logic for this may be seen from the block structure shown in Fig. A8.1. In general, any variable declared in the main program, function, or procedure will be global to all the blocks contained within that main program, function, or procedure. Thus, for example, a variable declared in the declarations identified at ⑥ in Fig. A8.1 will be available as a global variable for the procedure identified at ⑦. It will not, however, be defined for the main program or for the procedures identified at ② and ③. As a final observation on global variables, it should be noted that a variable defined by a declaration in both an outer block and an inner block using the same identifier in both will be treated by the compiler as two separate variables. It will, however, be available as a global variable to any other blocks contained within the outer block (assuming that there is no declaration of the variable in them).

As a general programming rule, especially in complex programs, the use of global variables should be minimized (if not eliminated). The use of pass-by-value and pass-by-reference variables to link program components in general produces programs that are far easier to understand, debug, and modify than those which use global variables.

Forward Declarations

One of the most useful features of functions and procedures is that they can include statements that call other functions and procedures.[10] In such a usage two situations arise. The first of these occurs when a function or procedure calls another function or procedure that has already been declared. For example, the statements at ⑧ in Fig. A8.1 which are part of the function block identified by ⑤ might call the procedure at ② or the one at ③. Similarly, the procedure declared at ⑦ might include statements that would call either the ② or ③ procedures. This situation causes no problems and is treated as normal by the compiler. The second situation that may occur when a function or a procedure calls another function or procedure is that the subprogram *which is called* is not declared until a later point in the overall program. Such a situation will generate a compiler error since the computer is unable to identify the called subprogram and thus considers its name as an undefined identifier. PASCAL provides a means of correcting this situation. Basically it makes it possible to give the compiler a message, in effect telling it that the missing identifier will be defined in a subsequent portion of the program. Such a message is called a *forward declaration*. It must be placed immediately preceding the function or procedure that does the calling. The forward declaration has a format almost identical with that which would be used for the heading of the subprogram that is called. The only difference is that the reserved word FORWARD is used as a suffix (separated by a semicolon). The only other change required in the program is that the actual subprogram is defined with an *abbreviated heading* containing (after FUNCTION or PROCEDURE) only its identifier. As an example of the use of a forward declaration, suppose the statements at ④ in Fig. A8.1 which are part of the procedure identified by ② include a call to the function at ⑤. The key portions of a typical program illustrating such a situation are shown in Fig. A8.2. The circled numbers shown in this figure correspond with those shown in Fig. A8.1. In the figure note the following:

1. The forward declaration for the function Five occurs in ①. It consists of the function heading followed by the reserved word FORWARD.
2. In the statement in ④ of procedure Two the function Five is called using the actual parameters j and x.
3. The actual statements defining the function Five start at ⑤ with an abbreviated function heading giving only the name of the function.
4. The output value of the function is determined in the statement in ⑧ which uses the formal parameters i and r defined in the forward declaration for the function given in ①.

At this point the reader might ask, "Why not simply put function Five in the main program declarations before procedure Two, and thus avoid the need for a forward declaration?" As long as function Five does not contain any calls to procedure Two, this is a logical and even preferred solution. However, if function Five does contain such calls, the interchange of the positions of the two subprograms will be of little help, since the forward declaration for function Five is simply replaced by one for procedure Two. To further illustrate the use of the FORWARD declaration, in Program A8.2 procedure SumRealImag is used to add two complex numbers, using the function Sum to perform the addition (obviously the addition could also have been done directly in the procedure by a simple arithmetic expression). In Program A8.2*a* the logical order is used; that is,

[10]As a matter of interest, in PASCAL a function or a procedure can even call itself. This is called *recursion*. It is a feature that is available in very few programming languages.

```
     PROGRAM . . . . . . . .;
①    VAR . . . . . . . .;
     {The next line gives the forward declaration for function Five}
     FUNCTION Five(i : integer; r : real) : real; FORWARD;
②    PROCEDURE Two . . . . . . . .;
     VAR x, y, : real; j : integer; . . . . . . . .;
③       PROCEDURE Three . . . .; VAR . . . .; BEGIN . . . .   END;
     BEGIN   {The statements defining Procedure Two follow} . . . . . . . .;
④       y := Five(j, x);   {An example of a call for function Five} . . . . . . .
     END;
      {The actual statements defining function Five follow}
⑤    FUNCTION Five;   {Note that an abbreviated heading is used}
⑥    VAR . . . . . . .;
⑦       PROCEDURE Seven . . . .; VAR . . . .; BEGIN . . . .   END;
⑧    BEGIN . . . . . . .;
        Five := sqr(i) + r;   {An example of the determination of Five} . . . . . . .;
     END;
     BEGIN   {Statements of main program}
     .

     .
     END.
```

Figure A8.2 An example of a forward declaration.

Program A8.2a

```
PROGRAM AddComplexNumbers (input, output);
VAR r, r1, r2, i, i1, i2 : real;
FUNCTION Sum (a, b : real) : real;
   BEGIN Sum := a + b   END;
PROCEDURE SumRealImag(VAR r, i : real; r1, r2, i1, i2 : real);
   BEGIN r := Sum(r1, r2); i := Sum(i1, i2)   END;
BEGIN
   read(r1, i1, r2, i2);
   SumRealImag(r, i, r1, r2, i1, i2);
   writeln('real part = ', r); writeln('imag part = ', i)
END.
```

Program A8.2b

```
PROGRAM AddComplexNumbers (input, output);
VAR r, r1, r2, i, i1, i2 : real;
FUNCTION Sum(a, b : real) : real; FORWARD;
PROCEDURE SumRealImag(VAR r, i : real; r1, r2, i1, i2 : real);
   BEGIN r := Sum(r1, r2); i := Sum(i1, i2)   END;
FUNCTION Sum;
   BEGIN Sum := a + b   END;
BEGIN
   read(r1, i1, r2, i2);
   SumRealImag(r, i, r1, r2, i1, i2);
   writeln('real part = ', r); writeln('imag part = ', i)
END.
```

the function is declared before the procedure that calls it. In Program A8.2*b* the procedure is declared first, and a forward declaration is used to indicate that the function will appear later.

There are several variations of the forward declaration which may be implemented as advanced programming techniques. One of these occurs when a function or procedure, rather than being supplied in the program, is to be taken from some library of subprograms contained in the computer. The details of how such a library is set up and accessed will vary in different computers. In the program, however, an *external declaration* is required. This is very similar to the forward declaration in that it gives the name and formal parameter list for the subprogram. The word FORWARD at the end of the declaration, however, is replaced with the word EXTERN. A variation of the external declaration occurs when the library program is written in FORTRAN rather than PASCAL. The PASCAL language (as implemented in some computers) permits accessing such a program. In this case the word EXTERN is replaced by FORTRAN. It should be noted that global variables cannot be used to communicate with external functions or procedures. Only formal parameters, either pass-by-value or pass-by-reference, are permitted.

The material presented in this appendix, namely, the block-structured characteristics of the PASCAL language, provides the key to writing programs that are easily understood and readily maintained and modified. In any programming project, before a single line of code is written, the programming tasks should be carefully delineated and organized into a hierarchical block structure. This should always be considered one of the first and most important parts of any programming task.

A.9 USER-DEFINED DATA TYPES

In Sec. A.4 we presented a discussion of the four standard data types used in PASCAL—integer, real, boolean, and character. In the design of the PASCAL language, however, it was recognized that in many programming situations having only these four types of data is overly restrictive for the user. To avoid this restriction, PASCAL allows programmers to define their own data types. This capability is one that is absent in many high-level languages. It is a valuable feature that contributes greatly to the efficiency and readability of PASCAL programs. We have already seen an example of a user-defined type in our discussion of arrays in Sec. A.6. In this section we generalize the concept and show how it may be applied.

Type Declaration

Before we introduce the user-defined type declaration, let us review some properties of a *standard data type,* for example, the integer one. This data type consists of a set of numbers (positive or negative integers) that are the allowed values that an integer variable can take on. Generalizing these properties to a *user-defined type,* we see that two pieces of information regarding such a type are required: (1) the name of the type (analogous to integer, real, etc.) and (2) the set of allowable values (analogous to the ''numbers'' of the integer type) that a variable of the user-defined type may have. These two pieces of information are continued in the *user-defined type declaration.* It has the form

TYPE type-name = (list-of-values) (40)

where TYPE is a reserved word. The components of this declaration are described below.

Type-name An identifier specifying a user-defined type whose allowable values are given in the following part of the declaration.

List-of-Values A list of the permitted values (separated by commas) that variables that are declared to be of the type given in type-name may have.

The values given in the list-of-values of the type declaration are considered to be ordered, the first one listed being the lowest and the last one being the highest. Like other declarations, TYPE is used only once in a block, and it is followed by as many entries (separated by semicolons) as are needed. An example of a type declaration follows:

```
TYPE daysOfWeek = (sun, mon, tue, wed, thu, fri, sat);
      employees = (bryant, clark, smith, jones);
         colors = (black, brown, red, orange, yellow, green, blue, violet,      (41)
                   gray, white);
         months = (jan, feb, mar, apr, may, jun, jul, aug, sep, oct, nov, dec);
```

Variables for each of these user-defined types may be declared in the VAR portion of the declarations. For example,

```
VAR day, holiday       : daysOfWeek;
    foreman, mechanic  : employees;
    bandcolor          : colors;                                             (42)
    vacation           : months;
```

The type declaration and the variable declaration can be combined using the following form of variable declaration:

```
VAR identifier      :  (list-of-values)                                      (43)
```

In this case, a type is not formally defined, only a variable and its allowed values. If more than one variable of a given type is to be declared, it is usually simpler to explicitly define a type rather than having to retype the list-of-values for each variable (and perhaps make an error). As an example, the items given in (41) and (42) could be rewritten as follows:

```
TYPE daysOfWeek        = (sun, mon, tue, wed, thu, fri, sat);
     employees         = (bryant, clark, smith, jones);
VAR day, holiday       : daysOfWeek;
    foreman, mechanic  : employees;
    bandcolor          : (black, brown, red, orange, yellow, green, blue,    (44)
                          violet, gray, white);
    vacation           : (jan, feb, mar, apr, may, jun, jul, aug, sep, oct,
                          nov, dec);
```

Note that, because of the ordering of the allowable values of a user-defined type, boolean expressions involving them can be written. For example, using the declarations of (44), day < wed would be true when day was equal to sun, mon, or tue and false when day was equal to wed, thu, fri, or sat. As another example of the use of ordering, a variable declared as a user-defined type may be used as an index between any of the values given in the list-of-values for that type. For example, using (44), we might write

```
FOR bandcolor  :=  black TO white DO statement                               (45)
```

The FOR statement would make 10 iterations of *statement*. As an example of the use of user-defined types, consider the determination of an employee's paycheck. Assume that he or she is paid $3 per hour on Monday to Friday and twice that on Saturday and Sunday. The computation of the amount of the paycheck is performed by Program A9.1. The input to the program will be the number of hours worked on each day from Sunday to Saturday.

Program A9.1

```
PROGRAM PayCheck (input, output);
VAR day:(sun, mon, tue, wed, thu, fri, sat);
      totalpay, dayhours, daypay:real;
BEGIN
   totalpay:= 0.0;
FOR day:= sun TO sat DO
   BEGIN
      read(dayhours);
      daypay:= dayhours * 3.00;
      IF (day = sun) OR (day = sat) THEN
         daypay:= daypay * 2.0;
      totalpay:= totalpay + daypay
   END;
   writeln('Paycheck =' totalpay)
END.
```

Subrange Type

Another user-defined type that finds frequent application is called a *subrange type*. This is established by a type declaration having the form

TYPE type-name = lowerlimit..upperlimit (46)

The components of the declaration are defined as follows.

Type-name An identifier specifying a scalar type (including user-defined ones) the range of whose allowed values is given in the following portion of the declaration. Note that for user-defined types, when a variable is declared to be of type-name (using a variable declaration), two constraints are imposed upon it: (1) it must be of a certain type, and (2) in that type it can have only certain allowed values.

Lowerlimit, Upperlimit The lowest and highest values permitted for a variable declared to be of type-name. Note that lowerlimit and upperlimit must themselves be of the same type. In addition, their type must either be the (standard) scalar types integer or char or must be a user-(previously) defined scalar type. Lowerlimit and upperlimit must also, of course, obey the ordering implicit in their type, that is, lowerlimit < upperlimit must be true. Finally, note that, for standard types, it is the type that lowerlimit and upperlimit have which determines the type of any variable declared to be of type-name.

As examples of declared subranges, for the standard types we might have

```
grades = 0..100;
 digits = '0'..'9';
letters = 'A'..'Z';                                    (47)
```

The first of these is a subrange of integer type while the last two are subranges of char. As examples of subranges for user-defined types, using (41) we might have

```
      weekdays = mon..fri;
  summermonths = jun..aug;                    (48)
```

Note that the notation defining subranges is one that we have already used. In our discussion of arrays in Sec. A.6 this notation was used to define the range of the array indices.

A.10 RECORDS AND SETS

Records

In Sec. A.6 we introduced the array, our first data structure, that is, a method for the storage of large quantities of data. An array was described as an ordered list of data items *of the same type*. In many programming situations, however, it is desirable to be able to store different types of data items in the same structure. For example, a payroll program might require storing an employee's name (type char), social security number (type integer), and pay rate (type real). PASCAL includes a provision for storing such a diversity of information using a data type called a *record*. This is declared in the type declarations section of a block using the following format:

```
TYPE record-identifier = RECORD
                         field-identifier1 : data type;
                         field-identifier2 : data type;
                                .                  .
                                .                  .
                         field-identifier  : data type
                         END;                                  (49)
```

where RECORD is a reserved word. Note that no semicolon appears after the data-type for the last field identifier. The components of this declaration are expressed as follows.

Record-identifier The identifier giving the name of the record and establishing it as a user-defined type.

Field-identifier The name of each of the components of the record. Note that each component can be a different type. The types can include not only the standard ones but also user-defined ones. They may even be arrays.

To access a certain element of a record, that is, to store information in it or retrieve information from it, we use the following format:

```
record-identifier.field-identifier                      (50)
```

Note that, in effect, we can consider a record as a new type of n-element one-dimensional array in which each of the elements may be of a different type.

As an example of a record, consider the payroll situation mentioned at the beginning of this section in which it is desired to store the name, social security number, and pay rate of an employee. In addition, let us add the employee's sex and department. The type declaration might appear as follows:

```
TYPE payroll = RECORD
                  name  : ARRAY[1..20] OF char;
                  ssn   : integer;
                  scx   : (male, female);
                  rate  : real;
                  dept  : (accounting, sales, production)
               END;                                              (51)
```

In this declaration note the following:

1. In the name field-identifier, since the standard data type char is only for a single character, we have used an array to store up to 20 characters to represent a name. Thus we see that an array is permitted as an element of a record.
2. In the sex field-identifier and in the dept field-identifier, we have utilized a user-defined type (see Eq. 43), thus, these components of the record can only take on the indicated values.

Once a record has been defined using a type declaration, we can use a variable declaration to define one or more variables each of which will be a record. The declaration will have the form

VAR identifier1, identifier2, . . . : record-type (52)

In addition, we can define an array (or arrays), each element of which is a record, using the usual form of variable declaration. For example, to store the information contained in the payroll record of (51) for 50 employees, we would use the following:

VAR employee : ARRAY[1..50] OF payroll; (53)

Any particular record for any employee can now be accessed as

employee[index].field-identifier (54)

As an example of the versatility of the PASCAL record type of data, using the declarations of (51) and (53) let us write a program to determine how many of the 50 employees are female. This is done in Program A10.1.

Program A10.1

```
PROGRAM female (input, output);
TYPE payroll . . .;  {Same as (51)}
VAR employee . . .;  {Same as (53)}
   i, count : integer;
BEGIN count := 0;
   FOR i := 1 TO 50 DO
     IF employee[i].sex = female
        THEN count := count + 1;
   writeln('There are', count, 'females')
END.
```

If we desire to find out how many females there are in the accounting department, all we need to do is to modify the condition in the IF statement as shown in Program A10.2.

```
Program A10.2
PROGRAM FemaleAccounting (input, output);
TYPE payroll . . .;  {Same as (51)}
VAR employee . . .;  {Same as (53)}
   i, count : integer;
BEGIN
   FOR i := 1 TO 50 DO
      IF (employee[i].sex = female) AND
            (employee[i].dept = accounting)
         THEN count := count 1;
   writeln('There are', count, 'females')
END.
```

When complex logic decisions are being made, representing the record identifier (in this case the array name and index) many times can be cumbersome. To avoid this, PASCAL provides a WITH statement that allows the record identifier to be given only once, immediately after DO in the IF statement. The general form of the WITH statement is

```
WITH   record identifier DO                                   (55)
```

where WITH is a reserved word. Note that there is no semicolon after DO. The specification given by the WITH applies to whatever statement (including structured and compound ones) follows its DO. The names of the field identifiers in the record are then used directly as variable names. Using this technique, we may rewrite Program A10.2 in the form shown in Program A10.3.

```
Program A10.3
PROGRAM FemaleAccounting (input, output);
TYPE payroll . . .;  {Same as (51)}
VAR employee . . .;  {Same as (53)}
   i, count : integer;
BEGIN
   FOR i := TO 50 DO
      WITH employee[i] DO
      IF (sex = female) AND (dept = accounting)
         THEN count := count + 1;
   writeln('There are', count, 'female')
END.
```

Sets

Another type of data structure available in PASCAL is the set. Like the array and the record, it is defined as a collection of elements used to store data. It has a quite different characteristic from the array and the record; however, in that, as we shall shortly see, the collection itself may be operated on as a single entity. A set is defined by a type declaration having the form

```
TYPE set-identifier = SET OF base-type                        (56)
```

where SET and OF are reserved words. The components of this declaration are now described.

Set-identifier This identifier gives the type name (analogous to integer, real, etc.) of the set and establishes it as a user-defined type.

Base-type The set of allowed values that variables may have when they are defined as being of set-identifier type. The allowable values may be given as a list (enclosed in brackets and separated by commas) or by using the subrange types defined in Sec. A.9. For example,

```
TYPE letters = SET OF 'A'..'Z';
     numbers = SET OF 0..10;                                    (57)
     choices = SET OF [5, 10, 15, 20];
```

Variables which are of the type set-identifier may be defined using a conventional variable declaration. For example, using (57) we might have the following variable declaration:

```
VAR vowels, first5letters, newletters : letters;
    evenNumbers, primeNumbers, otherNumbers : numbers;          (58)
```

Variables that are declared as being of a type defined as a set are themselves sets and are called *set variables*. They are quite different from the simple variables that we have previously encountered. In the statement part of a PASCAL block, set-variables may be assigned any number of elements, from 0 (the null set) to all of the ones listed in the set declaration. The general form of the assignment statement for set variables is

```
set-variable := [set-elements]                                  (59)
```

where set-variable is an identifier previously declared as being of the type given by set-identifier, and set-elements is a list (separated by commas) of the elements of the original set which are used to define set-variable. For example, using (57) and (58), we might have the following assignment statements:

```
vowels       := ['A','E','I','O','U'];
first5Letters := ['A','B','C','D','E'];                         (60)
evenNumbers  := [0, 2, 4, 6, 8, 10];
primeNumbers := [1, 2, 3, 5, 7];
```

Note that, in any set or set-variable, there is no ordering; that is, [7, 1, 2, 5, 3] is exactly the same as [1, 2, 3, 5, 7]. Set-variables can be operated on using any of three set operators. These have the property that when they operate on sets they produce new sets. They are expressed as follows: [In the following discussion A, B, and C are all assumed to be set-variables declared as being of the same set-identifier type. The examples used refer to (57), (58), and (60)].

Set Operators

+ Set Union The assignment statement A := B + C gives A as the *union* of B and C; that is, any element of the set will appear in A if and only if it is an element of B or an element of C (or an element of both). As examples, consider the statements

```
newLetters := vowels + first5Letters;
newNumbers := evenNumbers + primeNumbers;                       (61)
```

As a result of these statements we have

```
newLetters := ['A','E','I','O','U','B','C','D'];
newNumbers := [0, 2, 4, 6, 8, 10, 1, 3, 5, 7];                  (62)
```

– Set Difference The assignment statement A := B – C gives A as the *difference* B – C; that is, an element of the set will appear in A if and only if it is an element of B but not an element of C. For example, the assignment statements

```
  newLetters := vowels − first5Letters;
newNumbers := evenNumbers − primeNumbers;                    (63)
```

produce the results

```
  newLetters := ['I','O','U'];
newNumbers := [0, 4, 6, 8, 10];                               (64)
```

∗ **Set Intersection** The assigment statement A := B ∗ C gives A as the *intersection* of B and C; that is, an element of the set will appear in A if and only if it is an element of both B and C. For example, the assignment statements

```
  newLetters := vowels * first5Letters;
newNumbers := evenNumbers * primeNumbers;                     (65)
```

produce the results

```
  newLetters := ['A','E'];
newNumbers := [2];                                            (66)
```

The set operators described above have the function of producing sets. Another operation that is possible on sets is the use of the relational operators (Table A4.3) in boolean expressions to compare sets. The operations are

= **Set Equality** The boolean expression A = B is true if and only if every element of set A is an element of set B and every element of set B is an element of set A (otherwise it is false). The sets used in this expression or any other boolean expression may be set-variables or lists of set elements. For example,

```
       vowel = ['A','E','I','O','U'] is true
primeNumbers = [1, 2, 3, 5] is false (7 is missing)
```

<> **Set Inequality** The boolean expression A <> B is true if and only if A = B is false (and it is false only if A = B is true). As examples

```
      vowels <> ['A','E','I','O','U'] is false
primeNumbers <> [1, 2, 3, 5] is true
```

<= **Set Inclusion** The boolean expression A <= B is true if and only if every member of set A is also a member of B (otherwise it is false). Effectively, this says that A is contained in B. Note that B may have elements which are not in A. As examples,

```
      vowels <= letters is true
primeNumbers <= 1..6 is false
```

A boolean expression with a function similar to that of the set inclusion one is used as a test to determine whether a single (nonset) variable has its current value contained in a given set-variable. This set is performed by the boolean expression

```
variable IN set-variable
```

For example, declaring VAR c : char and using the assignment c := 'D', then c IN vowels is false while c IN first5Letters is true. Similarly, declaring VAR x : integer, and using the assignment x := 4, then x IN even is true while x IN primeNumbers is false. As an example of the use of set operators, consider Program A10.4.

Program A10.4

```
PROGRAM SetOperations (input, output);
TYPE numbers = SET OF 0..5;
VAR result, even, prime : numbers;
PROCEDURE TestSetMembers (x : numbers);
  BEGIN
    IF 0 IN x THEN write(' 0');
    IF 1 IN x THEN write(' 1');
    IF 2 IN x THEN write(' 2');
    IF 3 IN x THEN write(' 3');
    IF 4 IN x THEN write(' 4');
    IF 5 IN x THEN write(' 5');
    writeln
  END;  {End of testing procedure}
 BEGIN
   even := [0, 2, 4]; prime := [1, 2, 3, 5];
   result := even + prime; write('Union:');
   TestSetMembers(result);
   result := even − prime; write('Difference:')
   TestSetNumbers(result);
   result := even * prime; write('Intersection;');
   TestSetMembers(result)
 END.
```

In this program the variables even and prime are given the indicated elements of the set of integers from 0 to 5. The procedure TestSetMembers is used to output the results of applying the various set operators defined above. The output from this program will be

```
Union: 0 1 2 3 4 5
Difference: 0 4
Intersection: 2
```

As an example of the use of relational operators, consider Program A10.5.

Program A10.5

```
PROGRAM SetBoolean (input, output);
TYPE numbers = SET OF 0..5;
VAR test, even, prime : numbers; x : integer;
BEGIN
  even := [0, 2, 4]; prime := [1, 2, 3, 5];
  test := [0, 2, 4]; x := 4;
  IF test = even THEN writeln('Set equality with even');
  IF NOT test = even THEN writeln('Set inequality with even');
  IF test <> prime THEN writeln('Set inequality with prime');
  IF test <= prime THEN writeln('Set included in prime');
  IF NOT test <= prime THEN writeln('Set not included in prime');
  IF x IN even THEN writeln('x is in even');
  IF x IN prime THEN writeln('x is in prime')
 END.
```

The output from this program will be

```
Set equality with even
Set inequality with prime
Set not included in prime
x is in even
```

Appendix B

Description of Procedures

In this appendix, descriptions are given of three procedures that form a part of the software package presented in this book. In general, these procedures have been designed with the minimum level of sophistication necessary to effectively perform their function. The obvious advantage of such an approach is the minimization of computer memory requirements as well as the reduction of compilation and execution times. The user who wishes a higher level of performance will find that these programs form an excellent basis upon which embellishments may be readily added.

B.1 THE PROCEDURE Plot5

The procedure Plot5, which was originally introduced in Chap. 3, is used to provide a printer-constructed plot of one or more variables as defined by a sequence of values of data stored in a two-dimensional array. The plot is printed with the positive abscissa direction oriented vertically downward on the printed page, and with the positive ordinate direction going from left to right across the page. The identifying statement for this procedure is

```
PROCEDURE Plot5 (VAR y : plotArray; numPlots, numPoints, ordMx : integer);
```

A summary of the characteristics of the procedure is given in Table 3.1. Here we concentrate on giving a description of how the procedure operates. A flowchart of logic is given in Fig. B1.1. A listing of the statements of the procedure is given in Fig. B1.2. The most significant variables encountered in this subroutine are:

line	A one-dimensional array of 101 char variables in which numeric information is stored corresponding to the desired form of the line of the plot that is currently being printed.
numPlots	A variable specifying the number of functions that are to be plotted.
row	An index that is used internally in the program to indicate which line of the plot is currently being printed.
numPoints	A variable specifying the number of time values at which each of the functions is to be plotted, that is, the number of lines that the plot will have.
ordMx	An input variable specifying the maximum value desired for the ordinate scale. This variable is also used (ordMx = 999) to select the automatic scaling option.
y	The two-dimensional array of variables y[j,i] giving the *i*th value of the *j*th function that is to be plotted.

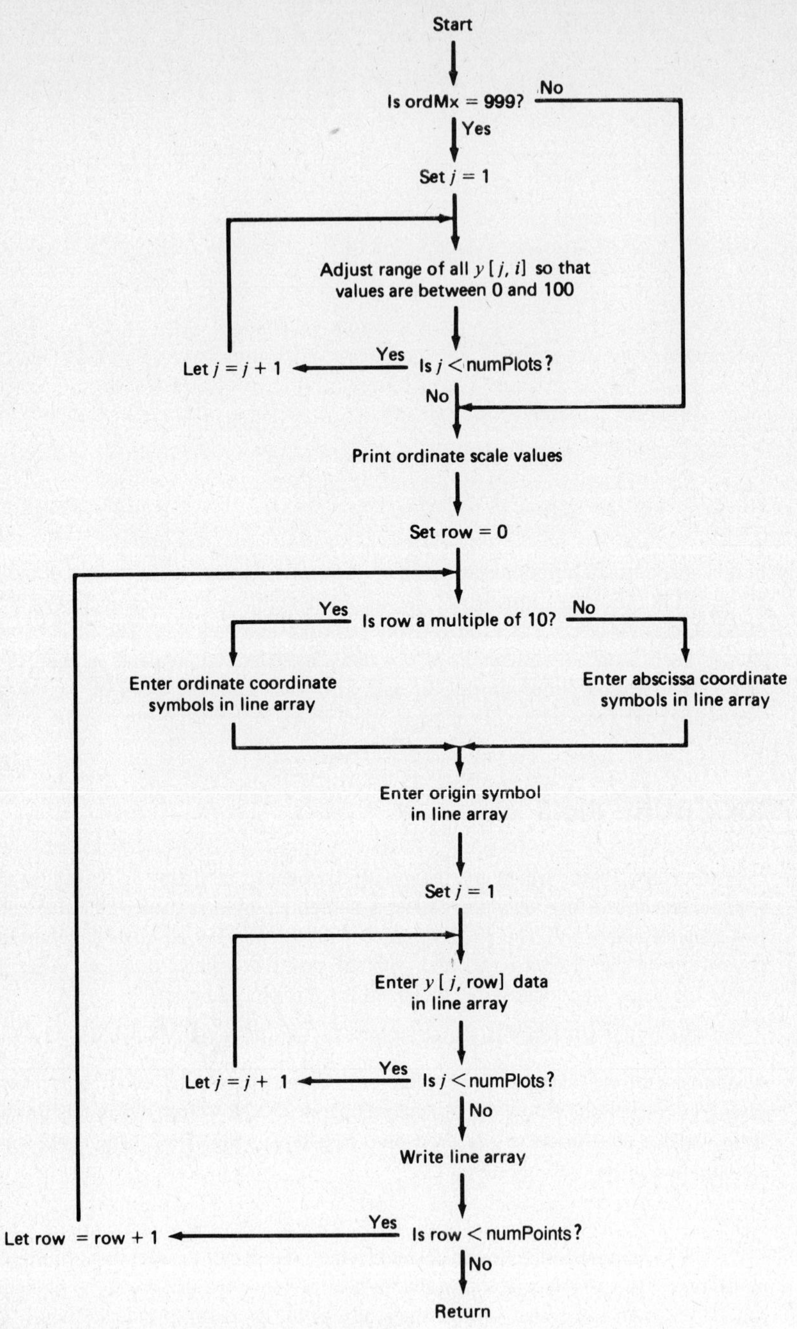

Figure B1.1 Flowchart of procedure Plot5.

The first operation of the procedure Plot5 is to determine whether a scaled or an unscaled plot is required. To this end, the variable ordMx is tested. If this variable has the value 999, the data representing the first function to be plotted, that is, the data stored as y[1,i] (i = 0,1,. . ., numPoints), is examined to find the maximum and minimum values. The data for this function is then scaled so that it covers a range of 0 to 100. The process is then repeated for the second function to be plotted, which consists of the data stored as y[2,i] (i = 0,1,. . ., numPoints), etc. The maximum ordinate variable ordMx is then set to 100. Finally, data on the maximum and minimum values and the scale factor of each function to be plotted are printed. From this point the operation of the procedure Plot5 is the same regardless of whether the automatic scaling option is used or not.

At this point in the program a set of ordinate values are computed and printed. These values range from ordMx-100 to ordMx. The program now sets the variable row, which indicates the line of the plot that is currently being printed, to 0 and sets the variables in the line array to the alphanumeric values corresponding to the plus and minus signs that are used to form the coordinate grid lines parallel to the ordinate. The numerical data of the first point of the function to be plotted is now converted to an index variable yData whose value corresponds to the plotted position of the data. The char value of the letter A is then stored as line[yData]. This process is then repeated for the first data point for any other functions that are to be plotted, using the char value of B for the second function, C for the third, etc. The characters A, B, C, etc., representing the values of the function, replace the char values of the plus and minus signs previously stored in the line array. The line array is now printed and the value of row is incremented. The process is repeated, with the symbol ''I'' being used to form the grid coordinate lines parallel to the abscissa for each value of row that is an even multiple of 10. The process is terminated when the line for which the value of row equals the value of numPoints is printed. At this point, control is returned to the main program.

B.2 THE PROCEDURE Root

The procedure Root introduced in Chap. 8 is used to find the roots of a given polynomial. It has the identifying statement

```
PROCEDURE Root (n:integer; a:array11; VAR p:carray10);
```

A summary of its characteristics is given in Table 8.3.[1] Here we present a description of how the procedure operates. A flowchart of the logic used in this subroutine is given in Fig. B2.1. A listing of the statements of the procedure may be found in Fig. B2.2. The procedure calls itself recursively until the boolean variable exit is true. It uses a Lin-Bairstow method to determine a quadratic factor of the given polynomial. It then removes this factor to obtain a polynomial of reduced order. The process is repeated until all the roots have been found.

The first operation of the procedure Root is to test the polynomial to determine if there is a zero root, that is, if the coefficient a_0 is zero. If there is such a root, it is stored in the p array, the coefficients in the a array are reordered, and the degree of the polynomial is lowered by 1. This process is continued until all the zero roots have been removed. The procedure then tests the degree of the remaining polynomial. If this degree is less than or equal to 2, a quadratic or first-order explicit formula is applied to find the root or roots. These are then stored in the complex

[1]The variable name b used in the table corresponds to the variable name a used in this discussion.

```
PROCEDURE Plot5 (VAR y : plotArray; numPlots, numPoints, ordMx : integer);

(* Given the parameters listed below, this procedure will
   create a plot of function(s) vs time.
   Special types required:
      plotArray = ARRAY[1..5,0..100] OF real.
   Definition of parameters:
      y        = The two-dimensional array containing the values
                 of each function.
      numPlots     = The number of functions to be plotted.
      numPoints    = The number of time increments to be plotted.
      ordMx = The maximum value of the function which can
                 be plotted.*)

TYPE cArray = ARRAY[0..100] OF char;
VAR i, j, ordLbl, row, yData, yzi : integer;
    yMax, yMin, yScl, yzr           : real;
    grid1, grid2, grid3             : char;
    line                            : cArray;

BEGIN
   writeln('1');
   yzr  := 0;
   yMin := 0;
   yScl := 1;
   IF ordMx = 999 THEN
      BEGIN
         yMax := -1.0E+50;
         yMin :=  1.0E+50;
         FOR j := 1 TO numPlots DO
            FOR i := 0 TO numPoints DO
               BEGIN
                  IF y[j,i] > yMax THEN
                     yMax := y[j,i];
                  IF y[j,i] < yMin THEN
                     yMin := y[j,i]
               END;
         yScl := 90 / (yMax - yMin);
         writeln(' ','       VALUES OF THE FUNCTION FROM',yMin:12,
               ' TO',yMax:12,' SCALED BY',yScl:12);
         yzr  := - yMin * yScl;
         yzi  := ROUND(yzr + (10 - ROUND(yzr) MOD 10));
         IF (yzi - yzr) >= 10 THEN
            yzi := yzi - 10;
         ordMx := 100 - yzi
      END;
   write('0','    ');
   FOR i := 0 TO 10 DO
      BEGIN
         ordLbl := 10 * i - 100 + ordMx;
         write(ordLbl:4,'      ')
      END;
   writeln('VALUES OF A');
   FOR row := 0 TO numPoints DO
      BEGIN
         IF (row MOD 10) = 0 THEN
            BEGIN
               grid1 := '-';
               grid2 := '+';
               write(' ','  ',row:4)
            END
         ELSE
            BEGIN
               grid1 := ' ';
               grid2 := 'I';
               write(' ','       ')
            END;
         FOR i := 0 TO 100 DO
            line[i] := grid1;
         FOR i := 0 TO 10 DO
            IF (10 * i) = (100 - ordMx) THEN
               line[10 * i] := '0'
```

Figure B1.2 Listing of procedure Plot5.

```
        ELSE
           line[10 * i] := grid2;
     grid3 := 'A';
     IF row <= numPoints THEN
        FOR j := 1 TO numPlots DO
           BEGIN
              yData := ROUND((y[j,row] - yMin) * yScl
                       - yzr - ordMx + 100);
              IF (yData >= 0) AND (yData <= 100) THEN
                 line[yData] := grid3
              ELSE
                 IF yData < 0 THEN
                    line[0] := '$'
                 ELSE
                    line[100] := '$';
              grid3 := SUCC(grid3)
           END;
     FOR i := 0 TO 100 DO
        write(line[i]);
     writeln(y[1,row])
   END
END;
```

Figure B1.2 *(Continued)*

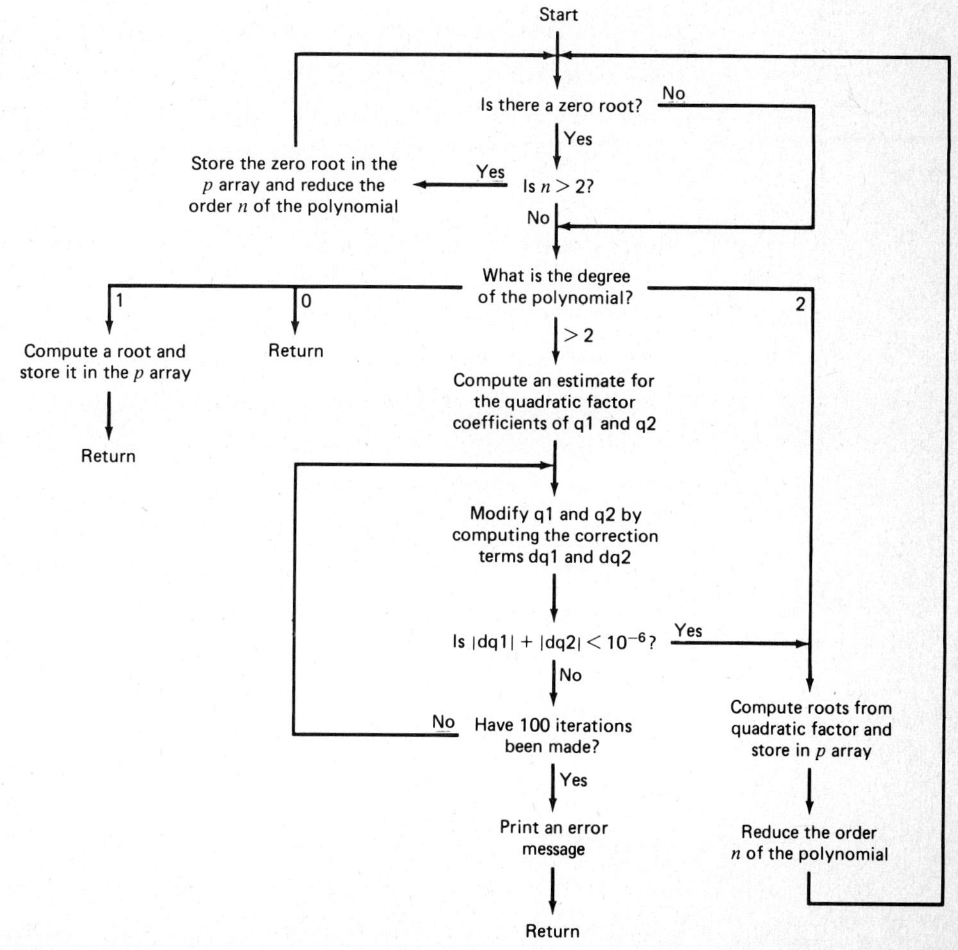

Figure B2.1 Flowchart of procedure Root.

```
PROCEDURE Root (n : integer; a : array11; VAR p : cArray10);

(* Given a polynomial of the form:
      a[n]*x**n + a(n-1)*x**(n-1) + ... + a[1]*x + a[0]
   this procedure will return its real and complex roots.
   Special types required:
      complex  = RECORD
                 re, im : real.
      array11  = ARRAY[0..10] OF real.
      cArray10 = ARRAY[1..10] OF complex.
   Definition of parameters:
      n = The degree of the polynomial.
      a = A one dimensional array containing coefficients
          of the polynomial.
      p = The complex array through which the roots are
          passed. *)

CONST errMax  = 1E-6;
      iterMax = 100;
VAR i, iter, sign                  : integer;
    det, discr, dq1, dq2, q1, q2   : real;
    exit                           : boolean;
    b, c                           : array11;
BEGIN
   exit := FALSE;
      WHILE (n > 2) AND (ABS(a[0]) < errMax) DO
         BEGIN
            p[n].re := 0;
            p[n].im := 0;
            FOR i := 0 TO (n - 1) DO
               a[i] := a[i + 1];
            n := n - 1
         END;
   IF n > 2 THEN
      BEGIN
         IF ABS(a[2]) < errMax THEN
            BEGIN
               q1 := 1;
               q2 := 1
            END
         ELSE
            BEGIN
               q1 := a[0] / a[2];
               q2 := a[1] / a[2]
            END;
         iter := 0;
         b[n] := 1;
         c[n] := 1;
            REPEAT
               b[n - 1] := a[n - 1] - q2;
               c[n - 1] := b[n - 1] - q2;
               FOR i := (n - 2) DOWNTO 0 DO
                  BEGIN
                     b[i] := a[i] - q2 * b[i + 1] - q1 * b[i + 2];
                     c[i] := b[i] - q2 * c[i + 1] - q1 * c[i + 2]
                  END;
               det := c[2] * c[2] + c[3] * b[1] - c[3] * c[1];
               IF det = 0 THEN
                  det := errMax;
               dq1 := (b[0] * c[2] + b[1] * (b[1] - c[1])) / det;
               dq2 := (b[1] * c[2] - b[0] * c[3]) / det;
               q1 := q1 + dq1;
               q2 := q2 + dq2;
               iter := iter + 1;
            UNTIL
               (iter = iterMax) OR ((ABS(dq1) + ABS(dq2)) < errMax);
         IF iter = iterMax THEN
            BEGIN
               write(' ','***** ROOT PROCEDURE ');
```

Figure B2.2 Listing of procedure Root.

```
                writeln('DOES NOT CONVERGE *****');
                exit := TRUE
             END
      END
   ELSE
      CASE n OF
         0 : exit := TRUE;
         1 : BEGIN
                p[1].im := 0;
                p[1].re := - a[0];
                exit    := TRUE
             END;
         2 : BEGIN
                q1 := a[0];
                q2 := a[1]
             END
      END;
   IF exit = FALSE THEN
      BEGIN
         discr := q2 * q2 - 4 * q1;
         sign  := 1;
         FOR i := n DOWNTO (n - 1)  DO
            WITH p[i] DO
               BEGIN
                  IF discr < 0 THEN
                     BEGIN
                        im := sign * SQRT(- discr) / 2;
                        re := - q2 / 2
                     END
                  ELSE
                     BEGIN
                        im := 0;
                        re := ( - q2 + sign * SQRT(discr) ) / 2
                     END;
                  sign := -1
               END;
         n := n - 2;
         FOR i := 0 TO n DO
            a[i] := b[i + 2];
         Root(n,a,p)
      END;
END;
```

Figure B2.2 (*Continued*)

p array. However, if the degree of the polynomial remaining after any zero roots have been removed is greater than 2, an estimate of a quadratic factor

$$q_1 + q_2 s + s^2$$

as represented by the procedure variables q1 and q2 is made using the relation

$$q_1 = \frac{a_0}{a_2} \qquad q_2 = \frac{a_1}{a_2}$$

An iterative process is then initiated to determine the quantities Δq_1 and Δq_2 (as represented by the variables *dq*1 and *dq*2) that are used to improve the initial estimate of the quantities q_1 and q_2 by the statements

$$\text{q1} := \text{q1} + \text{dq1} \qquad \text{q2} := \text{q2} + \text{dq2}$$

The terms Δq_1 and Δq_2 are computed by examining the first- and zero-order coefficients of the remainder obtained by dividing the original polynomial by the estimated quadratic factor and making a Taylor expansion of these coefficients with respect to q_1 and q_2. This produces a set of two equations in which the quantities Δq_1 and Δq_2 are the unknowns. These equations are then solved to find the quantities Δq_1 and Δq_2. This iterative process is repeated until convergence is reached, as defined by the sum of the magnitudes of Δq_1 and Δq_2 being less than 10^{-6}. A test of the number of iterations is also made, and the process is terminated if convergence is not reached within 100 iterations. In such a case, an error message is printed. When suitably accurate values for the quadratic-factor coefficients q_1 and q_2 are found, the roots of the quadratic factor are determined and stored in the p array. The quadratic factor is then divided into the polynomial and Root is called recursively to operate on the remaining polynomial. The process is repeated until all the roots of the polynomial have been found.

B.3 THE PROCEDURE Xyplt

The procedure Xyplt introduced in Chap. 12 is used to provide a printer-constructed *x-y* plot of a set of data points (x_i, y_i) stored in two one-dimensional arrays. The identifying statement for this procedure is

```
PROCEDURE Xyplt (nPoints : integer; VAR x, y : array200;
                 abRng, abMax, ordMax : integer);
```

The plot is printed with the positive abscissa direction oriented vertically downward on the page and the positive ordinate direction going across the page from left to right. A summary of the characteristics of the procedure is given in Table 12.1. Here we present a description of how the procedure operates. The discussion is supplemented by the flowchart given in Fig. B3.1 and the listing given in Fig. B3.2. The most significant variables used in the procedure are:

k	An internal program index that is used to determine which of the data points is currently being plotted.
line	A one-dimensional array of variables line[i] in which are stored the char values of the symbols used to represent the coordinate grid lines and the points that are to be plotted.
row	An internal program index that is used to indicate the line number of the plot that is currently being printed.
abRng	An input argument giving the total range of values of the abscissa scale desired for the plot. If this argument is set to the number 999, an automatic scaling option will be used.
abScl	An internal program constant giving the ratio of lines per inch to units per inch for the abscissa.
abMax	An input argument that specifies the maximum value of the abscissa scale that is to be used in the plot.
ordMax	An input argument that specifies the maximum value of the ordinate scale that is to be used in the plot (the minimum value is ordMax − 100).
xData	An internal program variable indicating the line, that is, the value of row corresponding to a given value of data stored in the x array.
yData	A variable indicating the position on a given line corresponding to a given value of the data stored in the y array.
xScl and yScl	Factors used for scaling the variables stored in the x and y arrays when the automatic scaling option is used.
x	The one-dimensional array of variables x[i] in which are stored the abscissa coordinate values of each of the points it is desired to plot.
y	The one-dimensional array of variables y[i] in which are stored the ordinate coordinate values of each of the points it is desired to plot.

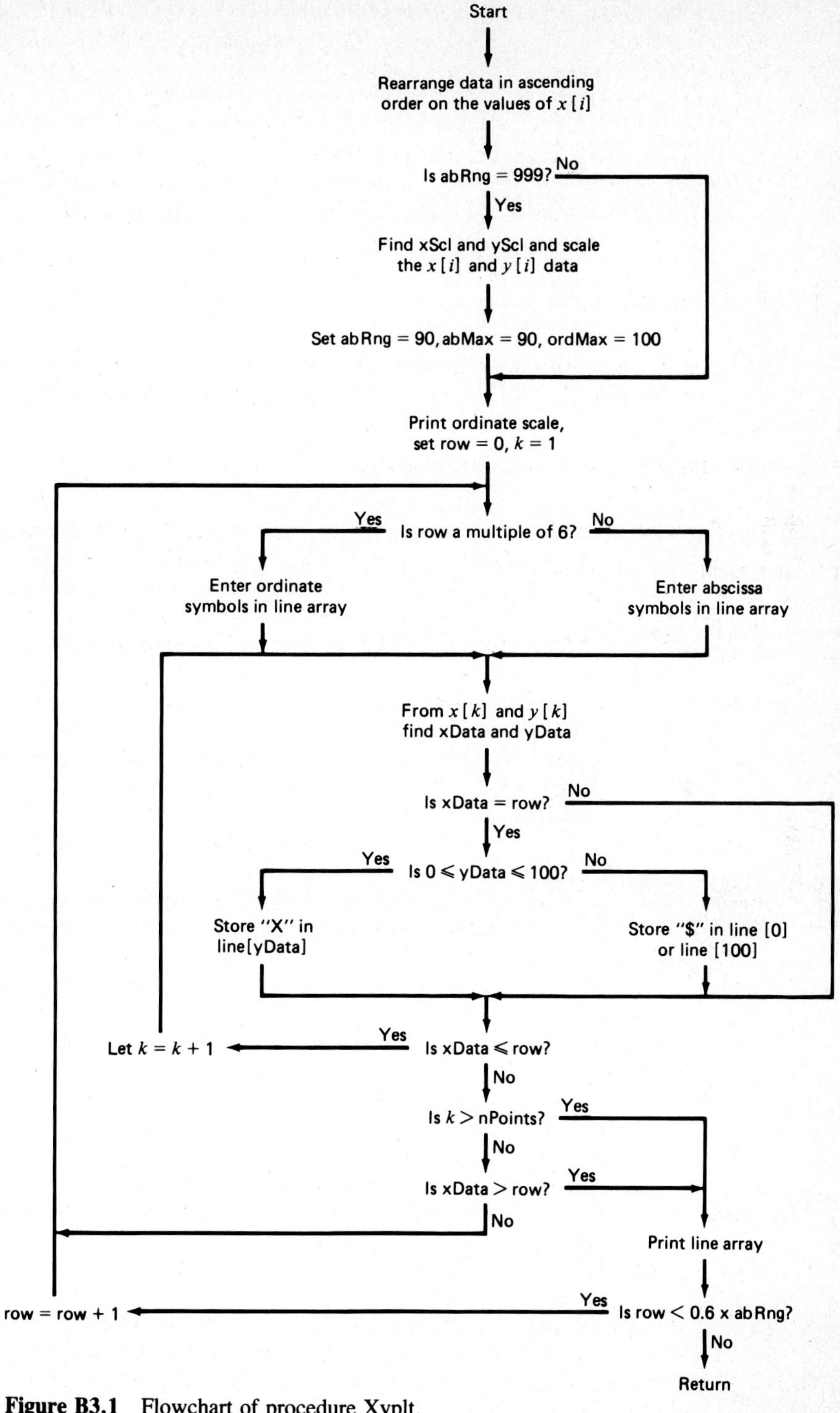

Figure B3.1 Flowchart of procedure Xyplt.

```
PROCEDURE Xyplt (nPoints : integer; VAR x, y : array200;

                  abRng, abMax, ordMax : integer);

(* Given the parameters listed below, this procedure
   will create a plot of y vs x  values.
   Special types required:
      array200 = ARRAY[1..200] OF real.
   Definition of parameters:
      nPoints    = The number of values of x and y.
      abRng    = The range of values of x which can be plotted
                along the abcissa.
      abMax   = The maximum value of x which can be plotted
                 along the abscissa.
      ordMax = The maximum value of y which can be plotted
                 along the ordinate.
      x       = The one-dimensional array of variables
                containing the values of x.
      y       = The one-dimensional array of variables
                containing the values of y. *)

CONST abScl = 0.600000;
TYPE cArray = ARRAY[0..100] OF CHAR;
VAR absLbl, i, j, k, ordLbl, row, xData, xzi, yData, yzi : integer;
    temp, xScl, yMax, yMin, yScl, yzr, xzr                : real;
    grid1, grid2                                          : char;
    line                                                  : cArray;
BEGIN
   writeln('1');
   yzi := 0;
   xzi := 0;
   FOR i := 1 TO (nPoints - 1) DO
      FOR j := (i + 1) TO nPoints DO
         IF x[i] > x[j] THEN
            BEGIN
               temp := x[i];
               x[i] := x[j];
               x[j] := temp;
               temp := y[i];
               y[i] := y[j];
               y[j] := temp
            END;
   IF abRng = 999 THEN
      BEGIN
         yMax := -1.0E+50;
         yMin :=  1.0E+50;
         FOR i := 1 TO nPoints DO
            BEGIN
               IF y[i] > yMax THEN
                  yMax := y[i];
               IF y[i] < yMin THEN
                  yMin := y[i]
            END;
         yScl := 90 / (yMax - yMin);
         xScl := 80 / (x[nPoints] - x[1]);
         writeln(' ','       VALUES OF X FROM',x[1]:12,' TO',x[nPoints]:12,
               ' SCALED BY',xScl:12);
         writeln(' ','       VALUES OF Y FROM',yMin:12,' TO',yMax:12,
               ' SCALED BY',yScl:12);
         yzr := - yMin * yScl;
         yzi := ROUND(yzr + (10 - ROUND(yzr) MOD 10));
         IF (yzi - yzr) >= 10 THEN
            yzi := yzi - 10;
         xzr := -x[1] * xScl * abScl;
         xzi := ROUND(xzr + (6 - ROUND(xzr) MOD 6));
         IF (xzi - xzr) >= 6 THEN
            xzi := xzi - 6;
         FOR i := nPoints DOWNTO 1 DO
```

Figure B3.2 Listing of procedure Xyplt.

```
                BEGIN
                   y[i] := (y[i] - yMin) * ySci + yzi - yzr;
                   x[i] := (x[i] - x[1]) * xScl + (xzi - xzr) / abScl
                END;
             abMax  := 90;
             ordMax := 100;
             abRng  := 90
          END;
       write('0',' ');
       FOR i := 0 TO 10 DO
          BEGIN
             ordLbl := 10 * i - 100 + ordMax - yzi;
             write(ordLbl:4,'      ')
          END;
       writeln;
       k   := 1;
       FOR row := 0 TO ROUND(abRng * abScl) DO
          BEGIN
             IF (row MOD 6) = 0 THEN
                BEGIN
                   grid1 := '-';
                   grid2 := '+';
                   absLbl := 10 * ((row - xzi) DIV 6) + abMax - abRng;
                   write(' ',' ',absLbl:4)
                END
             ELSE
                BEGIN
                   grid1 := ' ';
                   grid2 := 'I';
                   write('      ')
                END;
             FOR i := 0 TO 100 DO
                line[i] := grid1;
             FOR i := 0 TO 10 DO
                line[10 * i] := grid2;
             REPEAT
                IF k <= nPoints THEN
                   BEGIN
                      xData := ROUND(abScl * (x[k] - abMax + abRng));
                      yData := ROUND(y[k] - ordMax + 100)
                   END;
                IF xData = row THEN
                   IF (yData >= 0) AND (yData <= 100) THEN
                      line[yData] := 'X'
                   ELSE
                      IF yData < 0 THEN
                         line[0] := '$'
                      ELSE
                         line[100] := '$';
                IF xData <= row THEN
                   k := k + 1
             UNTIL
                (k > nPoints) or (xdata > row);
             FOR i := 0 TO 100 DO
                write(line[i]);
             writeln
          END
END;
```

Figure B3.2 *(Continued)*

The first operation of the procedure Xyplt is to rearrange the order of the data points so that they are in ascending order with respect to the values of the variables x[i]. Next, a test is made on the variable abRng to determine whether the automatic scaling option is desired. If it is (abRng = 999), the values of the variables x[i] and y[i] are examined to determine their maximum and minimum values and their ranges. The values of the points are rescaled so that the y[i] values range from 0 to 100 and the x[i] values range from 0 to 90. The maximum and minimum unscaled values and the scale factors used are printed as output. Finally, the variables abRng, abMax, and ordMax are set to 90, 90, and 100, respectively. From this point on, the operation of the program is the same whether or not the automatic scaling option has been selected.

At this point, the program starts plotting the data. Note that, to obtain a square grid, six printed lines (requiring a distance of 1 inch down the page) are equal to 10 units of the abscissa scale. This provides the same grid-line spacing as is used for the ordinate in which 10 spaces (requiring 1 inch across the page) corresponds to 10 units of the ordinate scale. Thus the y[i] data is evaluated on the basis of 10 divisions for 10 units of scale while the x[i] data is evaluated on the basis of 6 divisions for 10 units of scale.

In a manner similar to that done in the procedure Plot5, the program now enters a set of char values for + and − signs into the line array so as to construct an ordinate grid line. It then computes the x position of the variable x[k] and stores this as the index xData. The program now tests row to see if row is equal to xData. If not, it prints the line and increases row by 1. The operation is repeated until the correct line for x[k] is found. The program now examines the value of y[k] and computes an index yData indicating the position in the line array for this variable. The appropriate element of the array is changed to the char value of the symbol "X" to plot the point. K is then increased by 1 and x[k] is tested to see if it should also be plotted on the current line. If not, the line array is then printed, the index row is increased, and the process continues. Logic is provided to test the value of row so that coordinate grid lines parallel to the ordinate and constructed of + and − symbols are drawn every six values of row. In between these lines, coordinate grid lines parallel to the abscissa constructed of the symbol I are drawn every 10 spaces across the page. The process continues until all lines have been printed, at which time control is returned to the main program.

Appendix C

Microcomputer Implementations of the Programs

Among its many positive features, PASCAL has the advantage of requiring only a modest amount of computer resources as compared with many other high-level languages. As a result, it is an excellent choice as a language for microcomputers. As an example, one of the currently popular PASCAL compilers requires only 28K of memory for its operation.[1] Even in a 64K CP/M computer this leaves considerable memory space available for the program that is to be compiled. One of the features of PASCAL sometimes omitted in microcomputer implementations is the function parameter. The PASCAL procedures presented in this text that use this feature are readily modified to accommodate this omission. In the following pages figures are given that detail the modifications. A summary of these figures follows.

Figure	Example program from text	Procedure used
C.1	2.1	IntegTrpz
C.2	2.2	Romberg
C.3	3.2	Plot5
C.4	4.1	PiecewiseLin
C.5	5.1	DiffEqn
C.6	5.4	AdBash
C.7	6.1	MxDiffEqn
C.8	10.1	FourSeries
C.9	12.1	Xyplt

Since the actual details of using input and output files vary in different microcomputer implementations, only the standard PASCAL write statements are used in these programs. In addition, input is provided by assignment statements. Note that included in these listings (Figs. C.3 and C.9) are modified versions of the plotting subprograms Plot5 and Xyplt. These procedures have been altered so that their output is limited to 80 columns. This may be directly displayed on most microcomputer screens and printers. On all the listings, compiler ''Insert'' directives are used where necessary to load procedures that are not shown explicitly.

[1]TURBO PASCAL, available from Borland International, 4113 Scotts Valley Drive, Scotts Valley, CA 95066. This compiler has a built-in editor and sells (1985) for the bargain price of approximately $50.

```
PROGRAM Main (output);    (* Example 2.1 *)

(* Main program for IntegTrpz example
     n - Number of trapezoidal sections
     a - Value of integral *)

VAR
   a    : real;
   n, i : integer;

FUNCTION Y (t : real) : real;
   BEGIN
      y := abs(sin(t)) / (3.0*t + 1.0)
   END;

PROCEDURE IntegTrpz (timeA, timeB : real; n : integer; VAR area : real);

(* Trapezoidal integration procedure
      timeA - Initial value of independent variable t
      timeB - Final value of independent variable t
      n - Number of trapezoidal sections used
      area - Value of integral from timeA to timeB
   A function Y(t) must be used to define the integrand. *)

   VAR
      t, dt : real;
      i     : integer;

   BEGIN

      (*   Initialize the value of time t to the value timeA
           and calculate dt, the increment in time   *)

      t := timeA;
      dt := (timeB - timeA) / n;

      (*   Initialize the value of the integral   *)

      area := Y(timeA) / 2.0;

      (*   Compute the n - 1 terms in the summation   *)

      IF N <> 1 THEN
         BEGIN
            FOR i := 1 to n-1 DO
               BEGIN
                  t := t + dt;
                  area := area + Y(t)
               END
         END;

      (*   Finish the computation of the integral   *)

      area := (area + Y(timeB) / 2.0) * dt
   END;

BEGIN
   n := 0;
   FOR i := 1 TO 10 DO
      BEGIN
         n := n + 10;
         IntegTrpz(0.0, 3.0, n, a);
         writeln(n, a:12:9)
      END
END.
```

Figure C.1 Modification of Example 2.1 and procedure IntegTrpz.

```
PROGRAM Main (output);    (* Example 2.2 *)

(* Main program for Romberg example
     n - Number of trapezoidal sections
     a - Value of integral *)

VAR
   a : real;
   i : integer;

FUNCTION Y (t : real) : real;
   BEGIN
      y := abs(sin(t)) / (3.0*t + 1.0)
   END;

(*$I Itrpz.kp *)

PROCEDURE Romberg (timeA, timeB : real; nmax : integer; VAR a : real;

                   errmax : real; test : integer);

(* Romberg integration procedure
     timeA - Initial value of independent variable t
     timeB - Final value of independent variable t
     nmax - Maximum number of rows of R array
     a - Value of integral from timeA to timeB
     errmax - Maximum value of error used as limit
     test - Output parameter, set to 1 to output
        stopping conditions, otherwise set to 0
   A function Y(t) must be used to define the integrand.
   This procedure calls the procedure IntegTrpz.          *)

   VAR
      error, fac : real;
      i, j, m    : integer;
      r          : ARRAY[1..20,1..20] OF real;

   BEGIN

      (*   Use the procedure IntegTrpz to find the array element r11 *)

      IntegTrpz(timeA, timeB, 1, r[1,1]);
      i := 1;  m := 1;

      (*   Find the other rows of the r array  *)

      REPEAT
         i := i + 1;  m := m * 2;
         fac := 1.0;

         (*   Use IntegTrpz to find the first element in row *)

         IntegTrpz(timeA, timeB, m, r[i,1]);

         (*   Use extrapolation to find other elements in row *)

         FOR j := 2 TO i DO
            BEGIN
               fac := fac * 4.0;
               r[i,j] := (fac * r[i,j-1] - r[i-1,j-1]) / (fac - 1.0)
            END;
         error := abs(r[i,i] - r[i-1,i-1]);

      (* Check terminating conditions for REPEAT/UNTIL loop *)
```

Figure C.2 Modification of Example 2.2 and procedure Romberg.

```
   UNTIL (i >= nmax) OR (error < errmax);

   (* Check to see if terminating conditions are to be printed *)

      IF test > 0 THEN
         writeln(' Romberg, n =', i:3, '   error =', error);

      (* Store output from R array in variable 'a' *)

      a := r[i,i]
   END;

BEGIN
   FOR i := 2 TO 5 DO
      BEGIN
         writeln;
         Romberg(0.0, 3.0, i, a, 1.0E-04, 1);
         writeln('           n =', i:3, '     area =', a)
      END
END.
```

Figure C.2 (*Continued*)

```
PROGRAM Main (output);    (* Example 3.2 *)

(* Main program for example of plotting
      t - time
      g - value of inductor current
      a - plotting array   *)

CONST twoPi = 6.2831853;
TYPE plotArray = ARRAY[1..5,0..100] OF real;
VAR
   t, dt, g, dg : real;
   a            : plotArray;
   i            : integer;

FUNCTION Y (t : real) : real;
   BEGIN Y := 20.0 * (1.0 - cos(twoPi * t)) END;

(*$I Itrpz.kp *)

PROCEDURE Plot5 (VAR y : plotArray; nf, nti, ordMx : integer);

(* Given the parameters listed below, this procedure will
   create a plot of function(s) vs time.
   Special types required:
      plotArray = ARRAY[1..5,0..100] OF real.
   Definition of parameters:
      y     = The two-dimensional array containing the values
              of each function.
      nf    = The number of functions to be plotted.
      nti   = The number of time increments to be plotted.
      ordMx = The maximum value of the function which can
              be plotted.*)

TYPE cArray = ARRAY[0..70] OF char;
VAR i, j, ordLbl, row, yData, yzi : integer;
    yMax, yMin, yScl, yzr         : real;
    grid1, grid2, grid3           : char;
    line                          : cArray;

BEGIN
   writeln('1');
   yzr  := 0;
   yMin := 0;
   yScl := 1;
   IF ordMx = 999 THEN
      BEGIN
         yMax := -1.0E+30;
         yMin :=  1.0E+30;
         FOR j := 1 TO nf DO
            FOR i := 0 TO nti DO
               BEGIN
                  IF y[j,i] > yMax THEN
                     yMax := y[j,i];
                  IF y[j,i] < yMin THEN
                     yMin := y[j,i]
               END;
         yScl := 60 / (yMax - yMin);
         writeln(' ','        VALUES FROM ',yMin:12,
               ' TO ',yMax:12,' SCALED BY ',yScl:12);
         yzr  := - yMin * yScl;
         yzi  := ROUND(yzr + (10 - ROUND(yzr) MOD 10));
         IF (yzi - yzr) >= 10 THEN
            yzi := yzi - 10;
         ordMx := 70 - yzi
      END;
   write('0','    ');
   FOR i := 0 TO 7 DO
      BEGIN
         ordLbl := 10 * i - 70 + ordMx;
```

Figure C.3 Modification of Example 3.2 and procedure Plot5.

```
            write(ordLbl:4,'         ')
         END;
      writeln;
      FOR row := 0 TO nti DO
         BEGIN
            IF (row MOD 10) = 0 THEN
               BEGIN
                  grid1 := '-';
                  grid2 := '+';
                  write(' ',' ',row:4)
               END
            ELSE
               BEGIN
                  grid1 := ' ';
                  grid2 := 'I';
                  write(' ','       ')
               END;
            FOR i := 0 TO 70 DO
               line[i] := grid1;
            FOR i := 0 TO 7 DO
               IF (10 * i) = (70 - ordMx) THEN
                  line[10 * i] := '0'
               ELSE
                  line[10 * i] := grid2;
            grid3 := 'A';
            IF row <= nti THEN
               FOR j := 1 TO nf DO
                  BEGIN
                     yData := ROUND((y[j,row] - yMin) * yScl
                              - yzr - ordMx + 70);
                     IF (yData >= 0) AND (yData <= 70) THEN
                        line[yData] := grid3
                     ELSE
                        IF yData < 0 THEN
                           line[0] := '$'
                        ELSE
                           line[70] := '$';
                     grid3 := SUCC(grid3)
                  END;
            FOR i := 0 TO 70 DO
               write(line[i]);
            writeln
         END
END;

BEGIN
   g := 0.0; t := 0.0; dt := 0.1; a[1,0] := 0.0; a[2,0] := 0.0;
   FOR i := 1 TO 50 DO
      BEGIN
         IntegTrpz(t, t+dt, 2, dg);
         g := g + dg;
         t := t + dt;
         a[1,i] := g * 0.7;
         a[2,i] := Y(t)
      END;
   Plot5(a, 2, 50, 70)
END.
```

Figure C.3 (*Continued*)

```
PROGRAM Main (input,output);    (* Example 4.1 *)

(* Main program for PiecewiseLin example
     ndata - number of data points
     ydata - array of inductor voltage values
     tdata - array of time values
     t     - time
     amps  - inductor current
   Program requires IntegTrpz, PiecewiseLin (redefined
   as Y), and Plot5 subprograms *)

CONST ndata = 9;
TYPE
   plotArray = ARRAY[1..5,0..100] OF real;
   array50   = ARRAY[1..50] OF real;
VAR
   t, dt, amps, damps : real;
   tdata, ydata       : array50;
   a                  : plotArray;
   i                  : integer;

FUNCTION Y (t : real) : real;

VAR i, im     : integer;
    outRange : boolean;

BEGIN
   outRange := false;
   IF t = tdata[1] THEN Y := ydata[1]
      ELSE IF t < tdata[1] THEN outrange := true
         ELSE IF t = tdata[ndata] THEN Y := ydata[ndata]
            ELSE IF t > tdata[ndata] THEN outRange := true
               ELSE
                  BEGIN
                     i := 2;
                     WHILE t >= tdata[i]
                        DO i := i + 1;
                     im := i - 1;
                     Y := ydata[im] + (ydata[i] - ydata[im])
                        * (t - tdata[im]) / (tdata[i] - tdata[im])
                  END;
   IF outRange = true THEN
      BEGIN
         Y := 999999.0;
         writeln;
         writeln(' T = ', t,' OUT OF RANGE');
         writeln
      END
   END;

(*$I Itrpz.kp *)

BEGIN
   t := 0.0; dt := 0.1; amps := 0.0; a[1,0] := 0.0; a[2,0] := 0.0;
   tdata[1] := 0.0; ydata[1] := 0.0;
   tdata[2] := 0.5; ydata[2] := 36.0;
   tdata[3] := 0.9; ydata[3] := 41.4;
   tdata[4] := 1.3; ydata[4] := 42.6;
   tdata[5] := 1.8; ydata[5] := 38.3;
   tdata[6] := 2.4; ydata[6] := 25.1;
   tdata[7] := 3.0; ydata[7] := 11.0;
   tdata[8] := 4.0; ydata[8] := 3.0;
   tdata[9] := 5.0; ydata[9] := 1.0;
   FOR i := 1 TO 50 DO
      BEGIN
         IntegTrpz(t, t+dt, 2, damps);
         amps := amps + damps;
         t := t + dt;
         a[1,i] := amps * 0.7;
         a[2,i] := Y(t)
      END;
   Plot5(a, 2, 50, 70)
END.
```

Figure C.4 Modification of Example 4.1 and procedure PiecewiseLin.

```
PROGRAM Main (output);    (* Example 5.1 *)

(* Main program for linear time-varying DiffEqn example
     iter - Number of iterations between plotted points
     amps - Plotting array                                    *)

TYPE
   plotArray = ARRAY[1..5,0..100] OF real;

VAR
   i, iter : integer;
   t, dt   : real;
   amps       : plotArray;

FUNCTION G (t, amps : real) : real;
   BEGIN
      G := (4.0 - amps) / t
   END;

PROCEDURE DiffEqn (tInitial, yInitial, tstop : real; iter : integer;

                VAR y : real);

(* Runge-Kutta differential equation solving procedure
      tInitial - Initial value of time
      yInitial - Initial value of dependent variable y(t)
      tstop - Final value of time
      iter - Number of iterations from tInitial to tstop
      y - Value of y(t) at tstop
   A function G(t, y) must be used to define the derivative
   of y(t).                                                   *)

   VAR
      t, dt, g1, g2, g3, g4 : real;
      i                     : integer;

   BEGIN

      (*   Initialize the values of t and y and compute
           the value of dt                               *)

      t := tInitial;
      y := yInitial;
      dt := (tstop - tInitial) / iter;

      (*   Perform iterations to find the value of y at
           tstop                                         *)

      FOR i := 1 TO iter DO
         BEGIN
            g1 := G(t, y);
            g2 := G(t + dt/2.0, y + dt*g1/2.0);
            g3 := G(t + dt/2.0, y + dt*g2/2.0);
            g4 := G(t + dt, y + dt*g3);
            y := y + (g1 + 2.0*g2 + 2.0*g3 + g4) * dt / 6.0;
            t := t + dt
         END
   END;

(*$I Plot5.kp *)

BEGIN

   (* Initialize values of current and time at t = 0.05
      Set unused plotting array element to zero             *)

   amps[1,1] := 84.0 * 0.7;
   amps[1,0] := 0.0;
   t := 0.05;
```

Figure C.5 Modification of Example 5.1 and procedure DiffEqn.

```
    dt := 0.05;
    iter := 2;

    (* Compute ano plot the values of the inductor current *)

    FOR i := 2 TO 50 DO
       BEGIN
          DiffEqn(t, amps[1,i-1], t+dt, iter, amps[1,i]);
          t := t + dt
       END;
    Plot5(amps, 1, 50, 70)
END.
```

Figure C.5 (*Continued*)

```
PROGRAM Main (output);    (* Example 5.4 *)

(* Main program for linear time-varying Adams-Bashforth example
     a - Plotting array                                        *)

TYPE
   plotArray = ARRAY[1..5,0..100] OF real;

VAR
   a        : plotArray;

FUNCTION G (t, amps : real) : real;
   BEGIN
      G := (4.0 - amps) / t
   END;

(*$I Dferk.kp *)

PROCEDURE AdBash (tInitial, yInitial, tstop : real; n : integer; scale :

                  real; aPosition : integer; VAR a : plotArray;

                  test : integer);

(* Adams-Bashforth predictor-corrector differential-equation
   solving procedure
     tInitial - Initial value of time
     yInitial - Initial value of dependent variable y(t)
     tstop - Final value of time
     n - Number of intermediate steps used in going to tstop
     scale - Factor for multiplying output data for scaling
     aPosition - Index for position for storing data in
           plotting array
     a - Plotting array used to store intermediate values of y(t)
     test - Indicator used to check operation of corrector loop.
           Use test = 1 for output, test = 0 for no output
   Note: A function G(t, y) must be used to define the derivative
   of y(t). This procedure calls the procedure DiffEqn.    *)
```

Figure C.6 Modification of Example 5.4 and procedure AdBash.

```
CONST
   iter = 4;          (* Number of iterations for DiffEqn *)
   kmax = 10;         (* Maximum cycles for corrector *)
   errmax = 1.0e-8;   (* Maximum error for corrector *)

VAR
   t, dt, acor, ypcor, error : real;
   i, j, k : integer;
   yp      : ARRAY[0..4] OF real;

BEGIN

   (* Calculate variables, store initial values *)

   t := tInitial;
   dt := (tstop - tInitial) / n;
   j := aPosition;
   a[j,0] := yInitial;
   yp[0] := G(t, a[j,0]);

   (* Calculate three more values to start predictor *)

   FOR i := 1 TO 3 DO
      BEGIN
         DiffEqn(t, a[j,i-1], t+dt, iter, a[j,i]);
         t := t + dt;
         yp[i] := G(t, a[j,i])
      END;

   (* Enter the predictor-corrector section of program *)

   FOR i := 4 TO n DO
      BEGIN

         (* Apply the predictor equation *)

         t := t + dt;
         a[j,i] := a[j,i-1] + dt * (55.0*yp[3] - 59.0*yp[2]
                   + 37.0*yp[1] - 9.0*yp[0]) / 24.0;
         yp[4] := G(t, a[j,i]);
         k := 0;
         REPEAT

            (* Start the corrector loop *)

            k := k + 1;
            acor := a[j,i-1] + dt * (9.0*yp[4] + 19.0*yp[3]
                    - 5.0*yp[2] + yp[1]) / 24.0;
            ypcor := G(t, acor);
            error := abs(yp[4] - ypcor);
            yp[4] := ypcor
         UNTIL (k >= kmax) OR (error < errmax);
         a[j,i] := acor;
         IF test > 0 THEN
            writeln(' Ad/Bash Corrector, k =', k:3,
                    '  Error =', error);

         (* Update the derivative information *)

         FOR k := 1 TO 4 DO
            yp[k-1] := yp[k]
      END;  (* End of predictor-corrector section *)

   IF scale <> 1.0 THEN
      FOR i := 0 TO n DO
         a[j,i] := a[j,i] * scale
END;

BEGIN
   AdBash(0.05, 84.0, 2.5, 49, 1.0, 1, a, 0);
   Print5(a, 1, 49)
END.
```

Figure C.6 (*Continued*)

```
PROGRAM Main (output);     (* Example 6.1 *)

(* Main program for MxDiffEqn two-mesh resistor-inductor
   network example                                    *)

TYPE
   plotArray = ARRAY[1..5,0..100] OF real;
   array10   = ARRAY[1..10] OF real;

VAR
   t, dt            : real;
   i, iter          : integer;
   oldamps, amps    : array10;
   a                : plotArray;

PROCEDURE Gn (t : real; VAR amps, g : array10);
   BEGIN
      g[1] := 120.0 - 2.0*amps[1] + 2.0*amps[2];
      g[2] := 2.0*amps[1] - 5.0*amps[2]
   END;

PROCEDURE MxDiffEqn (tInitial : real; VAR yInitial : array10;

                  nVars : integer; tstop : real; iter : integer;

                  VAR y : array10);

(* Matrix Runge-Kutta differential equation solving procedure
      tInitial - Initial value of time
      yInitial - Array of initial values of independent variables
      nVars - Number of independent variables
      tstop - Final value of time
      iter - Number of iterations from tInitial to tstop
      y - Array of values of independent variables at tstop
   A procedure Gn (t, y, g) (where y and g are arrays) must
   be used to define the derivatives of the independent
   variables                                             *)

   VAR
      t, dt, dt2, dt6           : real;
      ytemp, g1, g2, g3, g4     : array10;
      i, j                      : integer;

   BEGIN

      (*   Initialize the values of t and the y array and
           compute the value of dt, dt/2, and dt/6       *)

      t := tInitial;
      dt := (tstop - tInitial) / iter;
      dt2 := dt / 2.0;
      dt6 := dt / 6.0;
      FOR i := 1 TO nVars DO
         y[i] := yInitial[i];

      (*   Begin outer loop to compute the nVars values of
           the independent variables at tstop          *)

      FOR i := 1 TO iter DO
         BEGIN

            (*   Compute the arrays g1, g2, g3, and g4    *)

            Gn(t, y, g1);
            FOR j := 1 TO nVars DO
               ytemp[j] := y[j] + g1[j]*dt2;
            t := t + dt2;
            Gn(t, ytemp, g2);
            FOR j := 1 TO nVars DO
```
Figure C.7 Modification of Example 6.1 and procedure MxDiffEqn.

```
            ytemp[j] := y[j] + g2[j]*dt2;
         Gn(t, ytemp, g3);
         FOR j := 1 TO nVars DO
            ytemp[j] := y[j] + g3[j]*dt2;
         t := t + dt2;
         Gn(t, ytemp, g4);

         (*   Combine the arrays g1, g2, g3, and g4
              to find the intermediate value of the
              independent variables                 *)

         FOR j := 1 TO nVars DO
            y[j] := y[j] + (g1[j] + 2.0*g2[j] + 2.0*g3[j]
                    + g4[j]) * dt6;
      END   (* End of outer loop with i index *)

   END;   (* End of MxDiffEqn procedure *)

(*$I Plot5.kp *)

BEGIN

   (* Initialize arrays and other variables *)

   t := 0.0;
   dt := 0.1;
   oldamps[1] := 0.0;
   oldamps[2] := 0.0;
   a[1,0] := 0.0;
   a[2,0] := 0.0;
   iter := 2;

   (* Compute and plot the values of the mesh currents *)

   FOR i := 1 TO 50 DO
      BEGIN
         MxDiffEqn(t, oldamps, 2, t+dt, iter, amps);
         a[1,i] := amps[1] * 0.7;
         a[2,i] := amps[2];
         oldamps[1] := amps[1];
         oldamps[2] := amps[2];
         t := t + dt
      END;
   Plot5(a, 2, 50, 70)
END.
```

Figure C.7 *(Continued)*

```
PROGRAM Main (output);   (* Example 10.1 *)

(* Main program for plotting the Fourier series
   of a square wave
     a0 - The Fourier series coefficient a0
     a - Array of Fourier series cosine coefficients
     b - Array of Fourier series sine coefficients
     c - Array of Fourier series magnitude coefficients
     phi - Array of Fourier series phase coefficients
     plt - Plotting array
   A function F(t) must be used to define the periodic
   function being analyzed.                              *)

TYPE
   array10 = ARRAY[1..10] OF real;
   plotArray = ARRAY[1..5,0..100] OF real;
   complex = RECORD
               re, im : real
               END;

VAR
   a, b, c, phi : array10;
   a0           : real;
   i            : integer;
   plt          : plotArray;
   k, h         : integer;
   omega        : real;

FUNCTION F (t : real) : real;
   BEGIN
      IF t > 1.0
         THEN F := -1.0
         ELSE F := 1.0
   END;

FUNCTION Y (t : real) : real;
      BEGIN
         IF k = 0
            THEN Y := F(t) * cos(omega * h * t);
         IF k = 1
            THEN Y := F(t) * sin(omega * h * t);
         IF k = 2
            THEN Y := F(t)
      END;

   END;

(*$I Itrpz.kp *)

PROCEDURE FourSeries (nHarmon : integer; period : real; iter : integer;

                VAR a0 : real; VAR a, b, c, phi : array10);

(* Procedure for finding the coefficients of the Fourier
   series for a periodic function of time f(t)
     nHarmon - Number of harmonic frequencies for which the
          coefficients are to be found
     period - The period T of the periodic function f(t)
     iter - The number of iterations used in the trapezoidal
          integration procedure IntegTrpz
     a0 - The Fourier series coefficient a0
     a - Array of Fourier series cosine coefficients
     b - Array of Fourier series sine coefficients
     c - Array of Fourier series magnitude coefficients
     phi - Array of Fourier series phase coefficients
   Note: A PASCAL function must be provided to define the
   periodic function f(t)
   Note: This procedure calls the function Arg and the
   procedure IntegTrpz                                   *)
```

Figure C.8 Modification of Example 10.1 and procedure FourSeries.

```
VAR
   (* omega - The fundamental frequency
      h - Number of harmonic currently being evaluated
          (both defined as main program variables)
      k - Index used to change integrand from cosine form
          to sine form, defined as a main program variable *)
   temp   : real;
   i      : integer;
   z      : complex;

BEGIN

   (* Integrate F(t) to find the coefficient a0 *)

   k := 2;
   IntegTrpz(0.0, period, iter, temp);
   a0 := temp / period;

   (* Find nHarmon coefficients and store in the arrays
      a, b, c, and phi                                  *)

   omega := 6.2831853 / period;
   FOR i := 1 TO nHarmon DO
      BEGIN
         h := i;

         (* Set k = 0 so integrand contains the
            product of f(t) and the cosine term  *)

         k := 0;
         IntegTrpz(0.0, period, iter, temp);
         a[i] := 2.0 * temp / period;

         (* Set k = 1 so integrand contains the
            product of f(t) and the sine term  *)

         k := 1;
         IntegTrpz(0.0, period, iter, temp);
         b[i] := 2.0 * temp / period;
         c[i] := sqrt(a[i]*a[i] + b[i]*b[i]);

         (* Define a[i] and b[i] as elements of a complex
            variable z so that the function Arg can be used
            to define phi[i] in all four quadrants       *)

         z.re := a[i];
         z.im := -b[i];
         phi[i] := Arg(z)
      END
   END;
BEGIN
   FourSeries(5, 2.0, 201, a0, a, b, c, phi);
   writeln;
   writeln(' a0 = ', a0);
   writeln;
   writeln('  i    a[i]     b[i]     c[i]    phi[i]');
   FOR i := 1 TO 5 DO
      writeln(i:3, a[i]:9:5, b[i]:9:5, c[i]:9:5, phi[i]:9:5)
END.
```

Figure C.8 *(Continued)*

```
PROGRAM Main (output);    (* Example 12.1 *)

(* Main program for constructing a Nyquist plot
   for a second-order low-pass network function
     num - Array of numerator coefficients
     den - Array of denominator coefficients
     x - Array of values of the real part of
          the network function
     y - Array of values of the imaginary part
          of the network function
     omega - Frequency (rad/sec)                *)

TYPE
   array200 = ARRAY[1..200] OF real;
   array11  = ARRAY[0..10] OF real;
   complex = RECORD
              re, im : real
              END;

VAR
   omega, vmag, vphase : real;
   x, y                : array200;
   num, den               : array11;
   i                   : integer;

(*$I 121xy.rps *)
PROCEDURE Xyplt (nPoints : integer; VAR x, y : array200;

                     abRng, aoMax, ordMax : integer);

(* Given the parameters listed below, this procedure
   will create a plot of y vs x  values.
   Special types required:
      array200 = ARRAY[1..200] OF real.
   Definition of parameters:
      nPoints     = The number of values of x and y.
      abRng   = The range of values of x which can be plotted
               along the abcissa.
      abMax   = The maximum value of x which can be plotted
                along the abscissa.
      ordMax = The maximum value of y which can be plotted
                along the ordinate.
      x      = The one-dimensional array of variables
               containing the values of x.
      y      = The one-dimensional array of variables
               containing the values of y. *)

CONST abScl = 0.600000;
TYPE cArray = ARRAY[0..70] OF CHAR;
VAR absLbl, i, j, k, ordLbl, row, xData, xzi, yData, yzi : integer;
     temp, xScl, yMax, yMin, yScl, yzr, xzr              : real;
     grid1, grid2                                        : char;
     line                                                : cArray;
BEGIN
   writeln('1');
   yzi := 0;
   xzi := 0;
   FOR i := 1 TO (nPoints - 1) DO
      FOR j := (i + 1) TO nPoints DO
         IF x[i] > x[j] THEN
            BEGIN
               temp := x[i];
               x[i] := x[j];
               x[j] := temp;
               temp := y[i];
               y[i] := y[j];
               y[j] := temp
            END;
   IF abRng = 999 THEN
```

Figure C.9 Modification of Example 12.1 and procedure Xyplt.

```
      BEGIN
         yMax := -1.0E+30;
         yMin :=  1.0E+30;
         FOR i := 1 TO nPoints DO
            BEGIN
               IF y[i] > yMax THEN
                  yMax := y[i];
               IF y[i] < yMin THEN
                  yMin := y[i]
            END;
         yScl := 60 / (yMax - yMin);
         xScl := 80 / (x[nPoints] - x[1]);
         writeln(' ','      VALUES OF X FROM ',x[1]:12,' TO ',x[nPoints]:12,
                 ' SCALED BY ',xScl:12);
         writeln(' ','      VALUES OF Y FROM ',yMin:12,' TO ',yMax:12,
                 ' SCALED BY ',yScl:12);
         yzr := - yMin * yScl;
         yzi := ROUND(yzr + (10 - ROUND(yzr) MOD 10));
         IF (yzi - yzr) >= 10 THEN
            yzi := yzi - 10;
         xzr := -x[1] * xScl * abScl;
         xzi := ROUND(xzr + (6 - ROUND(xzr) MOD 6));
         IF (xzi - xzr) >= 6 THEN
            xzi := xzi - 6;
         FOR i := nPoints DOWNTO 1 DO
            BEGIN
               y[i] := (y[i] - yMin) * yScl + yzi - yzr;
               x[i] := (x[i] - x[1]) * xScl + (xzi - xzr) / abScl
            END;
         abMax  := 90;
         ordMax := 70;
         abRng  := 90
      END;
   write('0','    ');
   FOR i := 0 TO 7 DO
      BEGIN
         ordLbl := 10 * i - 70 + ordMax - yzi;
         write(ordLbl:4,'      ')
      END;
   writeln;
   k   := 1;
   FOR row := 0 TO ROUND(abRng * abScl) DO
      BEGIN
         IF (row MOD 6) = 0 THEN
            BEGIN
               grid1 := '-';
               grid2 := '+';
               absLbl := 10 * ((row - xzi) DIV 6) + abMax - abRng;
               write(' ','  ',absLbl:4)
            END
         ELSE
            BEGIN
               grid1 := ' ';
               grid2 := 'I';
               write('       ')
            END;
         FOR i := 0 TO 70 DO
            line[i] := grid1;
         FOR i := 0 TO 7 DO
            line[10 * i] := grid2;
         REPEAT
            IF k <= nPoints THEN
               BEGIN
                  xData := ROUND(abScl * (x[k] - abMax + abRng));
                  yData := ROUND(y[k] - ordMax + 70)
               END;
            IF xData = row THEN
               IF (yData >= 0) AND (yData <= 70) THEN
                  line[yData] := 'X'
```

Figure C.9 *(Continued)*

```
               ELSE
                  IF yData < 0 THEN
                     line[0] := '$'
                  ELSE
                     line[70] := '$';
            IF xData <= row THEN
               k := k + 1
         UNTIL
            (k > nPoints) or (xdata > row);
         FOR i := 0 TO 70 DO
            write(line[i]);
         writeln
      END
END;
BEGIN
   (* Initialize the variables and the arrays for the
      numerator and the denominator coefficients of the
      network function                                  *)

   omega := 0.0;
   num[0] := 60.0;
   den[0] := 1.0;
   den[1] := 1.0;
   den[2] := 1.0;

   (* Calculate and store the values of the real and imaginary
      parts of the network function                             *)

   FOR i := 1 TO 100 DO
      BEGIN
         SinStdySt(0, num, 2, den, omega, x[i], y[i], vmag, vphase);
         omega := (omega + 0.03) * 1.008
      END;
   Xyplt(100, x, y, 90, 60, 0)
END.
```

Figure C.9 (*Continued*)

Bibliography

For additional information on the numerical methods presented in this text, consult any of the following:

Atkinson, K. E. *An Introduction to Numerical Analysis,* John Wiley & Sons, Inc., New York, 1978.

Burden, R. L., J. D. Faires, and A. C. Reynolds. *Numerical Analysis,* 2d ed., Prindle, Weber, and Schmidt, Boston, 1981.

Hamming, R. W. *Numerical Methods for Scientists and Engineers,* McGraw-Hill Book Company, New York, 1962.

Kreyzig, E. *Advanced Engineering Mathematics,* 4th ed., John Wiley & Sons, Inc., New York, 1979. Chap. 19.

Ralston, A. *A First Course in Numerical Analysis,* McGraw-Hill Book Company, New York, 1965.

Smith, W. A. *Elementary Numerical Analysis,* Harper and Row, Publishers, Inc., New York, 1979.

For additional information on PASCAL programming, consult any of the following:

Grogono, P. *Programming in Pascal,* rev. ed., Addison Wesley Publishing Co., Reading, Mass., 1980.

Jensen, K., and N. Wirth. *PASCAL User Manual and Report,* 2d ed., Springer-Verlag, New York, 1978.

Jones, W. B. *Programming Concepts, A Second Course,* Prentice-Hall, Inc., Englewood Cliffs, N.J., 1982.

Koffman, E. B. *Pascal: A Problem Solving Approach,* Addison Wesley Publishing Co., Reading, Mass., 1982.

Schneider, G. M., S. W. Weingart, and D. M. Perlman. *An Introduction to Programming and Problem Solving with Pascal,* John Wiley & Sons, Inc., New York, 1978.

Welsh, J., and J. Edler. *Introduction to Pascal,* 2d ed., Prentice-Hall, International, Inc., London, 1982.

For additional information on the network analysis techniques referred to in this text, consult any of the following:

Durney, C. N., L. D. Harris, and C. L. Alley. *Electric Circuits,* Holt, Rinehart, and Winston, New York, 1982.

Hayt, W. H., Jr., and J. E. Kemmerly. *Engineering Circuit Analysis,* 3d ed., McGraw-Hill Book Company, New York, 1978.

Huelsman, L. P. *Basic Circuit Theory,* 2d ed., Prentice-Hall, Inc., Englewood Cliffs, N.J., 1984.

Johnson, D. E., J. L. Hilburn, and J. R. Johnson. *Basic Electric Circuit Analysis,* Prentice-Hall, Inc., Englewood Cliffs, N.J., 1978.

Van Valkenburg, M. E. *Network Analysis,* 3d ed., Prentice-Hall, Inc., Englewood Cliffs, N.J., 1974.

For additional information on the use of electrical circuits to model other types of engineering systems, consult any of the following:

Blackwell, W. A. *Mathematical Modeling of Physical Networks,* The Macmillan Company, New York, 1968.

Harman, W. W., and D. W. Lytle. *Electrical and Mechanical Networks, An Introduction to Their Analysis,* McGraw-Hill Book Company, New York, 1962.

Sanford, Richard S. *Physical Networks,* Prentice-Hall, Inc., Englewood Cliffs, N.J., 1965.

Thorn, D. C. *An Introduction to Generalized Circuits,* Wadsworth Publishing Co., Belmont, Calif., 1963.

Index